Discovering Relativity for yourself

Discovering Relativity
for yourself

with some help from
SAM LILLEY

CAMBRIDGE UNIVERSITY PRESS
CAMBRIDGE
LONDON   NEW YORK   NEW ROCHELLE
MELBOURNE   SYDNEY

Published by the Press Syndicate of the University of Cambridge
The Pitt Building, Trumpington Street, Cambridge CB2 1RP
32 East 57th Street, New York, NY 10022, USA
296 Beaconsfield Parade, Middle Park, Melbourne 3206, Australia

First published 1981

Printed in Canada
Typeset by Huron Valley Graphics, Inc., Ann Arbor, Michigan
Printed and bound by The Hunter Rose Company Ltd, Toronto, Canada

British Library Cataloguing in Publication Data

Lilley, Sam
Discovering relativity for yourself
1. Relativity (Physics)
I.   Title
530.1'1   QC173.55   80-40263

ISBN 0 521 23038 1   hard covers
ISBN 0 521 29780 X   paperback

To my students at classes in
Boston
Derby
Lincoln
Loughborough
Mansfield
Nottingham
Skegness
Sleaford
and
Stamford,
who taught me how to teach Relativity

CONTENTS

PREFACE

Another book on Relativity for the layman! Why? What's different about it?

There have been many books on Einstein's theory, written by authors who are highly expert in this field and who have gone to an immense amount of trouble to explain it with great logical clarity, yet in simple terms that should be comprehensible to any reasonably intelligent person even if he has had no scientific education. And *still* these books have left most of their readers bewildered. Why? After discussions with dozens of students I think I can answer that question. The difficulties that really trouble the layman are not those which you would logically expect. The Relativity expert, no matter how diligent and sympathetic, is unlikely to discover these difficulties, even more unlikely to know how to cope with them.

I am definitely not one of these Relativity experts. What I know of the subject has been learnt laboriously from their works. But I think I can claim to be an expert in something different – in the art of teaching science to the non-scientist. That has been my job since 1950.

I introduced Relativity amongst my courses in 1958. And since then I have been teaching it to carpenters and clerks, housewives, miners and insurance agents – to all sorts of people who have no special qualifications for learning the subject (and others like teachers and professional engineers who have limited qualifications). At first I taught it badly. But the customs of the Adult Education world enabled me to learn by my mistakes. Discussion bulks large in my classes. Students are encouraged to raise their difficulties and argue about them – till I find how to get them past the barriers. They are encouraged to *argue back* – try to prove Einstein (and me) wrong. They must be genuinely convinced before we move on.

How much (or little) my students learned from the earlier classes I do not know. But I learned an immense amount. Amid these discussions and arguments I came to understand that their main difficulties were not the ones that might reasonably be anticipated. Nor were they ones that could be dealt with by mere logical explanations. They needed a more roundabout approach. So I developed the practice of tape-recording every class, listening to the playback, analysing students' problems, spotting my deficiencies, and keeping systematic notes of all this. From autumn 1963 to spring 1970 I did this for every Relativity course I took – a total of 650 hours of recordings, and maybe 2000 or more hours of playback.

Thus I came gradually to understand the unconscious habits of thought lying behind each difficulty. And step by step I worked out the appropriate teaching methods – which have (to use the jargon) been validated by the greater ease and expedition with which my students of recent years conquer their difficulties. So I now feel that some purpose would be served by making these methods more widely available in book form. Translating face-to-face teaching into print has led to some stylistic eccentricities. But these can surely be excused if the book does (as I hope) help more people to feel at home with this key sector of twentieth-century culture.

In order to facilitate the task of exposing and then coping with students' real problems, it was necessary to ensure that mere technical difficulties, especially mathematical ones, did not get in the way. So I have gradually evolved methods that employ a minimum of mathematical techniques, and a scheme in which even that minimum is taught point by point as we need it. A glance at the pages ahead will show how slowly mathematics comes into the picture.

And this mathematical simplification has allowed me to do something which I believe is entirely new – to obtain the main results of the *General* Theory of Relativity by elementary methods. Instead of the fearsome tensor calculus I use only a simplified version of the infinitesimal calculus (*the* calculus of school terminology), which the student learns as he goes. (One new point arises incidentally that might even interest Relativity experts – §§32.26–7, 32.31–3.)

My debts are many. First to Hermann Bondi, whose article in *Discovery,* December 1957, first convinced me that Relativity is a suitable subject for Adult Education, and whose 1963 television series and subsequent books gave further inspiration (even if I no longer make much use of his methods). Then to many other authors from whom I have borrowed – and among these I must specially mention L.D. Landau and G.B. Rumer (*What is Relativity?* (1960) Oliver and Boyd, Endinburgh and London) and J.T. Schwartz (*Relativity in Illustrations* (1962) New York University Press, U.S.A.). Also to Don Chapman for painstakingly searching out errors in typescript and proofs. But my biggest debt of all (except, perhaps, to my wife for patience) is revealed in the Dedication.

University of Nottingham   Sam Lilley
Department of Adult Education   1980

INTRODUCTION

You may well be wondering whether this book is going to be too difficult for you. You may have special worries about whether you can cope with the mathematics that features rather prominently in the later pages. So let me assure you that this is a book for people who, *when they start on it,* are acquainted with arithmetic and *nothing more.* I undertake to teach you all the mathematics you need as you need it. If my assurance is not enough (and why should it be?), please read at least to page 4 before deciding whether to carry on or not.

The Special and General Theories of Relativity

But first of all, what *is* this Theory of Relativity? It is divided into two parts. By far the more important of these is the Special Theory of Relativity, which is roughly speaking the theory of how the world would appear to people who were used to moving around at very high speeds. And it must be steady motion – no speeding up, slowing down or swerving is permitted.

This Special Theory starts from the very simple idea that there is no means of knowing whether you are really moving or not. Not much in that, you would think. But when you follow up the consequences of this apparently innocent beginning, they turn out to be shattering. The world, says Relativity, is decidedly different from what we have hitherto believed.

Suppose (to take the most staggering assertion of the lot) that a pair of twins separate, one staying on Earth, the other going on a long fast space journey and returning. The theory says that on reunion the travelling twin will have aged less than the stay-at-home. And even when it's not shocking us like that, Special Relativity always modifies and sometimes revolutionises the old theories. Some of its consequences are of great practical importance in electronic engineering or nuclear power production. And the whole is essential to any real understanding of the space and time we live in.

The other part, the General Theory of Relativity, begins by dropping the steady-motion restriction of the Special Theory and studying the experiences of observers moving in any manner whatsoever. What emerges in the long run, however, is a new theory of gravitation, slightly better than Newton's – probably very much better in extreme conditions. It does not have much practical significance as yet, and maybe never will have. Yet studying it is one of the most exciting experiences that can happen to anybody. And it leaves us feeling that we really understand the space and time we live in.

Is Relativity difficult?

Relativity has become a bogy word. It has the reputation of being fiendishly difficult. You may, as I've said, be wondering whether you can cope.

You can – I'm sure of that.

Certainly you are going to have to work hard. You have been warned! But the same could be said of gardening; and you would not call gardening difficult. Relativity is not difficult in the way that running a four-minute mile is difficult. That needs special aptitudes and abilities. Relativity makes no such demands. Very many, perhaps most, of us are capable of mastering it. Our latent abilities only need bringing out. True, there must be *some* minimum level of intelligence required – relativistic thinking has not yet been observed among apes. But I know from many years of teaching that large numbers of people with no special abilities or qualifications and with very ordinary educational backgrounds are capable of studying the subject successfully and getting a great deal of enjoyment in doing so.

Of course there are senses in which this theory really is difficult. When I tell you that a fast space traveller will age less than his stay-at-home twin, you may well answer 'I can't imagine that.' And neither can I! But if you can't imagine something, does that mean you're also entitled to say you don't believe it?

There are lots of things which we can't imagine – because they're too big or too small or in some other way too far outside our experience – and which we nevertheless believe, because the evidence convinces us that it must be so. Think of all the things you accept about stars and galaxies and quasars; or of your belief that this page consists mostly of empty space between tiny tiny atoms.

It's the same with Relativity. We can't imagine this space traveller business. But that's simply because we've no experience that is even remotely comparable – we haven't travelled anything like fast enough. And my imagination – or Einstein's – fails just as badly as yours.

Yet I believe, and Einstein believed – and I hope that in due course you will believe – that the traveller *will* age less than the homelover. Why? Because there's overwhelming evidence – a combination of experiment, observation, logic and mathematics – which forces us to agree that it must be so.

Again, changing long-held beliefs is always a problem. And so your difficulty in this study is going to lie just as much in *unlearning what you think you know* as in learning what you think you don't know. We've all got a number of ideas that we've held so long, without consciously thinking about them, that we've come to feel that they're 'obviously' true; that things could not conceivably be arranged otherwise. Deeper study shows that they are *not* true – they only appear to be so within the narrow range of our experience. But these old beliefs are so deeply embedded in our way of thinking that they keep on misleading our intuitions long after our intellects have proved them false.

So there are difficulties of unlearning and difficulties of imagination, which worry even the experts. But we do agree, I'm sure, that we have to unlearn what we can *prove* false, and we have to believe what we can *prove* true, even if imagination boggles. And the process of proving the results of Relativity Theory is, I do assure you, moderately easy. All of it is easy in principle. Some parts of it are technically difficult – in the sense that playing the Emperor Concerto is difficult. But it's all easy in principle, just as playing the piano is easy in principle. And some parts of Relativity are technically easy as well – as easy as 'Chopsticks' on the piano. It's these technically easy parts that we'll concentrate on, though we shall improve our techniques as we go.

But those techniques will include – in the later parts of the book – a fair bit of mathematics. And it's about ten to one that you are worried about that. You think you can't do mathematics. Actually you can. Most of us have been *taught* mathematics either very badly or not at all. And so at present we can *do* mathematics either very badly or not at all. We come to imagine that the fault lies in ourselves – that we have the wrong sort of mind for mathematics. But that's not true. We can all do mathematics well enough to cope with this book, though naturally some will find it easier than others.

If you look ahead, you'll see that we do a little arithmetic and simple geometry on pages 15–16 and 38–44. But we don't use those algebraic symbols that perhaps frighten you till page 102 (they get passing mention earlier). After that our use of mathematics gradually increases; and *I guarantee to teach you all you need as you need it.*

You doubt your ability to learn? Nonsense! That can only mean that your teachers approached the subject from the wrong angle. Mathematics (so far as we are concerned) is only a language. Now you have already demonstrated that you are good at learning languages, for you have mastered at least two of them. Two? Yes, you've learnt both spoken English and written English – two distinct languages that had to be learnt separately (or was it your experience that as soon as you could speak you could also read and write?). So you're not going to fight shy of learning another language now – especially as the gap between written English and mathematics takes less bridging than the gap between spoken and written English.

Mathematics is a sort of shorthand language, specially designed to make some types of statements very briefly and concisely. Anything you can say mathematically could also be said in ordinary English. But it might then be so long-winded that you'd get lost in it. Mathematical language is very brief and compact, enabling us to carry out trains of reasoning that would otherwise be too long and cumbersome.

Every activity has its own special language. If you don't believe that, just try explaining how to start a car and get up to top gear without using any specialist language like 'steering wheel' or 'clutch pedal'. Try to explain part of your job (or hobby) without using specialist jargon. Please *do* these exercises – I want you to be convinced.

Every activity has its own language. If you refuse to use it, life grows difficult. But anybody who takes a bit of trouble can learn the specialist language.

The same is true of mathematics. There are many reasoning processes which, if you tried to carry them through in ordinary English, would become so lengthy that you'd get lost. Put the same reasoning into mathematical language, and quite suddenly the problem is reduced to manageable proportions. And you can have all this in return for the trouble of learning a little bit of a new language – and not a very difficult language at that, much easier than French is to an English speaker, a language that we'll learn in easy stages as we need it. So you'll throw away preconceived fears of mathematics, won't you?

How to use this book
Please note: how to *use* it, not how to *read* it.

Passive reading is not enough. You will also have to do a lot of thinking for yourself, wrestling with difficulties, solving problems. And you ought to regard the book chiefly as a means of guiding and helping you in this more active learning.

With most of the things you normally read you can think out the meaning as fast as you can take in the words. So you get used to continuous reading at a steady pace. In face-to-face teaching I expand and repeat my material to produce the same effect. But in a book, that would be too costly, and so I've had to reduce my wordage by a factor of four or five. And you must be prepared to allow for that – by slower and more careful reading, by pausing to let a sentence sink in, by rereading a difficult paragraph, and so on.

Don't believe something because I say so. I want you to argue and object. Try to pick holes in my reasoning. Naturally you will make sure that your arguments are not silly ones. You must look for objections to your own objections. Only after some hard thinking on these lines should you seek further help from the book. I think I know from experience pretty well all the points that people raise. And I've tried to deal with the main ones where they are most likely to arise. But I can't cover everything; so be prepared to wrestle further on your own.

I shall frequently want to prod you into working things out for yourself. That's the function of the **passages in heavy type** (Bold or Clarendon in printer's jargon). Whenever that type face appears, there's something for you to *do*. Take it seriously.

Usually I shall follow up immediately with the answer. So don't spoil everything by letting your eyes wander down the page. I suggest you use a postcard as bookmarker; and whenever, on turning a page, you see some of this heavy type, you immediately cover the answer with your card – till you've had a serious go at whatever you're asked to do. But don't be discouraged if you fail. Quite often the task will really be too difficult, but I just want to make sure that you've done some serious thinking before we get down to tackling it together. (When a question is printed in ordinary type, it is largely rhetorical. Move on as soon as you've understood its significance.)

Even when I'm not pushing you, the more active you can be in your studies, the better you're likely to learn. For example, you'll meet many diagrams. And I'm sure you'll understand better if you redraw them step by step as you follow the description in the text (a freehand sketch, rather than an accurate drawing, will usually be good enough). And *keep your drawings,* suitably labelled – we'll often refer back to them, and this will save you the trouble of constantly turning back the pages.

Don't think of your studies as a sort of steady flow from beginning to end. Rather think of yourself as having a series of tasks to complete.

The numbered sections will give you a rough guide – at the end of each you could ask yourself what you were supposed to learn and whether you have taken it in properly and coped with any tasks involved. The end of a chapter is a place for more thorough consolidation.

But don't carry this too far. Insistence on fully mastering each section before going on to the next could bring you to a complete stop. If a difficulty is stubborn, make a note of it, carry on for a few pages (maybe more) and then come back to your problem. You've been stretching your mental muscles in the meantime, and very often you'll find that your former difficulty melts away.

You'll have realised that this is not a book to rush through. Don't try to do too much at a time – your mind won't cope. Do an hour or two-hour stint of concentrated work one day; then give your mind a chance to digest what it has eaten, and come back two or three days later for the next relativistic meal. (In the meantime a little thinking in the bath or while walking will probably help.) So you won't expect to get through this book in a week or two. You might manage it in a dozen weeks. More likely it will need months. And there would be no disgrace in taking two or three years to do the job properly.

I shall frequently have to refer back to something we've already done. To make this easier, each chapter is divided into numbered sections called *articles*. The word 'article' will usually be abbreviated to the sign '§', so that §3.5 means the fifth article of Chapter 3. Doubling the '§' sign indicates plural, so that the 2nd, 3rd and 4th articles of Chapter 7 would be §§7.2–4.

Except when I specifically ask you to reread something, references back to earlier articles are meant to help you with difficulties arising from forgetfulness and the like. So don't waste time looking them up unless you actually need help. Sometimes it's more convenient to explain a difficulty *after* it has arisen. So always look one article ahead before starting to worry about something that's not clear.

I've usually put the teaching of purely mathematical points (for those who must learn their mathematics as they go) into separate articles, which are distinguished by having the letter M appended to the article number – as §3.17M on page 40. If your mathematics is good, these may need only a glance. To help you recall things that slip your memory (if I've not given a reference back) the index will lead to (e.g.) definitions of elementary mathematical terms that would not usually be indexed – even words like 'proportional' and 'product'. If you forget your Greek letters, look them up under 'Greek'. And see the list of mathematical signs at the end of the index.

Forgive me if in the last few pages I've sounded like a mixture of schoolmaster and parson. I honestly think you may find it worthwhile to read this advice from time to time.

What can you hope to achieve?

At the end of all these efforts you will still be less technically competent in Relativity Theory than a recent university graduate whose course has included a few lectures on the subject – less able to do the calculations or apply Relativity to practical problems. On the other hand, you may well outstrip that university product in your understanding of what the Theory of Relativity tells us about the nature of the world we live in. Certainly you will have tackled some interesting philosophical problems that the professional usually finds it convenient to bypass.

Paradoxically, this situation arises because you will have worked clumsily and inefficiently. If somebody has a business manufacturing furniture, he will use the most sophisticated machine-tools available. You might decide to make a table and four chairs because you think it fun; and you would make do with a few simple hand tools. As a maker of furniture he would be far more efficient. But you would learn more about the properties of wood, nails and glue. The ability to deal with the idiosyncrasies of the material has been built into his machines. But if your chisel is not to slip and your joints are not to collapse, you must learn a great deal about how your tools and materials behave.

Similarly, anybody who is learning Relativity professionally is expected to employ sophisticated mathematical techniques in order to reach quickly the parts of the theory that concern him. He does not have to ask what goes on inside his mathematical 'machine'. We shall only use very simple 'mathematical hand tools'. We'll be less efficient at churning out relativistic results, and we'll have to work harder at it. This will force us to examine more closely what we are actually doing. And so, despite your technical inefficiency, you might well end up with a deeper understanding of what the Theory of Relativity is about than all but the best of experts.

Bon voyage!

1

Getting started

1.1 When you're in a train, you can't say whether it is really moving or not – that's an everyday experience from which our theory begins.

You can, of course, see the telephone poles flashing by. The naive interpretation of what you *see* would be that the poles are moving and you are not. Actually, all you can justifiably assert is the *relative* motion of the poles and yourself. And for further confirmation, think about all those films showing the inside of a railway compartment travelling at high speed – **think about how the illusion of motion is actually produced** (Remember: bold type means work for you – page 5).

From this sort of experience we've learnt that watching objects around us merely gives information about *relative* motion. It can never tell us whether they are moving or we are (or both).

1.2 Well then, would some test done *inside* the train tell us whether it is moving or not? For the present let's stick to trains in steady motion – at constant speed on a straight bit of track. I suggest that everything in this moving train happens in exactly the same way as if it were stationary. For example, the steady motion of the train doesn't make any difference to the problem of keeping your balance. You stand upright just as you would in a stationary train. **What happens if you drop something?**

It falls straight down *as you see it,* and lands right beside your feet – just as it would do if the train were motionless. **See if you can think of any experiment performed inside the train that would give different results when moving and when stationary.** Nobody has succeeded up till now.

Anyway, as you've been longing to point out, I've only been discussing motion relative to the Earth's surface. The Earth itself is moving far more rapidly relative to the Sun; and our experiments show no sign of that – nor of the Sun's own motion. It looks as if steady motion just cannot be detected.

1.3 Of course it's different if the motion is not a steady one – if the train is speeding up or slowing down, for example. To balance in a train that is braking hard, you must lean backwards. Anything you drop falls a little behind in a train that's speeding up, a little in front in one that's slowing down. There are, in fact, many experiments for distinguishing accelerated motion, like speeding up or slowing down, from steady motion. But I return to the point that, provided we stick to steady motion, *we do not know of any way of distinguishing between moving and stationary.* So it doesn't make sense to say that this thing is 'really' moving and that one is 'really' stationary. It doesn't make sense, because there is no way of testing whether it is true or not. All we can say is that one thing is moving or stationary *relative to another.*

1.4 A nineteenth-century scientist would have accepted those arguments, but with reservations. 'This applies to mechanical experiments,' he would have said, 'and I agree that *they* will never reveal whether you are really moving or not.' 'But,' he would have alleged, 'there are other types of experiments – particularly optical ones – that *will* do this.' He was wrong in saying so, but we shall understand the basis of Relativity better if we take a little while to discover what he actually thought and why he was wrong. And so we must study some properties of light.

   Light, as I'm sure you know, has some of the properties of a wave motion. Notice that I'm careful not to say that light *is* a wave motion – you'll soon know why. Light has *some of the properties* of wave motion – in the sense that you can solve problems about how it travels by treating it *as if* it consisted of waves, which (without prejudicing ourselves about the extent to which they actually exist) we can call *light waves.*

1.5 Any set of waves has a number attached to it, called its *frequency* – which is defined as the number of waves that pass a given point in one second.

   The frequency of light waves is associated with our sensation of colour. Low frequency gives red light and high frequency gives violet. And in between are the familiar colours of the rainbow.

   Of course, the colours are just our subjective response to the light. It's the frequency itself that is physically important.

1.6 But does it all stop with red light at the low frequency end and violet at the high? Certainly not. There are other radiations of the same general nature as light, having the same wave-like properties, but with higher or lower frequencies. It just happens that the eye is sensitive to a small range of frequencies, which we therefore single out for the name 'light'.

If we work up to higher frequencies beyond the violet, we come first to ultra-violet rays (which put the burn in sunburn), then X-rays and finally gamma-rays (given out, for example, by radioactive substances). And at lower frequencies than red we find infra-red (radiant heat), then microwaves and beyond that radio. In Relativity Theory the distinction between visible light, X-rays, radio and the rest has no significance. So to save words we adopt the convention of using 'light' for all or any of them, without distinction.

1.7 One very important point is that *in a vacuum* all these radiations travel at exactly the same speed, no matter what their frequency. By the convention of §1.6 we speak of this speed as 'the speed of light', even if we happen to be talking about radio or X-rays. It is very close to 300 000 kilometres per second or roughly 186 000 miles per second (300 000 is the accepted way nowadays of writing three hundred thousand – no commas, please).

Let me stress: this refers to the speed of light *when passing through a vacuum.* In a material medium (like air or glass) the speed is lower and varies with frequency. But we shall seldom be concerned with that. Unless I say otherwise, 'speed of light' always means its speed in vacuum – which does not vary.

1.8 It's often convenient to think of this enormous speed as 300 metres (984 feet) per *micro*second (millionth of a second). It is also 1080 million kilometres per hour (670 million miles per hour), so that a jet plane does about a millionth of the speed of light. Another useful comparison is that the Earth goes round the Sun at just under a ten-thousandth of the speed of light.

Finally the speed of light can also be expressed as about 9·5 million million kilometres per year. And this provides a convenient unit when we have to deal with very large distances: the *light year,* which is defined as the distance that light travels in one year. Think of it as nearly 10 million million kilometres (6 million million miles).

1.9 When we said in §1.7 that the speed of light is always the same, we were naturally thinking of light coming from a stationary source – stationary relative to whoever measures the speed. But suppose the source is moving. Will that effect the speed of light?

Well we've said that light behaves in many ways as if it were wave motion. And both theory and experiment (try it in your bath) show that for other wave motions the movement of the source does *not* affect the speed of the waves. So we might expect the same to be true of light. But we must not rely on arguing by analogy; we must test the matter experimentally.

One test a few years ago made use of certain sub-atomic particles (bits of matter smaller than atoms) called neutral pions, which are produced in violent collisions between other particles that have been speeded up in an accelerator ('atom smasher'). The neutral pion is a very short-lived particle. On average it lasts about a five-thousandth of a millionth of a millionth of a second. As it dies, it gives out two spurts of light – actually gamma-rays (§1.6).

The experiment was very simple in principle – merely to measure the speed of gamma-rays coming from the death of pions moving at 0·999 75 of the speed of light. The conditions were such that if even a ten-thousandth of the speed of the source had been added to the speed of the gamma-rays, the effect would have shown up. Yet there was no detectable difference. The conclusion is clear:

> The speed of light is not affected by the motion of the source.

1.10 What we've said about light so far is completely non-controversial – I mean as regards the controversy between Relativity and the older view of things. The men of the nineteenth century would have agreed with it all. But now comes the crucial point that distinguishes the modern relativistic outlook from all that preceded it. This arises when we ask another question, apparently as innocent as the last, namely: *How does the speed of light depend on the motion of the person who is measuring it?*

Thinking of light in terms of wave motion (§1.4) had suggested that the speed of light is unaffected by the motion of the source (§1.9) – a fruitful suggestion, since experimental test showed it to be correct. And in many other instances, some of them far more subtle, the wave-motion analogy has suggested the right answer. But analogies are untrustworthy: they can give the right answer time after time, and then – without warning – let you down. That's what happened in this case.

Nineteenth-century physicists, having got the right answers time and again from the wave theory of light, were naturally not as cautious as we know we have to be – after the event. They didn't carefully say 'Light has some of the properties of a wave motion.' They just said 'Light *is* a wave motion.'

1.11 But waves in what? Sea waves are waves in water. Sound is waves in air. But what is light waves in? Certainly not in any known material like water or air – for light travels through a vacuum. It must (they concluded) be waves in some subtler medium, which they agreed to call the *aether*.

They were only *assuming* the existence of this aether, of course – on the basis that there had to be *some* medium to carry the light waves. Their assumption was a fruitful one in that it led to theories that united the sciences of optics, electricity and magnetism into a single integrated whole, and to such practical results as radio.

But it also led to trouble.

1.12 The speed of light, we've said, is always the same. It doesn't vary with frequency (§1.7), nor with the motion of the source (§1.9).

Always the same relative to what? The waves-in-aether theory gave an obvious answer. In other wave motions the speed of the waves is constant relative to the medium in which they are travelling – think about waves on the moving surface of a river. Theory confirms that this should be true of all wave motions. So naturally nineteenth-century physicists went on to deduce that the speed of light is constant *relative to the aether* – a very plausible argument surely.

1.13 Now back to the question of §1.10: How does the speed of light depend on the motion of the observer? Assuming the conclusion of §1.12 to be correct, **how would you answer that question?**

In Figure 1.13 let the paper represent the aether. The thin arrows represent light travelling across this paper at 300 000 kilometres per second in any direction. If observer A (represented by a dot) is stationary on the paper, then no matter what direction the light is coming from, he will measure its speed as 300 000 kilometres per second.

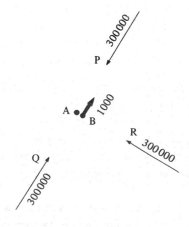

Figure 1.13

Observer B is moving at 1000 kilometres per second in the direction of the broad arrow. If he measures the speed (relative to himself) of light coming from direction P, he will get the answer 300 000 + 1000 = 301 000 kilometres per second. And for light from direction Q he will get 299 000 kilometres per second.

The interesting thing is that this gives a recipe for deciding whether you are moving through the aether or not. Measure the speed of light in various directions. If it is always the same, you are stationary in the aether. If it varies with direction, you are moving through it.

We could use this idea to measure the Earth's motion through the aether. Is it just the annual motion round the Sun? Or is there a further contribution from the motion of the Sun itself? That was the point of a famous experiment carried out by A. A. Michelson (in 1881) and repeated with more refinement in collaboration with E. W. Morley (in 1887).

1.14 The experiment cannot be done in that simple form – with the light coming from an outside source. I shall give only an outline description of the more complicated experiment that *was* done.

The apparatus (Figure 1.14) consists essentially of a light source S and two mirrors, M_1 and M_2, placed at equal distances from S and arranged to reflect the light back to S. The line from S to mirror M_1 lies in the direction of the Earth's motion, and SM_2 is perpendicular (at right angles) to that direction. The idea is to compare the times of the light's out-and-back journeys *in* the direction of the Earth's motion and

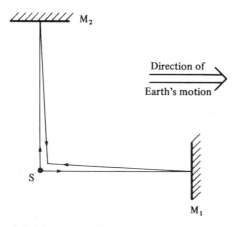

Figure 1.14

across it. The fact that the source is moving doesn't matter (§1.9). **But have you another worry?**

1.15 You may feel that any expected difference in the times would be cancelled out by the compensatory effect of outward and return journeys. But an analogy will show that this is not so.

Consider a 40-metre square raft (Figure 1.15) being towed at 3 metres per second through a calm, motionless sea (which it does not disturb). The raft corresponds to Michelson's apparatus; the sea to the aether. And the light beams are replaced by two well-trained dolphins, one of which swims from P to Q and back in and against the direction of the raft's motion, the other from P to R and back across the direction of motion. Each must keep close to its side of the raft, so that as seen by an observer on the raft it simply swims from one corner to another and back – exactly like the light beams of §1.14. They both swim at 5 metres per second relative to the water – corresponding to the constant speed of light through the aether.

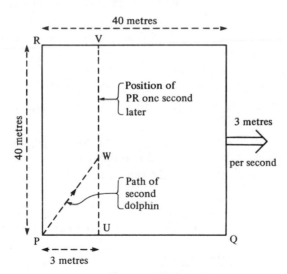

Figure 1.15

1.16 **Can you calculate the time of the journey from P to Q and back?**

In one second the dolphin swims 5 metres; but on the outward leg the raft goes 3 metres in the same direction. So, as seen by somebody on the raft, the swimmer has only covered 2 metres along the edge, and its effective speed relative to the raft is therefore $5 - 3 = 2$ metres per second. Thus it takes 20 seconds to get from P to Q. **You can check** that it does the return journey at $5 + 3 = 8$ metres per second relative to the raft and takes 5 seconds – giving a total time of 25 seconds for the round trip.

1.17 The calculation for the crosswise journey is more complicated. Consider what happens in the first second. In that time the edge moves to position UV, 3 metres to the right (this part of the figure is not to scale). To keep alongside the raft, the dolphin has to go diagonally along PW. And so the distance from P to W is 5 metres (1 second's swim at 5 metres per second).

So we know three things about the triangle PUW: (1) the angle at U is a right angle; (2) the side PU is 3 metres long; and (3) the side PW is 5 metres long. A piece of geometry which I expect you know (and if you don't, check my statement by a scale drawing) enables us to deduce that the side UW is 4 metres long.

So in this one second the dolphin has covered an effective distance of 4 metres along the edge of the raft. Clearly the same reasoning applies to each subsequent second, and so its effective speed relative to the edge of the raft is 4 metres per second. Thus the total journey of 80 metres takes 20 seconds.

The upshot is that – in spite of initial suspicions about cancelling effects – the double journey across the raft's direction of motion takes *less* time than the journey in and against the direction of motion.

1.18 To make sure you've followed the reasoning of §§1.16–7, do a similar calculation for the case in which the raft is 36 metres square and is towed at 4 metres per second, while the dolphins still cover 5 metres per second. Or, if you're O.K. in algebra, work out the general case, using c for the speed of light, v for the Earth's speed through the aether and l for the distance from light source to mirror. (Answers at end of chapter.)

1.19 According to nineteenth-century ideas, these calculations should apply to the Michelson–Morley experiment (§1.14) – except that different numbers would be involved, of course. The light should take longer for the journey SM_1S (Figure 1.14) than for SM_2S. The expected difference was *exceedingly* small. But if it had been even 100 times smaller, the apparatus could still have detected it. Yet when the experiment was performed, *no time difference at all was detected*. (Nor was this a freak result. The experiment was repeated several times down to 1930, ever more accurately, with the same negative outcome. An experiment with lasers in 1972 confirmed the result with yet more precision.)

To Michelson, hoping to measure the Earth's motion through the aether, this was a terrible disappointment. To others it was an absurdity to explain away. To Einstein it was the chance to create a new world outlook.

1.20 It may have occurred to you that if the Earth actually dragged a body of aether along with it (instead of just pushing its way through), then the Michelson–Morley result is exactly what we should expect (because the apparatus would be stationary in the *local* aether). However there is good evidence (chiefly from a phenomenon called the Aberration of Light) that this does not happen. To explain would involve developing a theory that we should never use again. So I'm going to ask you to take it on trust. Anyway, the nineteenth-century scientists agreed about this point.

1.21 Now recall that all this arises from the assumption that the speed of light is constant relative to the aether (§1.12) and the deduction that its speed as measured by any observer should depend on his own motion (§1.13). The Michelson–Morley experiment was supposed to detect this effect. And it didn't. Experiment showed that the light took the same time whether travelling *in* the direction of the Earth's motion or *across* it. **What does this seem to imply about the speed of light in these directions?**

The distance travelled is the same for both journeys (length SM_1 = length SM_2). So if the light takes the same time, it travels at the same speed. We seem to be reaching the conclusion that the speed of light (relative to the Earth) is the same in the direction of the Earth's motion as across it. Surprising! **Do think it over and consider how it relates to the line of thought we've been developing since §1.12.**

You'll have spotted that it directly contradicts the reasoning of §1.13, where we seemed to prove that one's measurement of the speed of light must depend on one's motion through the aether. It's as if the speed of both dolphins relative to the raft (§§1.15–17) remained 5 metres per second even with the raft moving. It does seem absurd, doesn't it? **Keep on thinking hard.**

The motion of the Earth, this experiment seems to be saying, has no effect on the speed of light relative to an observer on Earth. But there's nothing special about the Earth, surely. So we can generalise and say that the speed of light does *not* depend on the motion of the observer – *any* observer.

But how can one reconcile this with the reasoning of §1.13, which seems to prove that it does? The world of science was worried. All sorts of tricks were tried for getting round the difficulty. None of them worked. You can't get round it.

1.22 Eventually it was Einstein who took the bold step of saying: Let's stop trying to circumvent this difficulty; let's accept the experimental result at its face value, and see where it takes us. In other words, he proposed to work out a new theory based on the assumption that

> The speed of light (coming from any direction) relative to any observer (no matter how he is moving) is always the same.

(Remember that we're talking about the speed of light in a vacuum, and sticking to steady motion.) We can put that more concisely as

> The speed of light is *invariant*

– meaning that it stays the same, even when we change from one observer to another.

(Einstein was aware that nineteenth-century theories were in trouble in other ways. So he actually put his revolutionary proposal in a more general form that we'll come to later.)

1.23 I wonder if you've fully realised how this conflicts with our previously held notions about speeds. It makes the speed of light different from every other speed. If I am standing beside a railway and estimate the speed of a train as 80 kilometres per hour, and if your car passes me at 50 kilometres per hour in the same direction, then you will reckon the train's speed as 30 kilometres per hour relative to you. The speed of the train does depend very much on the motion of whoever is measuring it.

With everything that had been observed up till the 1880s the speed relative to an observer depended on the observer's motion. Naturally it was assumed that the same would be true of light. But when the experiment became possible, it was found *not* to be true of light. If we substitute light for that train, and if I measure its speed as 300 000 kilometres per second and you pass me at 100 000 kilometres per second, you too will find that this same light is going at 300 000 kilometres per second relative to you.

The speed of light is unique. Other speeds are relative to the observer who measures them. The speed of light is not relative to the observer. It is an absolute. That is the very startling idea that Einstein put forward. He might well have called his theory the 'Theory of the *Non*-Relativity of Light'.

1.24 You may find that idea difficult to accept, even unbelievable – because it conflicts with 'obvious commonsense'. You may feel like retorting 'I can't believe that the speed of light is the same for all observers, irrespective of how they are moving. And so I can't accept any theory based on that idea.'

Rest assured! You are not asked to believe this – or anything else – until you are convinced by evidence. Science does not work through acts of faith. The Michelson – Morley experiment certainly *suggests* that the speed of light is invariant. But it's very far from *proving* the point. So you're not being asked to believe this odd idea. But you are being asked to *take seriously the possibility that it might be so*, and to join with me in working out what *would* follow *if* it were so.

That's the job we start on now – working out the consequences of this seemingly absurd notion about the speed of light. Eventually we shall come to some consequences that can be tested experimentally. If the experimental results agree with the speed-of-light-the-same-for-all theory and disagree with theories based on earlier beliefs, then perhaps *your* beliefs will begin to change. This is standard scientific method – the normal procedure by which some theories come to be accepted, when so many are rejected.

But until that point is reached we're only doing a tentative exploration of what follows from this strange new idea. And we can be content with somewhat rough and ready methods, sacrificing thoroughness to simplicity. If experiment eventually confirms that the revolutionary idea about light is a good one, we can go back and refine our approach as necessary.

Answers to §1.18. The times are 40 seconds and 24 seconds. For the general case they are $2cl/(c^2 - v^2)$ and $2l/\sqrt{(c^2 - v^2)}$.

2

Rough and ready Relativity

2.1 I want now to show that *if* it's true that the speed of light is the same for all observers in all directions, *then* the world must be surprisingly different from what we've thought up till now. It's easy, for example, to deduce that

> It is impossible for an observer to move *at* the speed of light (relative to you, for example).

Take a light flash travelling in the same direction as this observer. If he were moving at the speed of light, he'd be travelling with this light flash – keeping alongside it, as it were. So its speed relative to him would be zero, which contradicts §1.22. To avoid the contradiction we are forced to conclude that he can't move with the speed of light.

2.2 Many parts of the theory can be worked out by reasoning that's not much more complicated than that. Consider the case of two observers called A and B observing two events P and Q. Observer A says that P and Q happened simultaneously (at the same time). Assuming that both observers are honest and accurate, **what will B say?**

We've always regarded it as obvious that two events either 'really' happen at the same time or 'really' happen at different times. There can be no difference of opinion on such a question. So B's answer *must* agree with A's. That's what we've believed till now. Well let's see.

To reduce the influence of past habits of thought, let's have it all happening in outer space in total darkness. The events can be explosions, and so the evidence is concerned with when A and B *saw* the light flashes coming from them.

2.3 Observer A has two facts to work from:

(1) The explosions happened at equal distances from him, but in opposite directions; and

(2) He *saw* them both at the same time – that is, the light flashes from the explosions reached him at the same time.

Try to follow my reasoning without a diagram. If that proves too difficult, Figure 2.5 will eventually come to the rescue. We are working on the assumption (§1.22) that the speed of light is invariant. So in A's view the two flashes travelled from the explosions to him at the same speed. As they had the same distance to go – fact (1) – they took the same time. And as they arrived together – fact (2) – they must have set out at the same instant. So we've deduced that

> According to A the explosions happened at the same time.

2.4 How (you may have asked) could A *know* that the explosions happened at equal distances from him? The answer is that he had a very long measuring rod, which stretched on either side of him to and beyond the explosions. These left burn marks on it, and A was able to verify later that the burns were equidistant from where he had been holding it.

2.5 This story is shown in Figure 2.5. In line (i) the 'stars' marked P and Q represent the explosions caught in the act of happening – at equal distances from observer A (represented by a dot). The thin arrows indicate the flashes setting out towards A. What line (i) shows is partly A's observation (1) of §2.3 and partly his conclusion that the explosions were simultaneous. Line (ii) shows A's second observation that the flashes reached him at the same time. If you had trouble with §2.3 run through it again now.

Figure 2.5

2.6 Just as the light flashes reach him, A passes (or is passed by) B. Note that I am not saying whether B is moving or not – just as I didn't say whether A is moving or not. If we insist – as we shall do – on steady motion only, then it doesn't make sense to say that either of them is, or is not, moving (§1.3). All we can do is to specify their *relative* motion.

So we'll say that B is moving, relative to A, in the same direction as the light from explosion P. In other words, if A thinks of himself as stationary, he will say that B, as he passes, is moving in the same direction as the light from P. (**How will B describe this relative motion?**)

Figure 2.6(ii) is simply Figure 2.5(ii) with the addition of B, who is shown passing A as the flashes arrive. Since the figure is drawn *as if* A were stationary, it must show B moving in the direction specified above. This is indicated by the thick arrow.

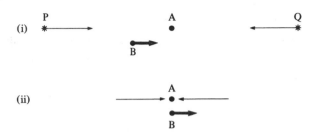

Figure 2.6

I have now given you all the available information about B. Bearing in mind the assumption of §1.22, **try to work out whether** B **will think the explosions happened simultaneously or not.**

2.7 Obviously B will also *see* the explosions at the same time. In A's view the two flashes had to travel the same distance to reach him. **Does B think the same?**

No. Since B is beside A when they both see the explosions – Figure 2.6(ii) – and is travelling towards the right in the diagram, it follows that when the explosions happened, B was nearer to P than to Q – as shown in Figure 2.6(i). So in B's opinion explosion Q happened further away from him than did explosion P. (If you feel doubtful, imagine B carrying a measuring rod like the one we gave A in §2.4. You'll easily convince yourself that the Q burn mark would be further than the P one from where he was holding the rod. So in his opinion . . .)

Now can you work out what B **thinks about this simultaneity question?**

The speed of light is the same in both directions according to B also – that is the revolutionary idea we're following up (§1.22). So to reach B the flash from Q had to travel farther than the one from P, and at the same speed. It therefore took longer. Since the flashes reached B together, the one from Q must have started earlier. All this is B's train of reasoning, and so we conclude that

According to B, explosion Q happened before explosion P.

2.8 In spite of what we worked out in §2.7, Figure 2.6 persists in telling us that the explosions occurred simultaneously. From that figure it also *looks as if* the flashes P and Q have the same distance to travel to reach either A or B (contradicting §2.7); the only difference seems to be that B *has to move* to reach the position where they pass, while A *is there already*. **Comment on that last please.**

We know (§1.3) that it's nonsense to talk about B having to *move* to reach the place where he passes A. Obviously the diagram is misleading us. The trouble is that it is drawn with A stationary and B moving, and is therefore biassed in favour of A's point of view.

Let's redress the balance by putting B's point of view in Figure 2.8. Here B is shown stationary, with A moving from right to left – so that the *relative* motion is the same as before. As B thinks of it, there are three instants to be considered. At stage (i) explosion Q occurs and its light flash starts off towards B. At (ii), P happens, nearer to B than Q was. And with the light from P starting later than from Q, but having a correspondingly shorter distance to travel at the same speed, the two reach B together – as shown in (iii). The diagram merely presents visually what we already know about B from §§2.6–7.

Though this diagram shows B's point of view, we can still use it to reason about A's opinion. The distance of Q from A in (i) is the same as the distance of P from A in (ii). So we can again use the argument of §2.3 to show that the explosions are simultaneous according to A – even if Figure 2.8 seems to say they're not.

Figures 2.6 and 2.8 are like two perspective views of the same scene. A perspective drawing only shows how things look from one point of view. Taken naively, it tells downright lies – the far side of the lake is higher than the near side! To extract sound information, you must know the rules and use them. Similarly, Figures 2.6 and 2.8 tell us lies

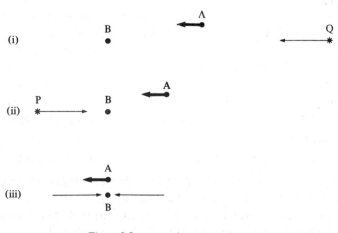

Figure 2.8

if we use them unintelligently. Logical thinking is needed to extract the truth from them.

Please make sure now that you've mastered the reasoning in §§2.3–8 –even if you still feel that it leads to impossible conclusions.

2.9 Let's sum up. We've deduced that these two explosions happened at the same time in A's opinion, but at different times in B's. You probably feel that one of them must be wrong. But there are plenty of questions on which people can have different opinions and yet all be right–people in different parts of the world rightly disagree about when noon occurs. I now suggest that it's the same in this case: A and B can disagree and yet both be right. The only important difference between them is in their motion. So I should want to draw the conclusion that:

> There is no unique correct answer to the question of whether two events happen at the same time or not. It depends on the motion of the observer. Observers moving relative to each other will give different answers, and yet be equally correct.

This conclusion is called the *Relativity of Simultaneity*.

2.10 You'll have noticed that our argument depends in a crucial way on the assumption that the speed of light is invariant (§1.22). If we assume instead that the speed of light is constant relative to the (supposed) aether, we end up with the conclusion that the old beliefs are true after all.

Take the case in which B is stationary in the aether (use Figure 2.8). Then A is moving through it. So arguing on the lines of §1.13, we see that according to A the speed of light coming from Q would be less than that from P. Then we should reason (in place of §2.3) that since the light has the same distance to cover from Q as from P, but at a slower speed, therefore . . . **You complete the argument to show that on the aether theory A agrees with** B.

So if you find the Relativity of Simultaneity hard to stomach, that's because your intellect may have agreed to accept the invariance of the speed of light as a basis for argument, but your intuition refuses to feel that way. Well, intuition must be taught its place. It's often good at suggesting ideas. But when logic and intuition conflict, it's logic (at least in this field) that we must trust.

2.11 **Why have we never noticed disagreements about simultaneity?**

Because at ordinary speeds the difference would be too small to detect. Between stages (i) and (ii) (Figure 2.6) B could only move a very short distance relative to A. So the light from Q has (in his view) only to go a tiny distance farther than that from P. And he infers that Q happens only a tiny, tiny fraction of a second before P – too small a time to notice.

2.12 Almost certainly you feel that this Relativity of Simultaneity business is all wrong. And I'm going to ask you in a moment to try to find faults in the arguments I've presented. You'll learn an enormous amount that way.

We all hold various beliefs which seem so obvious that we never question them. To examine them critically we must first bring them up to the level of consciousness. And one of the best ways of doing so is to argue very hard about such things as the Relativity of Simultaneity.

So now will you please make a sustained effort to find faults in §§2.2–9. When you think you've found one, don't just leave it hovering in your head. Try to formulate it clearly (in writing?) – for you'd be surprised how many seemingly inspired ideas turn out to be empty when you get down to the job of stating them precisely. One important hint: Make sure your arguments work equally well whether you refer to Figure 2.6 or Figure 2.8 – otherwise you'll be favouring one observer or the other.

When you feel you've got your objections clearly formulated, the next thing is to look for faults in your own fault-finding. You'll soon throw out some of your original ideas. And when you've got down to a hard core of seemingly unanswerable objections, only then have a look at §§2.13–18. And have another spell of thinking up objections between each article and the next. In any case, please study those articles carefully. If some of these worries have not occurred to you yet, they're going to give trouble later on. So be prepared to come back.

2.13 A common line of argument runs like this: 'A is midway between the explosions when the light flashes arrive. But B is right beside A at that moment, and so he is also halfway between the explosions . . .' and you can quickly complete an alleged proof that B agrees with A. **But do you see anything suspicious about the first sentence inside the quotation marks?**

The explosions happened at stage (i), and certainly A was halfway between them then. But when B passes A, the explosions are already part of the past. Now it's meaningless to say that A is halfway between two events that are over and done with – you'd laugh if I claimed at this moment to be halfway between the signing of the Magna Carta and the first H-bomb test. And if your reasoning is based on a meaningless statement, you can hardly expect anybody to accept it.

If you now want to modify your argument to suggest that at stage (ii) of Figure 2.6 A is halfway between 'the places where the explosions happened', would you please wait till you've studied §2.16, where I deal with this notion in another context.

2.14 People often argue that although B will disagree with A, nevertheless he can calculate how it will appear to A and make the necessary corrections. But equally, A can calculate how it will appear to B, and correct *his* ideas accordingly. Which is to make corrections to agree with which?

Of course there are plenty of cases where an observer is working in non-standard conditions and must make appropriate allowances (think of a round-the-world traveller calculating the date by counting sunrises). **So which of our pair is right and which must do the correcting?**

2.15 Maybe you suggest that the crucial question is whether this or that observer is moving or not. If one is stationary, his answer is correct and the other must make allowances. But we know from §1.3 that we can't just speak of an observer 'really' moving or not. We must specify motion relative to something. **Relative to what?**

Relative to the explosions, you perhaps suggest. **So what can we say about A's or B's motion relative to these?**

Perhaps you draw my attention to Figure 2.6, saying that it provides a good reason for preferring A's opinion. For it shows A stationary relative to the stars representing the explosions and B moving relative to them. But against that, consider Figure 2.8, which shows B stationary and A moving, relative to these stars. The question of which is moving and which is stationary seems to depend on which picture you use. Odd! **Are you not a bit suspicious of the idea of motion relative to the explosions? How would you detect this motion?**

Obviously, although I've spoken of 'explosions', we are really interested in ideal *events*, P and Q, which are instantaneous – occupy no time. **Now does it make sense to say that an observer is (or is not) moving relative to an instantaneous event?**

To decide whether one thing is moving relative to another, you must measure the distance between them on two separate occasions, and discover whether it changes or not. But the event only occurs *once* and so you can't measure the distance *twice*. Thus you can't decide whether the observer is or is not moving relative to the event. And if you can't decide one way or the other, then it's meaningless to say that he is doing so and equally meaningless to say that he is not. The words 'This observer is (or is not) moving relative to that event' take the *form* of a sentence, but actually they carry no meaning. And now we're back to the point made in §2.13 that we can place no faith in an argument that starts from a meaningless statement.

(Maybe you object that a real event is not instantaneous – it occupies a small interval of time. But likewise we cannot detect time-intervals shorter than a certain minimum. If the time-interval occupied by the event is less than the minimum we can detect, the practical effect is as if it were instantaneous – because we couldn't make the two distinct distance measurements – and you'll find that the above argument still holds good. Our explosions merely have to be very short and sharp.)

2.16 So motion relative to an event is out. But, you ask, couldn't we decide which observer is right and which must make corrections in terms of whether they are stationary or moving relative to *the place where an event* (say Q) *happened?*

But what is this 'place where an event happened'? **How would you identify it at some later time?**

Could you mark it with a flag, for example? In everyday life it seems easy. The place where Joe Smith was knocked down is taken to be the bit of street that was right below him at the moment of the accident. Ten minutes later I can still identify that spot and estimate that it was 30 metres from where I was standing. So I report to the police that 'Joe was hit 30 metres from here.'

But an observer who lived on the Sun would not agree. He says that during the 10 minutes since the accident I have been moving away from 'the place where it happened', carried along by the Earth's motion at about 30 kilometres per second (§1.8). So he asserts that I'm now nearly 18 000 kilometres from where Joe was knocked down.

You see, 'the place where the event happened' is a useful phrase if we all agree to use the same frame of reference – e.g., the Earth's surface. But as soon as we get away from such parochial conventions, we realise that there is no such thing as 'the place where the event happened' – there are no 'places' in empty space. And so there's no hope of establishing one observer right and the other wrong in terms of motion relative to 'the place where the event happened'.

Do, please, study this article carefully and think up more examples for yourself (e.g., in trains or ships). The place-where-the-event-happened mirage keeps on recurring. I'd give five to one that it will cause you trouble later. So do your best now to get it straight.

2.17 Articles 2.13, 2.15 and 2.16 exemplify something that we've got to watch all the time. We must keep examining the words we use to see if they refer to something real or are just meaningless phrases – like 'the place where the event happened'. To decide the point you usually have to ask how you would observe the thing that the words are alleged to refer to. If you can specify a clear procedure for making this observation, then that procedure defines the meaning of the words. (You will not insist that you could always carry out the procedure in practice, so long as it can clearly be followed in principle.) If you fail to specify such a procedure, the words are meaningless. And any reasoning that uses them is worthless.

That's the point of §2.4: to ensure that (1) of §2.3 is meaningful. And if you doubt my assertion that the 'halfway between' statement in §2.13 is meaningless, try to work out how you would test experimentally whether it is true or false.

2.18 Here's another argument. The events P and Q can't just happen in empty space. There has to be something to which they happen. Let's call it 'the vehicle' – I often think of P and Q as explosions aboard spaceships. It *is* meaningful, you say, to speak of A moving relative to the vehicle – because you could measure the distance between them on two occasions, and check that it changes. I agree. It *is* meaningful. **But is it relevant?**

No. Think of two spaceships, C and D (Figure 2.18) – C moving towards A, D away from him. As they pass, they both suffer explosions – either of which could be event Q. By §1.9 the flashes will travel at the same speed and – having covered the same distance – will reach A together. So nothing in the reasoning of §§2.3–7 is affected by the motion of either vehicle.

Figure 2.18

2.19 I've now dealt with the arguments and difficulties that most people raise. If you have a doubt that I haven't answered, please try to cross-examine yourself; almost certainly you will find that your trouble arises from modes of thought derived from our Earth-bound experience, for example our habit of taking the Earth's surface as frame of reference. If you still feel at the end that you've found an insuperable objection to my reasoning, I beg you to return to the subject from time to time and in the meanwhile to suspend judgement – but carry on with what follows as if I had proved my point.

The conclusion that *I* should like to draw from all this argument and counterargument is the one I put forward at the end of §2.9. More briefly:

> Observers in relative motion disagree about simultaneity.

2.20 Honest and accurate observers cannot disagree about reality – for the simple reason that the only sensible meaning for the word 'reality' (in a scientific context) is 'that which all honest and accurate observers agree about'.

So A and B cannot be disagreeing about reality, but only about some aspect of reality that appears differently in each observer's view of things. Thus we're driven to the conclusion that the notion of two events being simultaneous is not concerned with reality. It is only concerned with this or that observer's *view* of reality – with how reality *appears* to him.

An analogy may help. The statement 'The chimney of the house across the street is directly behind a telephone pole' is true for me, where I'm sitting now. It is false for my wife, a couple of metres away. And we are not surprised at this disagreement. For we do not believe that the statement 'The chimney is directly behind the telephone pole' is either true or false in itself. To make it true or false – and therefore meaningful – you have to add 'from so-and-so's point of view'.

And I'm suggesting that, in the same way, simultaneity is not concerned with a real relation between two events, but only with an observer's view of that relation. The bare statement that 'these two events happened at the same time' is meaningless. If one adds 'according to this observer', it becomes meaningful; but it is true for one observer and false for another.

2.21 Now I'm sure you won't find that convincing – yet! All your life you've been used to thinking that two events are either 'really' simultaneous or 'really' not simultaneous. And that belief has never landed you in trouble. So it's difficult to persuade yourself that you could perhaps be taking too narrow a view of things.

This is where we come up against those problems of imagination that we discussed in the Introduction. Here is the first important case where your logic has told you pretty firmly something that your imagination refuses to accept. And please don't think that I can suddenly make it easier for you to imagine things in a new way. You'll have to be as flexible as you can; and I'll keep prodding you in the right direction. Eventually your imagination will learn to cope better. Meanwhile let logic be your guide.

The basic trouble is the narrowness of our experience. Putting quantities into the reasoning of §2.11 (try this yourself if your mathematics is moderately good) we can easily calculate that at real life speeds explosion Q would, in B's opinion, precede P by something like a millionth of a millionth of a second – or less. No wonder we don't notice the discrepancy! But let us beware of assuming that what is true (for practical purposes) in our normal conditions must also be true in all circumstances. I think it was Eddington who remarked that a traveller who moved around the U.S.A. entirely by train would think that there is always a bell ringing at every level crossing. Our situation is rather like that. Restricted as we are to very slow relative motion, we never notice disagreements about simultaneity. And like Eddington's traveller we tend to conclude wrongly that our restricted experience has universal validity.

2.22 Now consider a rather different simultaneity problem. We'll have explosions P and Q as before. And observer C is related to them exactly as A was in §2.3. So by the same reasoning he says that the explosions happened at the same time. *At the moment when he says they happened* he is passed by observer D moving (relative to him) *at right angles* to the direction PQ. **What will D say?**

Figure 2.22(i) shows the situation as C and D pass (you must imagine D passing right over C, instead of just below him as drawn). By the time the light flashes reach C, D has moved (relative to him) to the position shown in line (ii). The light that reaches D has travelled in a different direction from that going to C – hence the sloping arrows. But it's clear from the symmetry of the situation that the two flashes will reach D at the same time; also that they had, in his opinion, the same distance to travel at the same speed (§1.22). Then the reasoning goes exactly as in §2.3 (**check the details**) to prove that in this case D *agrees* that the explosions were simultaneous.

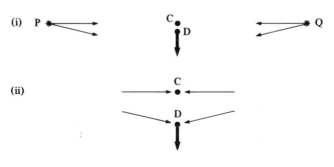

Figure 2.22

2.23 Turning to another topic, you probably feel that the length of a rod is a perfectly definite thing, about which – surely! – all observers will agree, no matter how they are moving. Don't be too sure.

First we must be clear what we mean by the length of a rod: we must specify the procedure for measuring it. Take (for simplicity) a rod less than a metre long; and consider how you would measure its length with the help of a metre rule, divided as usual into millimetres. If the rod is stationary relative to you and lying alongside the metre rule, you note that its ends lie opposite (say) the 23 millimetre and the 435 millimetre marks. And you conclude that the rod is $435 - 23 = 412$ millimetres long. Simple! But if the rod is moving – sliding past you lengthwise – **is this procedure adequate?**

Indeed not. One vital qualification must be added: you must observe which millimetre marks are opposite its ends *at the same time*. If you don't insist on that, the rod moves during the measuring process, and you get a false length measurement. **And now you see the trouble?**

Since observers disagree about 'at the same time', they are liable to disagree also about the lengths of rods.

2.24 To get a bit more insight, let's provide A and B with rods (which they carry with them). We'll define the length of each observer's rod by saying that one end of it is to be right beside explosion P when it occurs and the other right beside Q when *it* occurs.

Figure 2.24 is got by adding these rods to Figure 2.8 (continuous line for A's, broken for B's). In this picture, representing B's point of view, B and his rod are shown stationary – the same at each stage. And, as specified, the rod stretches from P at one end to Q at the other.

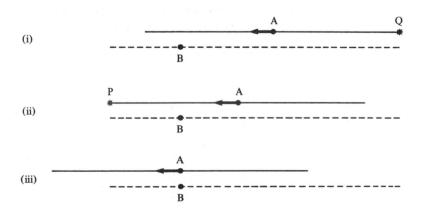

Figure 2.24

On the other hand, A is moving relative to B and carrying his rod with him – so that their positions shift between stages (i), (ii) and (iii). But, still as specified, one end of A's rod is beside Q when it happens and the other beside P when it happens.

Line (i) shows us the position of the rods at the same time – at the time of event Q – *according to* B. So it allows us to infer B's opinions about lengths. And it clearly shows B's rod covering the whole of A's and a bit more. Thus we conclude that:

According to B, A's rod is shorter than his own.

2.25 **But how does it seem to A?**

To get A's point of view we use Figure 2.25, showing the rods added to Figure 2.6. As before, each rod has one end beside P and the other beside Q, as they happen. But P and Q are simultaneous according to A. In his opinion, therefore, the two ends of B's rod coincide with the ends of his own *at the same time*. And so

According to A, the rods are the same length.

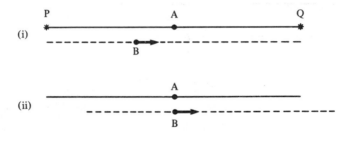

Figure 2.25

2.26 Comparing these conclusions, we see that

If the observers carry measuring rods extended in the direction of their relative motion, they will disagree about the comparative lengths of these rods.

We can summarise the disagreement like this: A says 'Of course they're the same length – the left ends coincided at P and the right ends at Q' (Figure 2.25); but B replies 'You cheated – you compared the right-hand ends and then shifted your rod before comparing the other ends' (Figure 2.24). If you want to argue about this conclusion, you must argue against yourself. I can't cover every objection.

2.27 **How does it work out if the rods are held at right angles to the direction of relative motion?**

This time use C and D of §2.22, and again let each observer's rod have one end beside P when it happens and the other beside Q when it happens – as in Figure 2.27. From §2.22 we know that C and D agree that P and Q are simultaneous. Thus each says that the ends of the other's rod coincided with the ends of his own *at the same time.* And so they agree that the rods are the same length. Thus:

> When the rods are held perpendicular to the direction of relative motion, the observers agree about their lengths.

And it's obvious (from the symmetry of the situation) that this conclusion would apply also to the lengths of the half rods stretching from the observers to event Q.

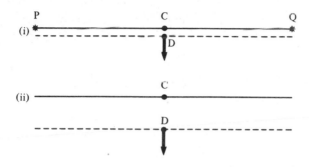

Figure 2.27

3

The dilation of time

3.1 Imagine two observers in spaceships moving relative to each other (steady motions as usual). Call them A and B, but don't identify them with A and B of Chapter 2. We shall draw our diagrams in such a way that if A is fixed on the paper, then B is moving across it horizontally from left to right.

Observer B is equipped with a flash-bulb, P, close beside where he is sitting, and a mirror, Q, some distance away – the line PQ being at right angles to B's direction of motion relative to A. When B fires his bulb, the flash travels from bulb to mirror and back, as indicated by the arrowed lines PQP in Figure 3.1. This shows how the experiment *looks to* B. If the spaceships are transparent, so that A can watch as he passes close to B, **how will it appear to him?**

Figure 3.1

3.2 As A sees it, B and his apparatus are moving from left to right. So Figure 3.2 shows how it looks to him. The position of the bulb just as it flashes is marked K. The mirror, by the time the light reaches it, will have moved to position L – so that the flash has gone diagonally along the line KL. And now it travels along LM back to the bulb, which has meanwhile moved to M – all this *as it appears to* A, of course.

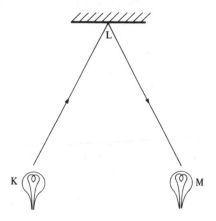

Figure 3.2

If this is hard to visualise, get a pal to act as B, moving his finger from navel to nose and back as he walks past you, to simulate the motion of the light. To him the finger goes straight up and down. But you will see it following a diagonal path like KLM.

3.3 Figure 3.3 combines the two previous diagrams (omitting some detail). **But am I right in assuming that the distances NL and PQ are equal?**

Give B a rod stretching from bulb to mirror, and give A a rod whose ends coincide with B's just as the light reaches the mirror (position NL). Now we have a situation just like that of the last paragraph of

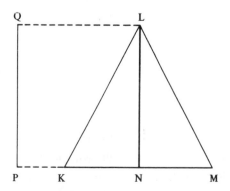

Figure 3.3

§2.27. So A and B will agree that their rods are equal in length – i.e., that distances NL and PQ are equal.

3.4 It is now geometrically obvious that the distance KLM which the flash has to travel according to A is greater than the distance PQP that it has to travel according to B. (You can demonstrate this by the device which ended §3.2.)

So far there is not a single point in all this that a classical, pre-Relativity physicist would disagree with. But now the disagreement begins. *He* would say that the time taken for the journey 'must' be the same no matter who is measuring it. To cover a greater distance in the same time, the light must travel faster according to A than according to B – that would be his conclusion. (The nose-and-navel model may help again.) **But does a Relativitist agree with him?**

3.5 On the relativistic assumption (§1.22) the speed of light is the *same* for A as for B. Having farther to go at the same speed, it must take longer according to A than according to B. Ponder!

Now concentrate on two *events:* (1) the lamp flashing; and (2) the flash getting back to the bulb again. **What surprising conclusion have we reached?**

> The time between these two events is greater according to A than according to B.

3.6 Startled? We've been accustomed to believing that there is a *definite time* between two events, which does not depend at all on who is measuring it. Yet now we've shown that if the assumption of §1.22 is correct, A's and B's measurements disagree. So (in this case at any rate)

> The time between two events depends on the motion of the observer who is measuring it.

3.7 Notice – this is important – that A's and B's measurements are not on a par. There's something special about B's timing. He could have a clock right beside the bulb as it flashes, and it will still be right beside the bulb when the flash comes back. So he can note the time of each event directly by his clock – without having to resort to any indirect processes.

The easiest way would be to work in the dark. The clock face will be illuminated as the flash occurs, and again by the returning flash. Mere subtraction would give B a *direct* measurement of the elapsed time. **But could A work in this direct way?**

No. His clock might be right beside the lamp when it flashes at K, but then it would be far away when the flash returns at M. At the best he can time one of the events directly; for the other he must do something more complicated.

He might use a second clock (present at M) to record when the flash returns. If so, he'll have to synchronise the clocks – by transmitting information between them and *inferring* just which tick of one corresponds to which tick of the other. Or he could arrange for a message to tell him that the light has arrived at M. But he would have to *allow for* its time of transmission. Or again he might measure the length PQ and the relative speed of the two ships, and thence *calculate* the time for the flash's journey.

So B can do his timing by direct observation. But whatever observations A uses, he is forced to supplement them with calculations or allowances or inferences.

3.8 It is convenient to have names for these direct and indirect modes of measurement:

> The *proper time* between two events is the time between them as measured by an observer in steady motion who is present at both, or by any observer who is stationary relative to this one.
>
> The time according to any other observer in steady motion is called an *improper time*.

A few comments may help. (1) We insist on steady motion; a time measurement by an observer whose motion is not steady is neither proper nor improper. (2) In §3.7 we spoke of B's *clock* as being present at the events. But we think of an observer and his clock as being in exactly the same place. Thus B is present at both events and his timing is the proper time. (3) Observers who are stationary relative to B will agree with B's timing; so their measurements are also counted as proper. (4) Any other observer will then be moving relative to B – like A in the case we've been considering. His time will be improper. (5) B is unique, but there can be any number of observers moving relative to him; hence we speak of *the* proper, but *an* improper time.

3.9 **Given any two events, is there always a proper time between them?**

We shall discover later that a pair of events can be so distant in space and so close together in time that no observer could possibly get from one to the other. In that case there would be no proper time. However, sticking to cases where proper time does exist, **can you make any general statement about the relations between proper and improper times?**

3.10 The observer present at both events is just like B. We can imagine him doing an experiment like B's, with the flash setting off at one of the events and returning at the other. Then the arguments of §§3.1–5 will apply – leading to the conclusion that:

> An improper time between two events is always greater than the proper time between them.

This effect is known as the *Dilation of Time*.

3.11 We should not, of course, expect the effect to be noticeable unless the observers' relative speed is a sizeable fraction of the speed of light. I'm sure you can **prove that for yourself** by showing that at low speeds distance KN will be tiny and so distances KLM and PQP will be very nearly equal.

If follows that time measurements concerning events happening on some object that is moving *slowly* relative to the observer can be taken for practical purposes to be proper times – even though they are strictly improper.

3.12 Now plenty of things in nature have built-in clocks and can be observed when moving very fast. So here, at last, is a chance to check the theory against experiment. The time between two ticks of the built-in clock is proper. Our version of the time between the same ticks is improper. And we have to check experimentally whether the two differ by the amount the theory predicts.

So now I'm going to take an actual case and work it out numerically. The calculation is fairly simple. But if you're weak in mathematics, you may need to take trouble with it. One word of advice. *Don't just read* what follows. *Write it out* for yourself – write the gist of it without the wordy explanations. And as you write each thing down, ask yourself: *Why* am I writing just this?

3.13 Let's take the speed of B relative to A to be 99·5 per cent of the speed of light. And for convenience take the distance from K to L to be 1000 units. In your own copy of Figure 3.3 you could mark the various distances as we come to them. **Now can you work out the distance from K to N?**

While the light goes from K to L the flash bulb goes from K to N – travelling 99·5 per cent as fast (relative to A). Hence the distance KN covered by the bulb is 99·5 per cent of the distance which the light travels in the same time – which makes it 995 units. (So Figure 3.3 needs stretching sideways.) If you can see how to calculate the distance NL, skip to §3.20.

3.14M (For the meaning of the 'M' see page 6.) The area of a square (as you know) is equal to the length of one side multiplied by itself. By analogy, any number or quantity multiplied by itself is called its *square*. For example, the square of 5 (or simply '5 squared') is 5×5, which comes to 25.

To save labour we use a standard abbreviation – writing 5^2 instead of 5×5. The little '2' up top means 'What you get by taking *two* 5s (or whatever number is written below) and multiplying them together' – i.e., the square of 5 (or whatever it is).

I shall shortly want to refer to the square of the distance from K to L. We write this as

      $(\text{distance KL})^2$.

The brackets indicate that the little '2' applies to the quantity described by the whole phrase enclosed between them.

If you are an absolute beginner, I advise you to read these abbreviations in full. Read 83^2 as '83 multiplied by itself' and $(\text{distance KL})^2$ as 'distance from K to L multiplied by itself'. When the ideas grow familiar, you can abbreviate to forms like 'the square of 83' and 'distance KL squared'. The process of finding the square of a number or quantity is called 'squaring' it.

3.15M The triangle KNL (Figure 3.3) is called a *right-angled triangle,* because the angle at N is a right angle. And a well-known geometrical proposition called *Pythagoras' Theorem,* states that

> In any right-angled triangle the square of the length of the longest side is equal to the sum of the squares of the lengths of the other two sides.

(The sum of two or more numbers or quantities is what you get by adding them.) Putting it less formally, multiply the length of each side by itself; add together the quantities so obtained for the two shorter sides; the result will be equal to the quantity so obtained for the longest side. Check that this is true for the triangle we met in §1.17. And if the Theorem is new to you, measure up a few right-angled triangles and check that it works for them.

In the case of triangle KNL, Pythagoras' Theorem tells us that

      $(\text{distance KL})^2 = (\text{distance NL})^2 + (\text{distance KN})^2$.

At first you should read that in full as 'The distance from K to L multiplied by itself *is equal to* the distance from N to L multiplied by itself *plus* the distance from K to N multiplied by itself.' Later you will learn naturally to abbreviate. But even then, *always read any mathematical statement as a complete sentence* – in this case 'Something-no-matter-how-much-abbreviated *is equal to* something-else.'

3.16M Any statement saying that one number or quantity is equal to another is called an *equation*.

A great deal of mathematics is concerned with manipulating equations. The methods of manipulation are very little more than obvious commonsense, as you will soon see.

3.17M It's clear, for example, that if the amount of beer in your glass is equal to the amount of beer in mine, then the amount of beer in my glass is equal to the amount of beer in yours. And every 'is equal to' relation is similarly reversible. In other words,

If you start with a true equation and interchange the sides, the result is still a true equation.

Doing this to the equation we met in §3.15, we get

$$(\text{distance NL})^2 + (\text{distance KN})^2 = (\text{distance KL})^2.$$

3.18M But from §3.13 we know that distance KL is 1000 units and distance KN is 995 units. So in that last equation we can write '1000' instead of 'distance KL' and '995' instead of 'distance KN'. And then it becomes

$$(\text{distance NL})^2 + (995)^2 = (1000)^2.$$

So (using the definition §3.14)

$$(\text{distance NL})^2 + 990\ 025 = 1\ 000\ 000.$$

3.19M We want to calculate the distance NL. So we need an equation that says

$$(\text{distance NL})^2 = \text{something-or-other}.$$

Can you see how to deduce such an equation from the one at the end of §3.18?

You probably answered

$$(\text{distance NL})^2 = 1\ 000\ 000 - 990\ 025.$$

Correct! But how did you get there?

If you said something about 'take it to the other side and change the sign', do forget it, *please*. Such mechanical rules get in the way of real mathematical thinking. If, however, you just feel that's the answer, but can't see why, that's very creditable. Every mathematician works like that at the frontiers of his competence. But then he demands to know the Why of it.

In fact it's commonsense. If the amount of beer in your glass is equal to the amount in mine, and if we both gulp down the same amount, then the amount remaining in your glass will be equal to the amount remaining in mine. It's always true that if two quantities are equal and you subtract the same quantity from both, then you end up with equal quantities. In other words,

> If you start with a true equation and subtract the same quantity from both sides, the result is another true equation.

Now in the equation at the end of §3.18 the left-hand side contains the quantity (distance NL)^2 that we're interested in. Unfortunately it also contains 990 025. So we subtract this last from the left-hand side – leaving (distance NL)^2. And to keep the balance we must subtract the same quantity from the right-hand side, giving 1 000 000 − 990 025, which is 9975.

3.20 If you jumped here from §3.13, you can presumably see how to apply Pythagoras' Theorem to prove the result the rest of us have just reached, namely

> (distance NL)^2 = 9975.

We can simplify the calculation at the cost of a mere $\frac{1}{4}$ per cent error by changing 9975 to 10 000. Then it follows that

> distance NL = 100

(since 100^2 = 10 000).

3.21 We started by taking distance KL as 1000 units, and we've deduced that distance NL is 100 units – $\frac{1}{10}$ as much. But our unit of distance could be any size we like. So no matter what the distance KL is, we could divide it into 1000 parts and take one of these parts as the distance unit. Thus the above reasoning shows that it will *always* be true (at the relative speed we've chosen) that

> distance NL = $\frac{1}{10}$ of distance KL;

and so (§3.3)

> distance PQ = $\frac{1}{10}$ of distance KL.

Applying the same reasoning to the light's return journey, we reach the conclusion that:

> The distance the light has to travel from lamp to mirror and back according to B's measurements is one-tenth of what it is according to A's measurements.

3.22 But we are working on the assumption that the light's speed is the same for these two observers (§1.22). So it takes one-tenth the time to cover one-tenth the distance; and we deduce that

> In the case where the relative speed of A and B is 99·5 per cent of the speed of light, the proper time (B's) is one-tenth of A's improper time.

3.23 Check your mastery of that calculation by doing a similar one. If you're weak on algebra, study §§3.13–22 again. Then (1) calculate the relation between A's and B's times when their relative speed is $\frac{3}{5}$ of the speed of light. Try it first without looking at the previous calculation, but if necessary work through the problem with the book open in front of you, changing the numbers as you go. *Hint:* take distance KL to be 5 units. If you need more practice, (2) do the case where the relative speed is $\frac{5}{13}$ of the speed of light. If you are algebraically competent, (3) work out the general case, using v for the relative speed expressed as a fraction of the speed of light. (Answers at end of chapter.)

3.24 As I said in §3.12, this prediction should be testable. One type of test makes use of sub-atomic particles (§1.9) called *muons*, which occur naturally in cosmic rays and can be produced artificially in atom-smashing experiments. The muon is a short-lived particle. After a short time it breaks up into other particles which don't concern us.

How short-lived? You can't actually say how long any particular muon will live. But you can make the statistical statement that

> Given a large number of muons in existence at one instant,
> then half of them will survive 1·5 microseconds (§1.8) later.

3.25 Notice that this statement makes no mention of how long the muons have already lived. No matter how long a muon's past life has been, it still has an even chance of dying within the next 1·5 microseconds.

So if you start at a certain time with a large bunch of muons, *every* 1·5 microseconds from then on the number of survivors will be halved. After 3·0 microseconds (1·5 microseconds twice over) the surviving fraction will be half of a half, which is a quarter. And carrying on similarly, at 4·5 microseconds, $\frac{1}{8}$ of the original bunch will survive; at 6·0 microseconds, $\frac{1}{16}$; at 7·5 microseconds, $\frac{1}{32}$.

3.26 For times *less* than 1·5 microseconds, clearly more than half will survive. In particular, the fraction surviving after 0·75 microseconds (half as long) is 0·707. You may know a neat way of getting that answer; if not, you must be content to check up that it's correct. In fact, if 0·707 survive at 0·75 microseconds, then the fraction surviving after a further 0·75 microseconds will be 0·707 of 0·707, i.e., $(0·707)^2 = 0·5$ (near enough) – which is the right answer for a total time of $0·75 + 0·75 = 1·5$ microseconds. So 0·707 is correct.

3.27 Cosmic ray muons are formed high in the atmosphere – as the outcome of a process that begins when the original cosmic rays collide with atmospheric atoms. They then rain down through the air at enormous speeds.

As they descend, they are continually dying off. So a standard muon detector will catch fewer of them at low altitudes than at high. And experiments comparing the numbers caught at various heights can be used to check the dilation of time. This amounts to using the muons as a fast-moving clock (§3.12).

3.28 One such experimental test was carried out by D.H. Frisch and J.H. Smith (*American Journal of Physics*, **31**(1963) 342). The apparatus was arranged to count muons hitting it *only* if they were travelling vertically downwards at between 99·5 and 99·54 per cent of the speed of light. They set it up on Mount Washington at a height of 1910 metres. In a typical run it counted 568 muons in an hour.

If a similar bunch of muons were allowed to continue their descent, how many should we expect to reach sea-level? To travel 1910 metres at 99·5 per cent of 300 metres per microsecond (§1.8) takes 6·4 microseconds. Now we have to complete the calculation on two alternative assumptions, to see which best fits the facts.

3.29 *Assumption 1: 'Commonsense' view of time.* The time to reach sea-level is 6·4 microseconds according to *all* observers (we assume). This is between 6·0 and 7·5 microseconds, and so (§3.25) between $\frac{1}{16}$ and $\frac{1}{32}$ of the original 568 should survive – i.e., between 18 and 35.

3.30 *Assumption 2: Relativistic prediction of Dilation of Time.* Now we have to distinguish between proper and improper time – that is, we must think about who is doing the measuring. The 1·5 microseconds in which half the muons die is clearly a proper time, measured by the muons' own clock – or by any clock moving with them. It would be absurd to imagine the muons committing suicide by our clock! And the laboratory measurements of their lives were carried out on slow-moving muons, so that they gave proper times (§3.11).

(Do you feel that 'fast-moving muons might behave differently'? **Careful!** They're only fast-moving *relative to us*. As measured by a clock moving with them, the muons' death rate must be the same, whether they're stationary in our laboratory or dashing down Mount Washington – otherwise we'd have found a way of picking out things that are 'really' moving, in contradiction to §1.3.)

3.31 **Is the journey time of 6·4 microseconds proper or improper?**

That figure was calculated (§3.28) from the height of the mountain and the speed of the muons, both measured by an observer fixed on the Earth's surface. So it's *his* version of the time – improper. The muon plays the part of B, while A is the Earth-fixed observer. And we already know that, with this relative speed, B's time is only a tenth of A's (§3.22). So the time of descent, as measured by the muons' own clock, is only 0·64 microseconds.

Now at 0·75 microseconds, 0·707 of the original muons would survive (§3.26). So after the shorter time of 0·64 microseconds rather more should survive – more than 402. On Assumption 2 that is how many we expect to count in an hour at sea-level.

3.32 And when Frisch and Smith set their apparatus to work at sea-level, the result was 412 in an hour. It is clear that this disagrees most violently with the prediction based on Assumption 1 and agrees as closely as one could reasonably expect with Assumption 2. **What conclusion would you draw?**

Obviously the 'commonsense' view about times must be wrong, since it leads to a prediction that is so completely at variance with the facts. **So can we infer that the dilation of time is an established truth?**

That would be going too far. In some future test *its* prediction might also disagree with the facts. *Both* theories might be wrong. But on present evidence the dilation idea looks pretty good.

3.33 We said in §1.24 that you were not required to *believe* that the speed of light is the same for all observers – only to work out what would follow *if* it were, and test these predictions against observed fact. Now the dilation of time prediction turns out to correspond more closely with the facts than does the familiar view. **So can we infer that the speed of light really is the same for all observers?**

Going too far again! That assumption certainly gives a prediction that fits the facts. But this could be sheer luck. The next prediction might be a complete failure. No matter how many tests a theory passes, it might fail the next one. And so we can never prove its truth with complete certainty.

But surely we'd say that this speed-of-light-the-same-for-all assumption seems, on present evidence, rather likely to be a good one.

There are no certain truths in science – only probabilities. Think of it in terms of betting – deciding on the basis of past form how to lay your future bets. In the last race the horse called Commonsense looked good at the starting post but didn't even finish; while outsider Relativity was a brilliant winner. If these two compete again, surely you'll put your money on Relativity – though it's still not a certainty.

A test like this one, where experimental evidence turns out to be in agreement with predictions, is called a *verification*. But the word 'verification' does not imply a claim to have proved the theory true – only a more modest claim to have produced evidence in its favour, evidence that it is likely to give good predictions when we use it again.

3.34 There's another way of putting this dilation of time business that is sometimes useful. Think of A watching B's clock through a telescope. Apply the conclusion of §3.10 to two ticks of B's clock, and try to work out: **What opinion would A form about the behaviour of B's clock?**

The time between these ticks, as recorded by B's clock, is *proper* (§3.8). But A's estimate of the time between the same two ticks is an improper time – which is therefore greater than what the clock actually shows (§3.10). He sees the clock recording less than he thinks it should. So he concludes that it is running slow – steadily losing time, not just lagging behind.

3.35 Maybe you objected that instead of A inferring that B's clock is running slow, it could be that B infers that A's clock is running fast. **Will that interpretation do?**

The reasoning of the previous article was concerned with an actual time recorded by B's clock, watched by A. But A does not do a direct time measurement by his clock. He only calculates what he thinks the time between those two ticks should be (§3.7). We have an *actual time* by B's clock, and A's *opinion* of what it *should be*. So we can deduce what A thinks about B's clock, but not what B thinks about A's.

3.36 **What will B actually think about A's clock?**

We could go through the reasoning of §3.34 again, changing every A into B and vice versa. And we should deduce that B also thinks that A's clock is running slow. And that leads to a very important conclusion:

> If two observers are moving relative to each other, *each* will think that the other's clock is running slow.

3.37 Do you feel there's a contradiction in *each* observer thinking that the other's clock is running slow? Try to clear this up for yourself, bearing in mind two points: (*a*) that when you judge that somebody else's clock is running slow, you are really judging that such-and-such ticks of his clock are simultaneous with such-and-such ticks of yours; and (*b*) that observers in relative motion disagree about simultaneity (§2.19). If you still can't see your way out of the difficulty, please be patient till we clear the matter up in Chapter 13.

3.38 **Did it occur to you to ask how the Frisch–Smith experiment looks to the muon?**

You may have reasoned thus: 'The muon, according to us, is travelling at almost the speed of light. But the same journey only takes a tenth as long by the muons' clock. So a muon must consider that it is moving, relative to us, at nearly ten times the speed of light!' **Anything wrong?**

Have you forgotten §2.26? **Try again.**

Think of Mount Washington as a measuring rod *carried by us*. We say it's 1910 metres long, but the muon disagrees.

Actually the observers must agree about their relative speed (since this is symmetrically related to them). So if the muon's version of the time is only a tenth of our version, then – to get the same speed – its estimate of the distance must also be a tenth of ours. The muon thinks that our Mount Washington measuring rod has contracted to 191 metres. You can easily generalise this into the conclusion that

> If observers carry measuring rods extended in the direction of their relative motion, then each thinks that the other's rod has *contracted* in the same proportion as he thinks the other's clock is running slow.

This *apparent* change in length is called the Fitzgerald Contraction. Unfortunately there is no method of testing it experimentally.

Answers to §3.23. (1) B's time is $\frac{4}{5}$ of A's. (2) $\frac{12}{13}$. (3) Ratio of times is $\sqrt{(1 - v^2)}$.

4

Three clocks and a pair of twins

4.1 I want now to consider a set-up involving *three* observers, A, B and C, in relative motion on one straight line (or just off it, so that they can pass without colliding). We'll stick to steady motion, and so we can't say that any one of them is or is not moving (§1.3). But Figure 4.1 shows their *relative* motions.

At stage (i) B and C are approaching A from opposite sides (the arrows are meant to suggest motion relative to A). At stage (ii) B passes A. At (iii) C, on his way towards A, passes B on his outward journey. At (iv) C passes A, while B speeds away. And (v) shows C also receding.

There are three particular events in this story: the stage (ii) event of B passing A, which we'll call event P; the stage (iii) event–call it O–of

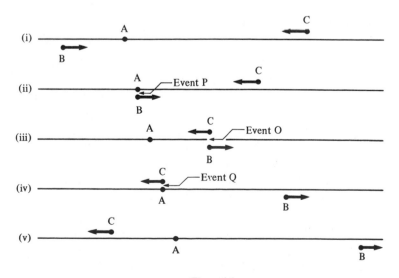

Figure 4.1

B and C passing each other; and finally Q, the event of C and A passing at stage (iv).

Do a mock-up by putting some object on the table in front of you to represent A, and using your left fist to reproduce B's motions and your right fist for C's. Use this model whenever you have difficulty in following a description. (Don't let the Old Adam mislead you into thinking you've now made A 'stationary'.)

4.2 Observer A is present at events P and Q. He can measure the time between them directly by his clock. But B is not present at Q, nor C at P. And so neither can make a direct measurement of the time from P to Q. (If this is not clear, reread §§3.7–8.)

However, B and C could co-operate to make a measurement of this time which involves only direct clock readings. For B can use his clock to measure the time from P to O; and C can do likewise from O to Q. And adding these measurements gives the time from P to Q as measured *jointly* by B and C.

But will B and C jointly get the same answer as A? This *three-clock problem* will be our concern for the next few pages.

4.3 In everyday life we assume that the answer is Yes. We assume that it doesn't matter how you move your clocks around or pass on the time from one clock to another (as B passes it on to C) – the time between two events will always be the same.

Obviously that assumption works in the slow-speed conditions of ordinary life. But would it work if high relative speeds were involved?

4.4 First consider the time from event P to event O. **Are A's and B's versions of this time equal?**

Observer B is present at both events, and so his version is proper (§3.8), whereas A's is improper. **And so...?**

From §3.10 it follows that the time from P to O is *greater* according to A than according to B. **How about the time from O to Q?**

It's greater according to A than according to C. **And now can you answer the question raised at the end of §4.2?**

The time from P to Q according to A can be got by adding his versions of the times P to O and O to Q. Since we've proved that these are respectively greater than B's version of the time from P to O and C's version of time O to Q, it follows that

> The time from event P to event Q according to A is greater than the time from P to O according to B *plus* the time from O to Q according to C.

I expect you worked that out almost entirely by yourself. You're getting quite good at Relativity!

4.5 Although we used improper times in our reasoning, the conclusion we've reached involves proper times only. It refers to direct timings by the clocks of the observers – a point that I'd like to emphasise by re-writing it in the form that

> The time between events P and Q is greater as measured by
> A's clock than as measured jointly by the clocks of B and C.

To put it at its baldest, A's clock ticks more times than B's and C's between them. You'd better do a numerical example. If B and C are both moving at $\frac{3}{5}$ of the speed of light relative to A, and if the time from P to Q measured jointly by B's and C's clocks is 8 minutes, **what is it by A's clock?** (Answer at end of chapter.)

4.6 The conflict with long-established belief becomes clear if we think of B saying, as he passes C, 'It's 4 minutes since I left A.' Then C adds B's 4 minutes to his own 4 minutes; and as he passes A, he calls out 'It's just 8 minutes since B passed you' – to which A replies 'You must be off your rocker: it's *ten* minutes.'

You feel it's all wrong, don't you? You somehow feel you *know* that the time from one event to another *must* be independent of the motions of the people who are measuring it. Yet what is this feeling based on? Experience?

No. To get a detectable time difference, the speeds would have to be enormously greater than anything you and I have experienced (**prove that by putting §§3.11 and 4.4 together**). So are you just generalising unjustifiably from a narrow range of experience – as in §2.21? Or can you see some other reason for believing that the time must be the same for A and the B–C pair?

To avoid seductive blind alleys, note that the time from P to Q according to B and C jointly is neither proper nor improper. The definitions (§3.8) demand measurement by *one* observer – not a team. So you'll have to try some other line. **Think it over before I discuss some of the arguments you might put forward.**

4.7 You may feel (in relation to §4.4) that since the times from P to O and O to Q according to A are *improper* (calculated or inferred), we are not justified in adding them to obtain A's proper (clock-measured) time from P to Q. **Think about that.**

When we speak of the time from P to O according to A, we really mean the time by A's clock from P to the moment when A thinks O happens. Similarly for the time from O to Q. So the point in question can be put as follows (all times being by A's clock): The time from P to the moment when A thinks O happens *and* the time from the moment when A thinks O happens to Q add up to make the time from P to Q. Put that way, surely the difficulty vanishes.

4.8 Did you suggest that the reasoning of §4.4 could be applied back-to-front, as it were, to give a result contradicting the one we've already got. Arguments like that often look plausible when you sketch them vaguely, but break down when you try to work out the detail. So I have to challenge you to write your proposed reasoning out on paper as a complete set of logical steps, like those of §4.4. There are several possible paths to follow. You'll find, I regret to say, that they all lead to dead ends. But you'll learn a lot from exploring them. **So have a go.**

You'll have found every time that you come up against an insuperable difficulty which turns on the following point. The reasoning of §4.4 works because A*'s improper times from* P *to* O *and from* O *to* Q *add up to make his* PROPER *time from* P *to* Q. And you can't find anything analogous in the proposed reverse argument. **But keep on trying.**

In the long run you'll have to admit that a situation with a *single* observer on one side and *two* on the other is not symmetrical; so every attempt to interchange the roles of A and the B–C team is bound to fail.

4.9 Another common argument is that A and the B–C pair are simply measuring the same thing in two different ways, and so must get the same result. **What do you think of that one?**

This is more an emotional response than a logical argument. It derives from the feeling, embedded deep in our culture, that we all share the same time ('time the ever flowing river' and all that). But we're trying to re-examine these long accepted beliefs. So to decide whether A and B–C are measuring the same thing or not, we must ask what they actually do.

They use measuring instruments called clocks. An instrument can only measure what is going on in the same place as itself. Instruments in different places are therefore measuring different things. (If I said that thermometers in different places were measuring the same thing, you'd doubt my sanity.) Now A's clock and the instrument consisting of B's and C's clocks used successively are *not* in the same place (except at events P and Q). Hence they can't be measuring the same thing. They're measuring the passage of time in different places – which is *not* the same thing. And if they're measuring different things, there's no surprise when they get different results.

4.10 You still have a feeling, I'm sure, that this is trick reasoning – an argument that is all right in appropriate cases but doesn't apply to time. Temperature, you say, does depend on place, but time doesn't. I know how you feel, but cross-examine yourself. Aren't you assuming that you already *know* the nature of time – when that's the very thing we have to investigate?

And having completed your self-examination, will you now make a mighty effort to trust the logic of our reasoning, accept the three-clocks result (with reservations if you wish) and see where it leads us.

4.11 Our old ideas receive a far worse blow from this three-clock stuff than from any of our earlier strange discoveries. If you look back at previous cases where relativistic theory conflicted with accepted beliefs, you'll find that they are not concerned with directly observed facts, but with observers' calculations or inferences – with their *considered opinions,* if you like. See the discussion of this point in §§3.7–8 and 3.10 – and similar things could be said about the Relativity of Simultaneity and the Fitzgerald Contraction. In all these cases, the unexpected assertion refers to the results of calculations or inferences about something which, by its very nature, *could not be observed directly.* So it would have been possible to believe that the ultimate reality was very much as we've always thought, but we'd met some unforeseen complication in the processes by which we make inferences and calculations about things that are not directly observable. That would have been very comforting. **Would that point of view be tenable after meeting the three-clocks affair?**

No. The three-clocks result, as we noted in §4.5, refers to direct measurements by observers who are present as the events happen and simply note times by their own clocks. And yet the outcome totally conflicts with our previous beliefs.

This means that time is very different from what we've thought till now. The time from one event to another is not a single unique thing, even when it's found by direct clock measurements alone. It depends on the motion of those who are doing the measuring.

4.12 So we've got to make radical changes in our ideas about the nature of time. Above all, we've got to get rid of the belief that there's a Public Time that we all share, which is the same for all observers no matter how they're moving. We have to admit that each observer has his own *private time* – the time that *he* measures by *his* clock – and that private times do not fit together to form a Public Time (though in slow-speed conditions they very nearly do).

4.13 You find it all puzzling? You reluctantly admit that things are different from what you thought, but you 'can't see why'? If 'see why' means explaining these unexpected discoveries in terms of familiar beliefs, then you're doomed to disappointment – the familiar beliefs have got to go. But if you simply mean that you want to replace the old beliefs by a new coherent theory of space and time and motion into which things like the three-clock affair and the Relativity of Simultaneity will fit naturally, then I promise you eventual satisfaction.

4.14 You'll have noticed that the three-clock affair is rather like the business of the space-travelling twin that I mentioned on page 1 – except that it shares the out-and-back journey between B and C. Let's make the two set-ups as closely comparable as we can. We'll consider a traveller D, and you'll have to imagine adding him to Figure 4.1 in accordance with the following description.

At the start, D is living with A. Shortly before stage (ii) he uses his rocket engines to accelerate away from A. He judges his acceleration so that he reaches B's speed just as B catches up with him shortly after stage (ii). At this moment he cuts out his engines and coasts alongside B. A little before stage (iii), D uses his motors in the reverse direction, and judges it so that he is coasting alongside C when he cuts out again. And yet another rocket manoeuvre, starting shortly before stage (iv), brings him back home to A.

Do a mock-up with your fists like that of §4.1. Now if A and D use their (identical) clocks to measure the time between parting and re-union, **do you think they will get the same result or not?**

4.15 If you accept the line I've been pushing, that clocks which move apart are measuring different things and would therefore be expected to give different results (§4.9) and that each observer has his own private time (§4.12), then you're bound to say that A and D have different motions and so you'd expect their clocks to record *different* time lapses between parting and rejoining – though naturally you'd be prepared to listen to some argument alleging a special reason why they should register the same time. **Do think it over carefully.**

You may feel that this conclusion is repugnant, but as a fair-minded enquirer after truth, I think you're bound to accept it as plausible. And in view of the close resemblance between the motions of D and the B–C team, you'd rather expect D's time to be less than A's.

Of course there are several points where this argument might be suspect. But I'm merely trying to suggest what would be the first reaction of an unprejudiced philosopher who had met the three-clock situation and was now confronted with the space-twin question.

4.16 Suppose A and D are not mere quasi-human abstractions (which is how I think of our observers), but actual flesh-and-blood twins. **Would you expect them at the end to have aged the same amount or not?**

Biological processes are clocks. After allowing for a margin of unreliability, they keep time with more conventional clocks. So if D's physical clock records less than A's, we'd expect his 'ageing clock' to do the same. This is the so-called *Space-twin Paradox*.

You may want to argue that an ageing clock will behave differently from other clocks. That seems to me like special pleading, but I don't want to discuss it at present. I only raised the question of ageing in order to make clear the full horror (as some think it) of where the theory is leading us. Now let's return to the less emotive formulation of §4.15: that we've got some reason to suspect that D's clock would record less than A's. We can call this version the *Clock Paradox*. As usual, **let's have your considered objections.**

4.17 You might try the same-reasoning-in-reverse type of argument that failed to upset the three-clock result (§4.8). This time, you suggest, the situation *is* symmetrical: there's one observer on each side of the comparison; all we can say is that A and D were together, moved apart, and came together again – but we can't say which of them moved. So in this case, you assert, the argument *could* be reversed to show that less time passes for A than for D – an absurd contradiction. **What about it?**

It won't do. Each time D uses his engines, his motion is accelerated, whereas for A it's steady motion all the time. And the difference can be detected (§1.3). So the relation between A and D is not symmetrical – you can't interchange their roles in order to run the reasoning in reverse.

(These accelerations also thwart any argument based on the idea that D's time is proper. See definition in §3.8, and compare §4.6.)

4.18 However, the accelerations raise another problem. We said (§4.15) that we'd be prepared to consider some alleged special reason why A and D should, after all, experience the same lapse of time. Some believe that the accelerations provide this reason. We derived the clock paradox, via the three-clocks problem, from the dilation of time. Now we established this last on the assumption that the observers are in steady motion. So, it is suggested, the accelerations introduce another time effect, which exactly compensates for the time dilation and makes the total time at the end the same for D as for A. **Do you consider that argument valid?**

The honest answer is 'I don't know'. All our reasoning up to the three-clocks result (§4.5) was concerned with observers in steady motion – not accelerating. The effects of accelerations would have to be investigated separately. Till that's done, we don't know.

If you feel like a little solo research, by the way, think about another traveller who uses the same accelerations as D at blast off, turn round and touch down, but goes farther (or less far) during the steady motion periods. What light does that throw on the question?

4.19 You'll have noticed that the last few articles contain more guesswork and much less solid reasoning than before. In fact we can't expect to make further progress until we get down to the hard grind of putting our foundations in order and developing the theory in a really systematic way.

I said in the Introduction that the real difficulties would be problems of imagination and of *un*learning (pages 2–3). The point of what we've done so far is largely to stretch your imagination and shake your former beliefs – so that your mind will be open to accept the more radical reconstruction of ideas that we'll start in the next chapter.

4.20 Meanwhile, as one last effort to liberate your imagination, I'd like to tell a story – about a middle-aged man who fell in love with a teenage girl. She refused him because of the disparity in their ages.

A nineteenth-century version of this yarn would tell how he voyaged round the world to forget her. (How your ancestors 3000 years ago would have boggled – especially at that unimaginable idea that if he kept on going in the same direction, he'd end up where he started! Let's learn from their failure of imagination.)

In a far distant future version, the woebegone lover sets out on a long fast space journey with speed and distance calculated to bring him back when he's just a year older than her – and then maybe she'll say Yes. But his rocket controls go wrong, and before he can repair them, he has travelled much farther than planned. So he arrives back in time to marry her daughter.

And now let's put our noses to the grindstone.

Answers to §4.5. Clearly the 8 minutes breaks up into 4 minutes from P to O by B's clock and 4 minutes from O to Q by C's. From §3.23, A's improper time from P to O is $\frac{4}{5}$ of B's – i.e., 5 minutes. Similarly from O to Q. So A's time from P to Q is 10 minutes.

5

Starting again

5.1 We turn now to ideas that look very simple compared to muons and space-travelling twins. But they are deceptively profound. Give them careful thought. First please reread §§1.1–3.

We can get a much simpler version of these things by asking what would happen in similar circumstances if there were no gravity to complicate matters. Let's imagine that while we are doing the experiments in trains, the Earth miraculously vanishes from under us, and the trains are left travelling in empty space. It's a reasonable guess that everything will happen as before, *except* that things will no longer fall.

In the case of the steadily moving train, when the Earth was still in position, we've seen (§1.2) that if you let go of a stone, it falls straight down (as you, in the train, see it). If you release the stone when the Earth has vanished, **what will it do?**

Same as before, with the falling cut out. So it just stays where you released it (as you see things, of course – somebody else might say that you and the stone move along together). Now take the case where the train's motion is not steady. If gravity ceases to act, **what happens to your stone when you let it go?**

Starting from §1.3, I'm sure you can work out that it moves away from you (forward if the train is braking, etc.).

This gives us the simplest possible way of distinguishing between steady and non-steady motion in no-gravity conditions. You have only to let your stone go. If it stays right beside you, you're in steady motion. If it drifts away, you're not.

5.2 'Letting go of a stone' is beneath the dignity of a theoretical physicist – he 'releases a test particle' (without giving it a shove, of course – that goes without saying in future). Observers whose test particles stay with them play a special role and are given an appropriate title: *inertial observers*. Let's have a formal definition:

Let an observer release a test particle. Then:
(1) If the particle remains stationary relative to the observer, he is called an *inertial observer;*
(2) If it moves away from him, he is a *non-inertial observer.*

(Naturally we must make sure that the test particle is not acted on by electrical or magnetic force, or pulled by a string, or anything like that. We shall always assume that no such forces are at work unless they are mentioned explicitly.)

5.3 At last we are in a position to clear up a certain vagueness on the subject of 'steady motion' that has been with us since Chapter 1. We've kept on using that phrase as if we knew what it meant. But did we? For our train at the start we took it to mean motion at constant speed in a straight line. But constant speed in a straight line *relative to whom?* We unconsciously accepted the Earth as our standard of reference. But relative to a young man in a sports car with his foot hard down, the train's speed would be far from constant. An inhabitant of one of Jupiter's moons would see it sometimes approaching, sometimes receding, sometimes moving transversely – nothing like straight-line-constant-speed motion.

Approached in that way, the idea of steady motion turns out to be elusive. You think you know what you're talking about, but when you look a bit deeper, you find you can't define your meaning.

Now, however, we're going to cut loose from all these difficulties. We'll not talk any more about steady motion, but only about *inertial motion* – the sort of motion that an inertial observer has. And that is precisely defined by the particle test of §5.2 – which is self-contained, not needing the help of any outside observer or frame of reference.

The straight-line-constant-speed approach helped us to get started. We'll also come back to it shortly and see that, in some conditions and suitably qualified, it still holds good. But I must stress: *inertial motion is defined by the particle test* (§5.2), not in straight-line-constant-speed terms.

5.4 You can do the particle test even when there is gravitation. **Please oblige by testing whether you are inertial or not.**

You held something up and released it. Did it stay put? No, it fell – moved away. So you are non-inertial. **How could you become inertial?**

Jump off a cliff. Under the action of gravity all things fall at the same rate (apart from air resistance effects, which you can eliminate by jumping in vacuum or minimise by using something dense as test particle). As you see it, the test particle stays with you. So you are inertial.

A spaceman in orbit provides another example of an observer who is inertial even when there is gravity. You've seen him on TV, releasing a camera spool, which hangs in the air beside him till he wants it again.

These examples should dispose of the idea that inertial motion always means motion at constant speed in a straight line. **Think about that point.**

5.5 I brought in gravity just now to prevent you from getting over-simplified ideas about inertial motion. But gravity does not concern us for a long time to come, because the *Special* Theory of Relativity only applies when gravitational effects can be ignored. In fact, we can now (at last!) say what this Special Theory is:

> The Special Theory of Relativity is a theory for describing and comparing the experiences of *inertial* observers in the *absence of gravitation.*

Please note the italicised limitations. (We can never actually get away from gravity. So in practice the theory is used in conditions where gravitational effects are small enough to ignore compared with other factors.)

5.6 That definition prompts us to ask: What is special about inertial observers in the absence of gravitation?

Well, when I wanted to demonstrate that inertial motion is not always in a straight line at constant speed, I had to invoke the aid of gravity (§5.4). Doesn't this suggest that perhaps when there's no gravity, an inertial observer *would* have straight-line-constant-speed motion? **Any criticisms of that suggestion?**

It's inadmissible as it stands. You must not speak of motion unless you specify *relative to what* (§1.3). **So complete the suggested statement.**

Make it 'relative to another inertial observer'. We're suggesting that

> In the absence of gravitation, any inertial observer moves in a straight line at constant speed relative to any other inertial observer. (It could be constant zero speed, of course.)

We couldn't use the straight-line-constant-speed idea to *define* inertial motion – because we couldn't answer the 'relative to whom?' question (§5.3). But when inertial observers have been defined by the particle test, this question *can* be answered, leading to the above precise statement.

5.7 The indented statement is now precise. But is it correct?

We can't test its truth directly, since gravity is always with us. The statement is an abstraction about what *would* be true *if* there were no gravity. Several lines of thought render it probable. (1) If it's false, how *does* an inertial observer move? If his path is not straight, why should it swerve this way rather than that? And so on. (2) Start with two trains at different places on Earth, each moving in a straight line at constant speed relative to the Earth's surface. Now make the Earth vanish as in §5.1 – and **you work out where this argument leads.** (3) When spacemen or astronomical bodies are separated by great distances, so that gravity is weak, we observe that their behaviour approximates to straight-line-constant-speed relative to each other.

Arguments like these convinced Newton that the statement was likely to be correct. And so he adopted it (using different language) as his First Law of Motion – one of the basic assumptions from which his theories of mechanics and gravitation were derived. The great success of those theories in making predictions that closely agreed with experiment and observation convinced Newton and his successors that the assumption was a good one (compare §§3.32–3). So we also adopt the indented statement of §5.6 as one of the assumptions on which our theory will be built. And we shall test its worth in terms of how good the resulting theory is at making predictions that agree with observed reality.

5.8 Let's get back to that steady-motion train of §1.2. We suggested that every internal experiment gives the same result whether the train is moving or stationary (relative to the Earth). An obvious extension is that any experiment will give the same result no matter what speed or direction the train is moving in.

Suppose, then, that we have many trains moving at different speeds in different directions. Observers in all of them do the same experiment, and so get the same result. Once more we make the Earth vanish. All these observers become inertial observers in the absence of gravity. And surely it's clear that their experiments will *still* all give the same result.

So the suggestion is that if a lot of inertial observers in the absence of gravitation all do the same experiment, they'll all get the same result. Or, as we shall usually put it for brevity, they are all *equivalent*. (Incidentally, '*same* experiment' means simply that if each completely describes his experiment as he sees it, all the descriptions will be the same.)

5.9 This conclusion is so important that we give it a name and state it formally:

> *Newton's Principle of Relativity*
> In the absence of gravitation, all inertial observers are
> equivalent, so far as mechanical experiments are concerned.

(The reason for that last restriction will soon appear.)

We can't, of course, know for certain that Newton's Principle of Relativity holds good. Somebody might find an exception tomorrow. We can only assume it as a hypothesis and go through the usual process of testing whether its consequences agree with observed facts.

5.10 Why the restriction 'In the absence of gravitation'? Well we shall see eventually that when there *is* gravitation, it's not quite true that all inertial observers are equivalent. Let's leave that till we need it, but there's no harm in adding the proviso now, since the Special Theory excludes gravity anyway.

5.11 This Principle, with the restriction to *mechanical* experiments very definitely included, was known to Newton (though he put it differently). And nineteenth-century physics wholeheartedly accepted it. But would it be possible to remove the restriction and restate the Principle for experiments of any type? Please reread §§1.10–13 and **tell me what a nineteenth-century scientist would have said.**

He would have insisted on the restriction. For he thought that optical experiments would give results depending on the observer's motion relative to the (supposed) aether. **Do we agree?**

Of course not. The Michelson–Morley experiment (§§1.14–19) was meant to exhibit this effect. And it failed to do so. The recipe that was going to demonstrate the non-equivalence of inertial observers in regard to optical experiments turned out to be a flop–as did other attempts on somewhat different lines. No experiment of any sort was ever successful in differentiating one group of inertial observers from another. Surely we must drop the 'mechanical' restriction from Newton's Principle of Relativity (§5.9) and so transform it into

Einstein's Principle of Relativity
In the absence of gravity, all inertial observers are equivalent.

5.12 In stating this Principle, we have at last done something approaching justice to Einstein–for this was the essential novelty that he introduced (though he stated it differently), far more important than the invariance of the speed of light (§1.22). Indeed the latter is just a special case of the former–if all inertial observers do the experiment of measuring the speed of light, they'll all get the same result.

Einstein's Principle of Relativity looks very simple, and provokes the 'Why didn't *I* think of that?' type of reaction. Yet it carries the most profound consequences. The Special Theory of Relativity is little more than a working out of these consequences.

We can't, of course, be sure about Einstein's Principle of Relativity, any more than of Newton's (§5.9). It is only a plausible assumption that we put into the making of our theory and eventually test in the usual manner. And just as for the invariance of the speed of light, you are not required to *believe* it – only to take it seriously, work out its consequences and test them experimentally (see §1.24).

5.13 It must be obvious by now that the concept of an aether (§1.11) is completely untenable. And maybe you ask: 'If light is not waves in the aether, what *is* it waves in?'. I now point triumphantly to the care with which I introduced the subject in §1.4, saying only that light behaves in some respects *as if it were* wave motion. You surely can't insist that these fictitious waves must be waves *in* something.

If you find that unsatisfying, I think it's because you feel that you ought to be able to explain light in terms of something else. I don't want to get involved in what could be a complicated argument. Can we just agree that for the purposes of this book we don't try to explain light? We merely describe its behaviour, and then take it as one of the basic things in terms of which we explain everything else.

5.14 Let's turn to a completely different line of thought. Chapters 2–4 were largely destructive in character. Under their impact our old world has fallen to pieces. You've got hold of a few bits of the new world, but they don't fit together. Things no longer make sense.

It's no use hoping we can simply tinker around with our old ideas till the new discoveries fit in place. We have to set about creating a completely new theory of space and time and motion. And in doing so, we'll take notice of the warnings of Chapters 2–4 about what we must not assume.

Before you can build a theory you need to be able to describe the phenomena you're dealing with, and so you have to decide what categories you'll use in your description. In everyday terms, if you're asked to describe a car accident, you might begin by saying that it happened half a kilometre down the road at 5.35 p.m. And in founding a physical theory, you'd be inclined to use similar categories – saying, for example, that a certain event happened at such-and-such a time at such-and-such a distance from here in such-and-such a direction. Such specifications have worked well in our slow-speed conditions. **But would they do when really high relative speeds are involved?**

5.15 No. For in the matter of specifying an event that happens somewhere else, other than where you are – for future brevity I'll speak of *an event 'over there'* – these ideas of time and distance turn out to be much less definite than we thought.

'The time of that event over there' might mean 'the tick of my clock which is simultaneous with the event.' But knowing that observers can disagree about simultaneity, we see that this meaning cannot be taken on trust. Alternatively it might mean 'the time recorded by a clock that is right beside the event as it happens'. But to compare that clock with mine I must bring them together – which involves relative motion and so introduces time-dilation complications. Thus the precise meaning of 'the time of that event over there' is far from obvious. Similar difficulties arise about distance measurements (§§2.26 and 3.38).

It's easy to see why 'the time (or distance) of that event over there' is less reliable than we thought. These are things that you can't know directly – because you are not over there! You can only know about them by means of transmitted information of some sort. We used to think it was obvious how to interpret the messages. Now we know it isn't.

5.16 So in building up our theory we must avoid categories like 'the time of that event over there'. **Then what sort of categories can we use?**

We do something that is very characteristic of twentieth-century science. We go back to the observations themselves; and we enquire what exactly it is that the observer observes, and what he merely infers from his observations. And then we base our theory on the actual observations, and leave the inferences to be worked out afterwards.

Now I hope it is obvious that an actual observation is something that can only take place where the observer is – *'on the spot'* as I shall call it for brevity. An explosion happens 'over there'. I can't observe it, because I'm not over there myself. **What can I observe?**

Well I can observe that the light flash reaches me – in other words, I *see* it. And the seeing happens *here*, where I am – it's an on-the-spot observation. I can't safely speak of *when the explosion happened* (§5.15). But if I say I *saw* it at such-and-such a time, then that's a hard fact, a direct observation.

And so our theory must start from on-the-spot observations – from the observer's observations of what *happens to him* (or what *he does*) *on the spot, where he is.*

Then we can imagine many observers, moving relative to each other, each making his own on-the-spot observations of the same set of events. They could compare and collate their results. And it is from these comparisons and collations that our new theory must emerge.

5.17 I'm sure you realise that our traditional 'commonsense' view of the nature of things was arrived at in just this way. All any human being knows directly is what happens *to him,* on the spot – you see in your own eyes, hear in your own ears. But people's observations can be compared, and thus we work out our ideas about the nature of the world. Each of us does part of the job for himself (particularly in the first year or two of life that we can't remember); more was worked out by our ancestors and passed down to us.

Unfortunately neither we nor they have had experience of high-speed motion. So we've worked out answers that do very well in low-speed conditions, but fail at high speeds. And that's why we're starting again and working out the general view of things that emerges from comparing and collating the on-the-spot observations of observers in very rapid relative motion.

5.18 But of course we can't actually do this in practice, and so we'll have to reason out what theories *would* emerge if it were done. We'll imagine many observers moving very fast relative to one another, making on-the-spot observations only. From a very cautious use of what we've learnt by experiment and observation (Chapter 1) we'll have to work out what sort of observational results these chaps would get. Then we must imagine them starting without preconceptions, comparing their observations, and developing a new theory of space and time to re-place the one we're having to abandon.

5.19 In one important way our task can be simplified. All the really interest-ing things in the Special Theory of Relativity can be worked out per-fectly well by sticking to one straight line – one dimension of space.

We consider observers moving (relative to each other) on this line. For theoretical purposes we'll have to think of them as having no size – as points. And they must be able to pass each other without colliding (so you may find it easier to think of them as moving on close parallel lines, as do A, B and C of Figure 4.1 – which gives a good idea of the sort of thing I mean). To help minimise the influence of ancient habit, we'd better think of them as spacemen, far removed from stars and planets, playing some space-age game that involves relative motion on a line.

Let's speak of this line as the *one-dimensional universe* – or the *1D universe* for short. Of course it's not really the Universe – the rest of space does exist – but it counts as the universe for our purposes, since we refuse to notice anything that's not on this line.

5.20 What sorts of on-the-spot observations are permitted? We'll have to be restrictive if our theory is to be generally applicable. For example, listening for the bang from an explosion can't be allowed – since sound is only transmitted in special conditions. Distance measurements are out, since an observer can't be simultaneously at both ends of the distance he's measuring.

We can allow the observer to notice that he's passing another observer – for at that moment the other *is* on the spot. And he could do the particle test (§5.2). He releases his test particle; if it's still 'here' later on, he's inertial; otherwise, he's not.

Again, he can observe light signals (flashes) and note the time *by his own clock* of receiving them (i.e., he may time when he *sees* an event).

We can also let him *send* light signals. And by agreements with other observers he can make this a useful part of his observations. It can be agreed, for instance, that another observer will send a reply the very moment he receives a signal from the first. If the first one notes the times of sending and receiving, he will have useful information about the relation of this other observer to himself.

You've probably recognised in the last paragraph a crude description of radar. This consists in sending radio signals to a target and getting replies (echoes). The apparatus times the moments of sending and receiving, and from these it automatically makes deductions about the position and motion of the target.

I hope you will spend a little while convincing yourself that the observations described in this article are *the only possible ones,* and in particular that an observer's only possible measurements are times by his own clock of the on-the-spot events that we've just listed.

5.21 Standardised clocks are essential if the comparison of various observers' times is to give useful results. And the only feasible way of standardising them is to insist that they must all be built to the same specification – if each observer gives a complete description of his own clock, these descriptions must be identical. (This is not as difficult as it might seem. A complete description of what is called an atomic clock can be given by merely stating a handful of whole numbers.)

The device of specifying identical clocks gets round a problem that worries many people: How do you know that an observer's clock is telling the right time? **Have you thought about that?**

We have no means of getting at time (as a quantity) except by using a clock. So we have to *define* time as what a standard clock measures. And each observer's private time (§4.12) is what *his* standard clock measures. So it's the right time by definition. If the clocks are identical, they are all measuring time in exactly the same way – and we can't ask for more. Questions about 'right' and 'wrong' time can only be questions as to whether the clock is or is not standard.

6

Space–time diagrams

6.1 You know how much a diagram can help your thinking. We desperately need such help now. But diagrams like Figure 4.1 are limited in usefulness, partly because they can only show a few snapshots out of a continuous story, mainly because they can only represent one observer's point of view on things like simultaneity – that's why we needed both Figure 2.6 and Figure 2.8.

6.2 As an introduction to a much more helpful type of diagram, consider some happenings on a straight bit of railway. At zero time the driver, A, of a train which is travelling at 60 kilometres per hour passes a starting post, from which we measure distances. At time 12 seconds he is passed by B, the driver of a second train travelling in the opposite direction at 30 kilometres per hour.

Figure 6.2 retells this little story by exhibiting graphically how the positions of A and B change in space and time. It is therefore called

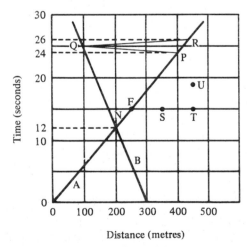

Distance (metres)

Figure 6.2

a *space–time diagram*. The motions of these engine drivers are represented by the two thick lines marked A and B. Any point on A's line represents an event happening to him. For example, the point N (where A's and B's lines cross) represents the event of A being passed by B. Following the broken line horizontally till it meets the time scale at the 12-second mark, we see that event N happened 12 seconds after zero time. And going vertically down from N we see that A was then 200 metres from the starting point. Point F might represent the event of A getting a fly in his eye. **When and where did it happen?**

Going horizontally to the time scale, it occurred at time 15 seconds (halfway between the 10 and 20 second marks). And running your eye vertically down, the position was 250 metres from the start.

(By an obvious convention, which you probably took for granted, we use the letter N for both the *point* in the diagram and the *event* that this represents.)

6.3 At time 24 seconds A gives a short toot on his whistle. The moment B hears this sound, he gives an answering toot, which A hears at 26 seconds. The point P (opposite 24 seconds on the time scale) represents the event of A sending the signal, and the thin line PQ shows how the sound travels till it reaches B at Q. (*Your own diagram* – see page 5 – will be clearer if you use red lines where I use thin ones. And there will be many similar diagrams. So get yourself a red ballpoint!) The second thin line shows B's reply travelling back to be heard by A at event R.

6.4 Diagrams like this are often useful – in working out railway timetables, for instance. But they won't do for us. It's most important that our diagrams shall not prejudge our conclusions. They must not, as it were, have their minds made up on questions that we want to keep open. **Can you see how Figure 6.2 fails in this respect?**

It assumes, for instance, that A and B (and any other observers) agree about times. This assumption is embodied in the single time scale on the left and in the horizontal grid lines, which are essentially lines of simultaneity. We can't let the diagram prejudge this question (§2.19). Again, the single horizontal distance scale assumes universal agreement about distance. And we can't have that either (§§2.26, 3.38).

In fact this diagram exhibits our old beliefs in pictorial form. It's what I call an *old-fashioned* space–time diagram. And we can't afford to use a diagram that reinforces the very ideas we are trying to question.

6.5 But all is not lost. Remember our resolution (§§5.16, 5.20) to base our theory on the times of on-the-spot observations obtained by each observer's own clock.

The point at the bottom left of Figure 6.2 represents zero time by A's clock – which information I've added to a new version of this diagram in Figure 6.5. And how do we know that the time of event N was 12 seconds? Why, A looked at his watch as B passed – so that's another time by A's clock, which we mark at N. Similarly F, P and R were timed by A as they happened, and are so marked. These are the observations that we've agreed to regard as firmly knowable facts.

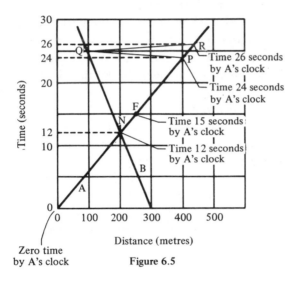

Figure 6.5

6.6 If A has timed any other events, we can mark these similarly in Figure 6.5. And the sum total of direct observational data concerned with A consists of this set of times of his on-the-spot events by his own clock. If we similarly marked B's line with *his* timings of his on-the-spot events, the diagram would then contain all the available information. The scales at the bottom and side offer nothing further. They are now quite useless. **Ponder on that.**

So why not wipe out these scales completely, and the graph-paper grid that goes with them? Then we've got Figure 6.6. And all the firm information is still there.

6.7 When using such a 'stripped down' diagram, it's easy to forget that it represents motion and change; one finds oneself thinking of it as a piece of geometry (two railway lines that cross at N!). So here's a way of bringing it to life. Get yourself a plain postcard (or the like). **This is**

a MUST! Across the middle of it cut out a slot about 100 millimetres (4 inches) long and a millimetre ($\frac{1}{25}$ inch) wide. Place the card on top of Figure 6.6, in such a way that the slot lies across the bottom of the diagram, just above the lower ends of A's and B's lines. Looking through the slot, you will now see short portions of these two lines, which you can easily think of as dots. The slot now corresponds to the railway line of §6.2, and the dots mark the positions of A and B just after zero time.

Now hold the card firmly with one hand, and with the other take hold of the paper that carries your version of the diagram, and move it steadily downwards behind the card. **What do you see?**

You see two dots moving closer together, passing each other (as point N appears in the slot) and then moving apart. (If you can't easily see what I say you should, widen the slot a bit.) You've got a moving picture, in which the slot is the railway line, and the two dots show the motions of drivers A and B (§6.2). As you continue pulling the diagram down behind the slot, **what else do you see?**

When P appears in the slot a *red* dot sets off from the black A dot and travels leftward till it meets the B dot. At that moment another red dot leaves B and travels back to meet the A dot at R. Our 'home movie' has enabled us to 'see' the whistle signal and reply of §6.3.

Notice that A's dot moves twice as fast as B's, and the whistle dots go faster still. The shallower the slope of a line, the faster the motion it represents. *Please keep your slotted postcard:* it will grow increasingly useful.

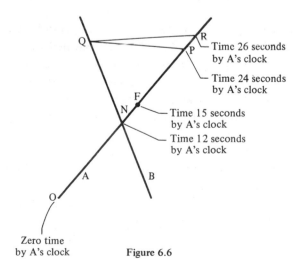

Figure 6.6

6.8 The essential point about the 'stripped down' diagram of Figure 6.6 is that it shows what observers can know directly, without committing us to anything beyond that. It says nothing about times of over-there events, nor about distances. These are matters that will have to be inferred – when we discover how – from the on-the-spot times that the diagram does record.

When we throw away the vertical and horizontal scales and the graph-paper grid to transform Figure 6.5 into Figure 6.6, it's surprising how much else we discard. **I'd like you to spend a few minutes thinking about what has been left out of the new diagram and how the omissions free us from prejudgements on questions we've got to investigate.**

I can only spare room for a checklist of prejudgements that go overboard: horizontals represent simultaneous events; higher on the paper always means later; verticals represent things happening at the 'same place' (cf. §2.16); there are fixed scales that apply, respectively, to the time and distance measurements of all observers. If you missed any of these, think again.

6.9 We can, of course, interpret Figure 6.6 as applying to happenings on the 1D universe (§5.19) instead of the railway. Drivers A and B become *observers* exchanging *light signals* (instead of sound). In these terms, sticking to A's experience, **what story does Figure 6.6 tell now?**

I hope you made good use of the slot technique (§6.7), which now gives a moving picture of happenings on the 1D universe. If necessary, **try again.**

At event O, A's clock reads zero; 12 seconds later *by A's clock* he is passed by B, and 3 seconds after that he gets a fly in his eye. At time 24 seconds by his own clock (event P) he sends a light signal. And B's answering signal reaches him at 26 seconds by his own clock.

6.10 The line labelled A now gives an account of as many of the on-the-spot events of A's life as we care to mark. We could even mark every tick of his clock; or we could selectively mark the ticks for (say) 10, 20, 30, . . . seconds – thus providing A with *his own private scale* for representing *his own private time* (replacing the vertical time scale that we've thrown away).

This line is called A's 'world line' (an ill-chosen word, I think, but we must bow to convention). Defining it more generally,

> The *world line* of an observer is a straight line or curve whose points represent all the events at which he is present – all the things that happen to him (or he does) on the spot, where he is.

And the times of these events, by the observer's own clock, are indicated on his world line (by marking them individually or by means of a time scale).

(Were you worried about curved world lines? Use the slot to investigate what sort of motion they represent.)

6.11 Figure 6.6 also uses another sort of line – PQ for example. This tells us that a light signal was sent at event P and received at event Q. So we call it a 'signal line'. And in general,

> A *signal line* is a line connecting points that represent events at which a particular light signal is sent or received.

Here, you'll notice, we say nothing about times. **Why not?**

Does a light signal carry a clock? Even if it does, we can't know what that clock reads. So we've no means of marking times on a signal line.

In my diagrams world lines will always be thick and heavy, signal lines thin and light. Your own diagrams will be clearer if you use red for signal lines.

By the way, the phrase 'light signal' (often just 'signal') will always mean an *instantaneous* pulse of light, unless I say otherwise. (On 'instantaneous', see end of §2.15.)

6.12 A minor difficulty arises concerning the slot technique (§6.7). What shows through the slot in any one position must represent events happening at the same time. So the procedure I've described can show only one observer's version of the story (§2.19).

Don't let this worry you. Before such questions become important, we shall have made improvements that will eliminate the difficulty.

6.13 We have stripped the space–time diagram of those features which embody old exploded beliefs. It is now as neutral as we can make it. To give it positive content we have to find ways of incorporating the key assumptions of Relativity Theory. And we must start by specifying precisely the new type of diagram that we are going to build up and use. We do so partly by using analogies from the old diagram, and partly by asking what would be appropriate ways of representing known properties of observers and signals. For example, from the diagrams we've been using so far we can take over the obvious idea that

> An event happening on the 1D universe (§5.19) will be represented by a point of a plane

– that plane being, of course, the sheet of paper on which you are drawing the diagram. (Check that a dot does behave like an instantaneous event when you use the slot in the manner of §6.7.)

6.14 There are two directions along the 1D universe. We can call them the *positive* and *negative* directions, and picture them as right and left in diagrams (Figure 6.14). A light signal will be called *positive-going* or *negative-going* according as it travels from the negative to the positive direction or the reverse.

Figure 6.14

6.15 On the analogy of Figure 6.6, we might try using straight lines for signal lines (§6.11), and distinguishing the two types by the way the line slopes. A positive-going signal line will slope upwards (i.e., towards the top of the page) from left to right, as in Figure 6.15; and a negative-going one will slope the opposite way. Check that the slot technique now shows these signals moving in the appropriate directions.

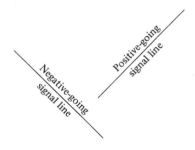

Figure 6.15

Two signal lines of opposite types can only cross once – as the figure illustrates. That is the diagram's way of saying that a positive- and a negative-going signal can only pass each other once – as we'd expect.

6.16 Can two positive-going signal lines meet? To settle that question, we recall that the speed of light, as viewed by any one observer, is always the same (§§1.7, 1.9). Actually we can't use speeds in our arguments, since speed is not an on-the-spot measurement – it is calculated from a distance and at least one over-there time. But we can put the previous statement in a form that could be verified by on-the-spot observations only: one light signal cannot catch up on another. **So what's the answer to the question at the start of this article?**

If they did meet, that would represent one signal catching up on the

other – use the slot on Figure 6.16 to check this point. So the answer is
No.

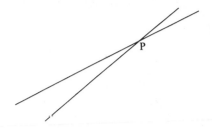

Figure 6.16

Now two lines that don't meet must have the same slope (in relation
to a horizontal drawn across the page). So all positive-going signals
must have the same slope; and similarly all negative-going ones must
have the same slope, but in the opposite direction (§6.15).

6.17 We can decide for ourselves what slope to use. And obviously we'll
pick the simplest one – 45° to the horizontal, as in Figure 6.17. So
summarising,

> For signal lines we choose straight lines making an angle of
> 45° with the horizontal, sloping upwards towards the right for

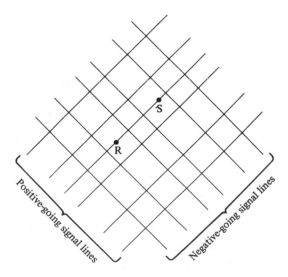

Figure 6.17

positive-going, and towards the left for negative-going, signals. And the direction of a signal line is upwards

– in the sense that if point S is higher than point R on the same signal
line (Figure 6.17), then a signal sent at event R will be received at
event S, but one sent at S will never be received at R. Now use the slot
technique to verify that the (red) dots representing light signals do
behave as they should.

6.18 Now what shall we use for the world line (§6.10) of an inertial observer
(§5.2)? Figure 6.6 would suggest a straight line. Obviously it can't be at
45° to the horizontal (**Why?** That would make it coincide with a signal
line, and the observer would be travelling *with* the light signal – which
is impossible, §2.1). **So is it to be steeper or less steep?**

By using the slot, you'll see that a world line less steep than 45°
represents something moving faster than light. And we don't know
whether that is possible. So it would be playing safe to *start* with a
world line steeper than 45° and investigate later whether the other case
is possible or not.

6.19 That argument (as you're probably eager to point out) is unsound –
because we mustn't talk about speeds (§6.16). However, it has given
us an idea, which we can now justify in permissible on-the-spot
terms.

First let's *try* a world line less steep than 45° (Figure 6.19). Let the
point Q, below this line, represent the event of a lamp flashing. Signal
lines QP and QR represent two light pulses travelling away from Q in
opposite directions. Check that with the slot, and **then finish off the
story that this diagram tells.**

The observer receives both signals – one at P and one at R (slot
again!). So he *sees* this event *twice*. **Is that possible?**

Again we don't know. You and I have never seen the same event
twice in this way. But we can't be sure it's impossible.

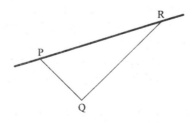

Figure 6.19

6.20 On the other hand, when the world line is steeper than 45° (Figure 6.20), only one of the signal lines starting from Q meets it. **What's the story this time?**

Slot! The observer sees the flash *once only*. And that *is* possible – it's everyday experience. So the sensible thing (as suggested in §6.18) is to give our first observer a world line steeper than 45°. In summary,

> We choose one inertial observer, A, and take as his world line a straight line making an angle of more than 45° with the horizontal.

In Figure 6.20, A's world line has been labelled with his name near the foot. That will be our usual custom.

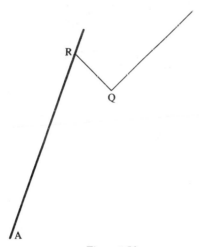

Figure 6.20

Having fixed up the world line of this one inertial observer, we'll have to do a lot of work investigating how the diagram represents his view of space and time, before we can think of putting in even one more observer.

6.21 You may want to argue that A's world line must be vertical. Your reasons are probably not very clear, but they may include a feeling that a sloping world line would make the speed of light (according to A) greater in one direction than the other. But you can't start making deductions like that before you've worked out how the diagram represents speed. In the meantime you'll have to be content with this argument:

You assert that A's world line *must* be vertical. All inertial observers are equivalent (§5.11), and so it would follow that *all* their world lines must be vertical. Then none of these world lines could meet. No two inertial observers could ever pass each other! Your suggestion has led to an absurdity, and so must be wrong.

6.22 The only permissible measurements are the observer's on-the-spot times (§5.20). So we have to put a time scale on A's world line (compare §6.10). Thus, as illustrated in Figure 6.22,

> We choose a point O on A's world line to represent the event of A's clock reading zero.
>
> Times of A's on-the-spot events, as measured by A's clock, will be represented to scale along his world line

−i.e., equal intervals of time as measured by A's clock will be represented by equal lengths of his world line. For future reference we can call this 'A's time scale'.

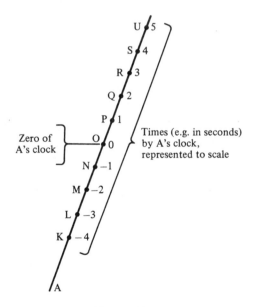

Figure 6.22

6.23M Even if your mathematics is weak, you probably weren't surprised to see times marked −1, −2, −3, −4 at the bottom of Figure 6.22. We all make some use nowadays of what are called *negative quantities* even if it's only to describe sub-zero temperatures.

  We have to be prepared to cope with times *before* zero time – represented in Figure 6.22 by points *below* O. Times before zero are essentially *times measured backwards* from event O. By convention we mark them −1, −2, . . . , meaning 1 second (or whatever unit we're using) measured backwards, 2 seconds measured backwards, . . .

It is common to read '−2 seconds' as 'minus 2 seconds', and if you are used to that, don't try to change. But even while you use that form, remind yourself that you actually mean 'measured backwards'. And if negative quantities are really new to you, I suggest you use the 'measured backwards' language until they become familiar – and when they *do* become familiar 'negative 2' is better than 'minus 2'.

6.24M There is an obvious relation between negative quantities and the process of subtraction. For example, '2 hours measured backwards from 10 p.m.' brings you to 8 p.m., which is also what you get by subtracting 2 from 10. That is why the same sign is used for negative quantities and subtraction.

Having invented the adjective 'negative' for quantities measured backwards, we now speak of quantities measured forward (usually written without any special sign) as *positive*.

6.25M Clearly the 'negative' idea can be applied to other things besides time. Temporarily think of Figure 6.22 as just a line with distances marked along it, in units of one pace. For somebody starting at O, an instruction to go 2 paces would take him to Q, while an order to go −3 paces would bring him to L.

In this case 'measuring backwards' can have a literal meaning – pacing in the opposite direction. Time can't be literally measured backwards like this. We have to rely on some clock that was in use before zero time. And if this records the time from M to O as 2 seconds, then clearly the time from O to M is 2 seconds measured backwards – that is, −2 seconds.

7

Time and distance 'over there'

7.1 Suppose we have some event Q at which A is *not* present – an over-there event, as I've been calling it. To us it seems natural to describe Q in terms of its time and its distance from A, according to A. But that implies that we shall have to find a way of expressing these two quantities in terms of A's on-the-spot measurements (since the latter are the only ones that we regard as directly knowable, §5.20).

Thinking of A as a spaceman, as suggested in §5.19, may stop us being misled by old habits of referring everything to 'fixed' reference points on Earth. If he is interested in an event occurring outside his own ship, **what method would a spaceman use for measuring its time and distance?**

Radar (§5.20). **That gives a clue worth thinking about.**

7.2 Figure 7.2 shows A's world line and the point representing event Q. (It is usual to mark the zero of A's clock at O, but leave the rest of his

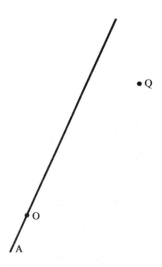

Figure 7.2

time scale – §6.22 – to the imagination.) **What information can A have about event Q?**

He can receive a signal from Q – at event R, say. **Draw the corresponding signal line on your own diagram** (hint, if needed, from Figure 6.20). The time of arrival of this signal is one piece of on-the-spot information. **Anything else?**

He can also *send* – at event P, say – a signal so timed that it arrives just as Q happens. **Draw the signal line.**

7.3 Your diagram should now look like Figure 7.3 (on page 78). Use the slot to verify that it tells the correct story.

But how could A know when to send this signal that is to arrive as Q happens?

He can't know directly. But he could send a series of coded signals; and somebody present at Q could tell him which of them arrived at that moment (or, if he's using radar, this information could be conveyed automatically in the echo). There's no difficulty in principle in assuming that A can know when a signal has to be sent to get to Q.

So here are A's two hard facts – the times of sending a signal to Q and of getting one from Q. **Please satisfy yourself that this is the *only* firm information that A can have about event Q** (§§5.15–16, 5.20). Now we seek rules by which we could calculate the time and distance of Q according to A from the times of P and R.

7.4 But from Figure 7.3 it looks as if there are *two* distances to consider: the distance that one light signal has to travel from A to Q, and the distance the other has to cover to get back from Q to A – put otherwise, the distance between events P and Q, and the distance between events Q and R. **Are these two distances equal or unequal according to A?**

My answer would be 'Equal'. Very likely you disagree. And this is another of those places where you should **carefully work out and formulate your objections, and then criticise yourself, before reading on.**

7.5 'They're obviously unequal', you (perhaps) exclaim, 'Figure 7.3 shows clearly that the distance from P to Q is greater than that from Q to R.'

That argument assumes that the lengths of the lines PQ and QR on the diagram (or perhaps the horizontal distances P to Q and Q to R) can be interpreted as the distances the signals have to go. But we haven't yet worked out how distances are shown on the diagram, and so we can't use it to deduce anything about distances. Besides, we made it clear (§§6.20–21) that A's world line can be given *any* slope (steeper than 45°). So you can't deduce anything from a property of the diagram that depends on this slope.

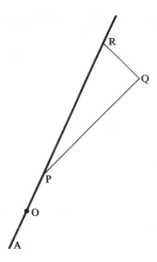

Figure 7.3

7.6 Maybe you objected: 'It all depends on A's motion relative to Q. If (e.g.) he's moving towards Q, then the light has further to go on the outward than on the return journey.' **Any comment?**

An observer moving relative to an *event?* We've been through this before – and further misleading thoughts that often follow. Please try to recall our previous discussion. And if necessary study again §§2.15–18.

I'm afraid you're still thinking in terms of some fixed frame of reference, though you're intellectually convinced that no such thing exists. Bad habits (derived from everyday usage) keep recurring. Thinking in terms of spacemen may help (§§5.19, 7.1). And I'll try one more line of persuasion.

7.7 Let me describe the sort of Earth-bound situation that may be misleading you. I am climbing a very long straight staircase. My enthusiastic young grandson runs up ahead of me, turns immediately he reaches the top, and runs down again. It's clear that he has farther to travel from me to the top than from the top back down to me. That's the sort of thing you're thinking of, isn't it? **But think critically, please.**

It does seem obvious that the boy's downward trip is shorter than his upward one. Concretely, he covers *fewer steps* of the staircase when going down than when going up.

And those 'fewer steps' are the crux of the problem. They're steps *of the staircase.* You're unthinkingly using the staircase as frame of reference. But this staircase is moving relative to me; and so you've established that the two distances are unequal, not according to me, but according to somebody who is stationary on the stairs – somebody for whom the staircase *is* the correct frame of reference.

If we want to measure distances according to *me,* we must use a frame of reference that is stationary relative to me. The staircase (if that's what we use) must move with me. In other words, it must be an ascending escalator, with me standing on one step. The boy's journeys are exactly as before – from me to the top and back. But the measurements in terms of steps trodden will be changed. **On this escalator does the lad have to cover the same or a different numbers of steps from me to the top and from the top back to me?**

This time it will be the same number of steps either way – the number of steps between the one I'm standing on and the one that reaches the top at the same time as he does. The distances are equal – *according to me.* And if I used light signals instead of my grandson, this would still be true. (If doubts persist, imagine this happening in dense fog, so that we don't know whether the escalator is working or not. The boy can still count steps.)

7.8 More generally, distances according to *any* observer must be measured in a frame of reference that is stationary relative to him. All the measuring apparatus must be stationary relative to him. Then we shall find that the distance from him to any event is equal to the distance from the event back to him. And that applies to the distances the signal has to cover from A to Q and from Q back to A in the situation of §7.2 and Figure 7.3.

7.9 Let's get back to our main business. Now that we're discussing over-there events, we are allowed to use what we know about the speed of light. So, assuming that we know the times of events P and R, **can you see how to work out the time of event Q according to** A?

In his opinion the light has the same distance to go on its outward journey to Q as on its return from Q. And (§1.22) it travels at the same speed both ways. Therefore it takes the same time from event P to event Q as from Q to R (still according to A). And so we conclude that:

> The time of event Q according to A is halfway between the times of events P and R by A's clock

– halfway between the times of sending a signal to the event and receiving one from it.

7.10 Now let's consider distances. Taking the radar approach, A will measure the distance of Q from himself in terms of – **in terms of what?**

In terms of the time between P and R – for the longer the signal takes to come back, the farther it has travelled. So when A wants to know a distance, what he actually measures is a time. Then surely the simplest thing would be to agree that *distance is measured by the time that light takes to cover it.*

7.11 There's economy in this, for now the unit of time will do for distance as well. If a distance is such that light takes 3½ seconds to cover it, we'll call it a *distance of 3½ seconds.* (Maybe you suggest 'light seconds' by analogy with the light year of §1.8. But the extra word adds nothing.)

Of course we can easily convert to more familiar units if we want. But normally our distances will be given in terms of whatever time unit we are using (second, year, etc.). I'll call this the *natural unit* of distance.

7.12 Using natural units, **what is the speed of light?**

Speed is distance-travelled divided by time-taken. In these units, distance-travelled by light *is* time-taken, and so its speed is time-taken divided by itself – which is 1.

And that's a handy size to have for this very important speed. Note also that it doesn't matter what unit we choose for time; the speed of light is always 1. So to express the speed of light in this system we don't have to name the units (as we must in other systems – 300 000 *kilometres* per *second,* 670 million *miles* per *hour*). The speed of light is the *pure number* 1, without any statement of units.

7.13 **Now can you see how to express the distance of event Q according to A in terms of the times of P and R by A's clock?**

According to A, in the time between events P and R light travelled from him to Q and back – covering twice the distance between himself and Q. And, using natural units (§7.11), this is the same as saying that twice the distance of Q from A is the time between P and R. So

> The distance of event Q from A according to A, expressed in natural units, is half of the time by A's clock between events P and R

– half of the time that elapses between sending a signal to the event and receiving one from it.

7.14 Re-examining the recipes of §§7.9 and 7.13, an interesting point emerges. We may speak of 'the time of Q according to A', but what we actually refer to is the time of one of A's on-the-spot events – namely the tick of A's clock halfway in time between sending the signal and getting the answer. And what we call 'the distance of Q according to A' is actually a time-interval in A's on-the-spot experience – half of the time that elapses between sending the signal and getting the reply. In spite of the language we use, we're still really concerned with on-the-spot measurements. (The event itself *is* 'over there' of course – since several observers can observe it.)

Let M be the tick of A's clock halfway in time between P and R (Figure 7.14). Then what we call the time of event Q is really the time of on-the-spot event M (§7.9). And the distance of Q from A (in natural units) is really the on-the-spot time-interval from P to M or from M to R (§7.13).

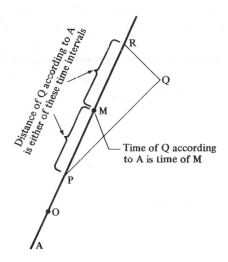

Figure 7.14

As we go on, we shall normally use the ordinary language of over-there times and distances. And that can sometimes lead to conceptual difficulties, which will be greatly eased if you keep reminding yourself of the present article. Make a note to do so.

7.15 The rest of this chapter contains some rather abstruse ideas. You may have to leave them half digested and come back later.

I expect you've realised that we have not been *discovering* recipes for the time and distance of Q according to A. We have no means of knowing these quantities, except through the times of P and R. So all we've been doing – all we *can* do – is to *decide* which particular combination of the times of events P and R we are going to call 'the time of Q according to A', and which other combination we are going to call 'the distance of Q according to A'. We've been *deciding what we shall mean* when we talk of these things – *defining* them, in other words. (You may think it odd to talk of 'defining' something with the humdrum familiarity of distance. But wait a page or two!)

Now a definition is not something that you deduce or discover. It's something you *decide on*. But we want our definitions to be useful in the real world, and we let that guide our decisions. That's why we've kept on appealing to the invariance of the speed of light.

7.16 You may have been worried by an apparently circular argument – using the properties of the speed of light to decide what we mean by time-over-there and distance, when we can't know anything about speed till after we've decided what we mean by time-over-there and distance. But now I'm sure you see that what we were really doing was choosing our definitions in such a way as to make sure in advance that they will lead to the conclusion that the speed of light is the same in all circumstances. And that precaution had to be taken – deciding on definitions which lead to any other conclusion would simply produce a theory that would be completely useless, since it would conflict with one of the most fundamental things we know about the real world.

7.17 In view of what we've said in §§7.15–16, we obviously must ask whether these definitions are consistent with everyday notions of time and distance (within the limited range of conditions where the latter work well). The easiest demonstration of this consistency is that radar is a successful technique – for *it* uses the definitions of §§7.9 and 7.13, and gives results agreeing with our ordinary methods of measurement.

Yet there is more to be said. Our ordinary notion of distance is based on the rigid measuring rod. At first sight this seems to be very different from the definition we've just adopted – so different that it looks like a remarkable discovery when we find that the two give the same result.

But an allegedly rigid measuring rod is in fact a very complex structure, built from a multitude of atoms. If it manages to stay the same length, that is because its atoms stay the same distance apart. But what enables them to maintain their constant separation?

Atomic physics answers that the spacing between atoms is determined by their natural periods of oscillation, which are in effect built-in clocks. The effect is *as if* (though the reality is more complicated) atoms kept exchanging light signals with their neighbours and these had to be sent and received in time with the ticks of the in-built clocks; that implies definite distances between them, and any deviation from these distances would be automatically corrected.

So the distance-keeping is done essentially according to the radar-type definition of §7.13. Thus the rigid rod does not give an alternative way of measuring distance – only a more complicated version of the §7.13 recipe, using billions of radar stations instead of just one.

7.18 And now the statement that the speed of light is always the same begins to look very odd indeed. How do we measure it?

Our theoretical observers (or real spacemen) couldn't measure it at all. Speed is distance-travelled divided by time; and they have no means of measuring distance except by the radar recipe of §7.13, which *assumes* that the speed of light is invariant.

In terrestrial speed-of-light measurements, the distances are found by rigid body methods. So it seems as if we're really discovering something important when our experiments tell us that the speed of light is invariant. **But think again.**

We measure distance by yardsticks or metre rules. But Dame Nature, in arranging for these to have a constant length, did so by using the recipe of §7.13. So she cunningly inserted the invariance of the speed of light into our measuring apparatus; and then – of course! – that apparatus tells us that the speed of light is invariant.

If you're lucky, one bit of mystery starts to make sense. It seemed strange that the speed of light should be so unlike all other speeds (§1.23). But now we see that asserting the invariance of the speed of light is only another way of saying that the quantity we call distance is defined as in §7.13.

This almost amounts to saying that the Michelson–Morley experiment is a fraud. At any rate the conclusion to be drawn from it is different from what we thought. What the experiment really does is to verify that Dame Nature fixes the length of the arms in the apparatus by means of the radar definition. This experiment is not concerned with the speed of light, but with the behaviour of the atoms in the apparatus!

8

Co-ordinate systems

8.1 This is the dullest chapter in the book. But do persevere–we'll be building the diagrammatic equivalent of a piece of mathematical machinery that will enable us to churn out exciting results for a long time to come.

We have to work out how A's ideas about times and distances will be represented on the space–time diagram that we started constructing in §§6.13–22. You probably know how we would have represented them on the old-fashioned diagram of Figure 6.2. To find the time of event S, we should go horizontally across to the time scale and read off '15 seconds'. And to find the distance of S from A, we should measure the horizontal length FS and use the distance scale at the bottom to interpret this as 100 metres.

These can't be the correct answers on the new diagram, but it would be desirable to find something analogous.

8.2 We'll start with the situation described in §7.2 and portrayed in Figure 7.3 – observer A; over-there event Q; light signal sent by A at event P, which arrives as Q happens; return signal from Q reaching A at R. For once you might draw your own version of the figure *carefully;* it will build up into something impressive. It's easier to get a good drawing if you put in the signal lines at 45° slope before doing the world line.

8.3 If A flashed a lamp at event P, he would send out light signals both ways along the 1D universe. The signal line for the positive-going signal is PQ (Figure 7.3). And the negative-going signal line is added as PS in Figure 8.3. Similarly, the signal SR represents the positive-going signal that A would receive (along with negative-going QR) at event R. **Use the slot to check up the story the diagram now tells.** (Events Q and S are related to A in the same way but on opposite sides. **What does this signify?** I'll answer later.)

Draw the line joining S and Q, and let M be the point at which it crosses A's world line. The figure PQRS is a rectangle (since its sides are all at 45° to the horizontal). So the lengths PM, MR, SM and MQ are all equal. And M is the same point as M of Figure 7.14.

8.4 We saw in §7.14 that the distance (in natural units, §7.11) of Q from A according to A is equal to the time-interval from P to M. And this is represented on the diagram by the length PM, using A's time scale (§6.22). But to represent the distance *from* A *to* Q it would be better to have some length stretching *from* A's world line *to* point Q – analogous to length FS of §8.1. **Any suggestions?**
   Length MQ, being equal to length PM (§8.3), will do nicely. So

> The distance (in natural units) of event Q from observer A
> according to A is represented, on A's time scale, by the
> length MQ in the diagram.

   Do make sure you've followed the *logic* of this argument – from the definition of §7.13, through §§7.14 and 8.2–4. That way lies confidence.

8.5 To avoid possible confusions, I want to adopt the convention that the word 'distance' shall always refer to the 1D universe and the word 'length' to the space–time diagram. Thus I can write (as I did in §8.4) of a certain *length* (on the diagram) representing a certain *distance* (in the 1D universe).
   Another confusion can be avoided if we agree that 'direction' always refers to one of the two directions in the 1D universe (§6.14); and that when we wish to describe the way a line in the diagram points, we'll

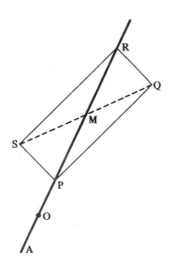

Figure 8.3

speak of its 'slope'. Thus §6.15 describes how signals travelling in opposite *directions* are represented by lines that *slope* different ways.

8.6 The length MQ (representing Q's distance according to A) is not measured horizontally (as it was in the old-fashioned diagram, §8.1). If you draw two or three versions of Figure 8.3, giving A's world line different slopes, you'll find that the slope of MQ depends on that of the world line. **Can you spot a relation between them?**

Through M draw a signal line, MU (Figure 8.6). **Can you answer now?**

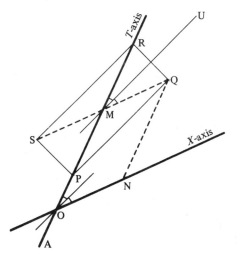

Figure 8.6

Since PQRS is a rectangle, with two of its sides parallel to MU, it follows that angles QMU and UMR are equal.† So

> The slope of the line MQ is such that it makes the same angle with a signal line as does A's world line, but on the other side.

And that's distinctly different from the old-fashioned diagram, where distances are always represented by horizontal lengths.

8.7 No matter what point we take as Q, the rule we've just worked out will always give *the same* slope for MQ. There's a conventional way of indicating this. Through O (representing the zero of A's clock) we draw a line which makes the same angle with a signal line as does A's world line, but on the other side (Figure 8.6). Then, of course, dis-

---

† Their equality is indicated by the two little circular arcs near M. We specify an angle by naming successively a point on one arm, the vertex, and a point on the other arm.

tances according to A are represented by lengths (like MQ) parallel to this new line – which is therefore called A's *distance axis,* and labelled (for reasons that can wait) '*X*-axis'.

Thus the distance axis specifies the slope to be used when measuring lengths that represent distances according to A.

8.8 Observer A's distances are represented according to the same scale as is used on his world line for times (§8.4). If the time scale (specified in §6.22) is marked along his world line (Figure 8.8), we can also mark the same scale along his distance axis. This replaces the horizontal distance scale of old-fashioned Figure 6.2.

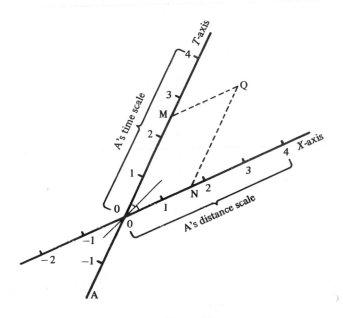

Figure 8.8

And then we have an easy recipe for reading A's distances from the diagram. If Q is *any* event, draw a line through the corresponding point Q of the diagram, parallel to A's distance axis, meeting his world line at M. Then in natural units the distance of event Q according to A is given by length MQ interpreted according to A's distance scale. (We could just as well use his time scale, of course; but normally we shall not mark either of them – §7.2.)

8.9 **What length in Figure 8.6 represents the time of event Q according to A?**

The time of Q is the same as that of M (§7.14). And by §6.22, the time of M (starting from the zero of A's clock) is represented on A's time scale by length OM. So

> The time of event Q according to A is represented on A's time scale by the length OM – or equally by length NQ (parallel to A's world line).

8.10 Thus A's times are represented on the diagram by lengths measured along or parallel to his world line. And so we say that his world line is also his *time axis* (cf.§8.7). And we label it '*T*-axis'.

Notice again the contrast with the old-fashioned diagram, where every observer used the vertical time axis (Figure 6.2). By identifying the time axis with the world line, our new diagram emphasises that each observer has his own private time (§4.12).

8.11 And now we have a recipe (like that of §8.8) for reading A's times from the diagram. Through Q (Figure 8.8) draw a line parallel to A's time axis, meeting his distance axis at N. The time of event Q is given by the length NQ, interpreted according to A's time scale.

8.12 **Can you prove the following propositions?**

> All the events which A regards as happening at the same distance from himself (in one direction) are represented by the points of a line parallel to his time axis.
>
> All the events which A regards as happening at the same time are represented by the points of a line parallel to his distance axis.

A line starting from any point of MQ, parallel to the *T*-axis and ending on the *X*-axis will be the same length as OM and NQ. Applying the recipe of 8.11 to this line (instead of to NQ) finishes the proof of the second statement. Use §8.8 for the first.

8.13 We can now cope with the difficulty that I mentioned in §6.12. Make a mark against the slot, roughly one third of the way along from the left-hand end, to represent the zero from which distances are measured. Now place the card on top of Figure 8.6 in such a way that: (1) the slot lies parallel to A's distance axis; and (2) the zero-distance mark lies over A's world line.

Then because of (1), all the points showing through the slot represent events that happen at the same time *according to* A – which circumvents the trouble of §6.12. And (2) arranges that the slot will give a correct representation of distance from A according to A.

Now move the diagram downwards (as in §6.7) but in such a way as to keep (1) and (2) always true. Then you will get a moving picture of happenings on the 1D universe *as they appear to* A. **Get some practice by applying this improved slot technique to Figure 8.6.**

You should see things like all of SMQ appearing in the slot at the same moment (implying that the corresponding events are simultaneous), the movement of red dots showing signals from A to events Q and S and back to A, and so on – all as they appear to A. **And now can you answer the question of §8.3?**

The new slot technique shows S and Q happening simultaneously and at equal distances from A in opposite directions. **Prove this formally, using the propositions of §§8.4 and 8.9.**

8.14M The time of Q is the same as that of M (§7.14), which is positive or negative (§§6.23–5) according as M, and therefore Q, lies above or below the distance axis. Also, event Q happens after or before O according as point Q lies above or below the distance axis (check with slot). Putting these statements together, events which (in A's opinion) happen after O have positive times (according to A), and those happening before O have negative times – which is just the sort of arrangement we should want. Similarly we can regard the distance of an event from A according to A as a positive or negative quantity according as it is on the positive or negative side of him (in the sense of §6.14 – a minor inconsistency will be cleared up in §16.2).

We shall normally arrange our diagrams in such a way that most of the times and distances that concern us are positive. But our reasoning would still work if any of them were negative. If you don't feel at home with negative quantities yet, forget the point for the present. I'll remind you when necessary.

8.15M If A has to specify an event like Q – locate it uniquely – he must give *two* measurements: its time and its distance. Correspondingly, the position of Q in the diagram is fixed by means of the two lengths that represent its time and distance. These two lengths are called the *co-ordinates* of Q – A's co-ordinates of Q. The length OM or NQ (Figure 8.6), which represents the time of Q according to A, is known naturally enough as the *time co-ordinate*. And length ON or MQ is the *distance co-ordinate*. Alternatively, if it better suits our purpose, we can think of the actual time and distance measurements as being the co-ordinates. Of course there's nothing special about Q, and so every point (or event) has its time and distance co-ordinates.

The whole system that we've been developing for laying out the diagram and measuring on it is called A's *co-ordinate system*. It consists essentially of A's time and distance axes, the co-ordinates themselves, and the rules for using them. Finally, the point O is called the *origin*.

8.16 Omitting the signal lines from Figure 8.6 gives Figure 8.16, which exhibits the essentials of A's co-ordinate system. The origin and axes are marked. It would be a useful exercise to read through §§8.2–15 again, leaving out anything that does not appear in Figure 8.16. This will give you a fuller description of the co-ordinate system. Use the slot, of course.

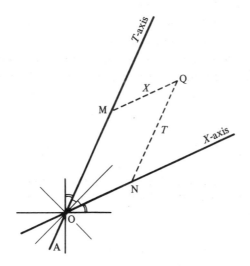

Figure 8.16

The markings at O show that, instead of the equal-angles-with-signal-line rule of §8.7, we could say that the distance axis makes the same angle with the horizontal as the time axis makes with the vertical.

8.17 Signal lines have disappeared from our diagram, and they will be absent from many others in future. As a result, it's easy to find yourself looking at things as if the time and distance of an event can be directly known without using transmitted information. And that can lead to serious misapprehension.

Somewhere or other you have a note reminding you to return occasionally to §7.14. Would you please add the present article – as a reminder that almost all knowledge of the over-there world is obtained by light signals, even when text and diagrams fail to say so.

8.18 To complete the labelling of Figure 8.16, it would be handy if we could mark the lengths NQ and MQ with 'the time of event Q according to A' and 'the distance. . .'. But such long-winded phrases would clutter the diagram. So we abbreviate. We agree to use the symbol T to stand for the phrase 'time of event Q according to A'.

Please note the exact way I've put it: the symbol T is to *stand for* the longer phrase. It is simply a shorthand way of writing 'the time of event Q according to A'.

And now we can, of course, mark T against the length NQ in Figure 8.16. This T is just a space- and labour-saving device – nothing more mysterious than that. If you tend to think of algebraic symbols like T and x and n in some other way, *not* as abbreviations for phrases, then I beg you please to read the 'M' article that follows.

19M It's most important that we shall approach these algebraic symbols (*a*) without fear, and (*b*) in a way that maximises our command over them. Whatever you may have been taught in the past, *don't* think of that T as something for which you can, in a particular case, substitute '3 seconds' or some other actual time. Of course you *can* do that. But it's rather a side issue. Think of T primarily as a short way of writing the longer phrase.

And for the present I'd like you to read it that way – in full. If one of my colleagues gets a note signed 'SL', he does not think that it comes from somebody called 'Ess Ell'. He knows it's short for 'Sam Lilley'. Similarly, when you meet T, don't say 'tee' (or think 'tee' inside your head); say (or think) the whole phrase 'the time of event Q according to A'.

Naturally, as with any other phrase you use frequently, you'll tend to abbreviate it as time goes on. Almost immediately you'll start leaving out the word 'event' – knowing quite well that Q must be an event. Later you may simply say 'the time according to A' or 'A's time' (if the context makes it clear that it's the time of Q you're concerned with). In suitably unambiguous circumstances, you may even reduce it to 'the time'. And maybe one day, when you're fully competent in these things, you'll just say 'tee'. But even then I hope the thought 'time of event Q according to A' will be at the back of your mind.

8.20 Similarly we'll use X to *stand for* 'the distance of event Q from A according to A'. And then we can mark X against the length MQ in Figure 8.16. (In reading that last sentence did you follow the advice of §8.19?) Alternatively we could mark T against OM and X against ON. (*Did* you read that correctly?)

Now you can see why I labelled A's axes as 'T-axis' and 'X-axis'. If T stands for the time (of Q) according to A, or more briefly 'A's time', then 'T-axis' is a handy way of writing 'A's time axis'.

21M To avoid confusion between two quite different uses of letters, we employ italic (sloping) type for letters – like T and X (or x or y) – used as abbreviations for phrases that refer to numbers or quantities; and Roman (upright) type when a letter is used for anything else – Q and M (or even m) for points and events, A and B for observers.

8.22 Now it's time to introduce other inertial observers. We can't just assign them what world lines we please (as we did for A in §6.20), because once A is in the diagram the world lines of other observers must (as it were) be what A says they should be. We might hope that other *inertial* observers would, like A, have straight world lines (experiments with the slot are indicated), but we need more than hope! Let's investigate.

8.23 Consider another observer, B, whose world line is straight (Figure 8.23). We can reset A's clock to read zero as B passes him, so that the crossing point of the two world lines becomes O (§6.22).

Draw the line HK parallel to A's distance axis, meeting A's world line at H and B's at K. We can allow K to take over (for a little while) the role formerly played by Q. Then H will similarly replace M (since HK, like MQ, is parallel to the X-axis). The propositions of §§8.4 and 8.9 remain true whatever letters we use to label the events. And in this case they tell us that the time and distance of K according to A are represented (on the same scale) by lengths OH and HK. So in A's

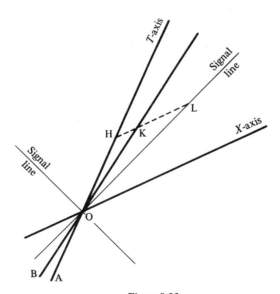

Figure 8.23

opinion, B travels the distance represented by length HK in the time represented by length OH.

Now imagine the line HK moving up the page – representing the passage of time. Then length HK increases proportionally to length OH. (That simply means that the two lengths increase together in such a way that when one of them is multiplied by a certain number, the other is also multiplied by the same number.) Thus, using the end of the previous paragraph, in A's opinion the distance travelled by B increases proportionally to the time of travel. In other words, the speed of B relative to A is constant; and then §5.6 tells us he's inertial. So

> The world line of any inertial observer, B, who passes A at event O is a straight line through point O.

(If you know enough to be worried about the 'conversely' question, you'll also know how to deal with it.) I've drawn B's world line steeper than 45°. We haven't proved that it must be so, but that can be investigated later (cf. §§6.18–20).

8.24 Will B's on-the-spot times be represented to scale along his world line, as A's are (§6.22)? It would seem probable. But a proper proof at the moment would be a big bore; so I'm postponing it to §15.7, where it will follow easily from something else we'll be doing. In the meantime let's *assume* that

> Times of B's on-the-spot events, as timed by his own clock, are represented to scale along his world line

– and that this will hold good for *any* inertial observer. (We don't know how B's scale will be related to A's, but you'll be surprised how far we can go without suffering from this ignorance.)

8.25 We originally specified only two things about A's representation on the diagram: (i) a straight world line, steeper than 45° (§6.20), and (ii) his on-the-spot times to be represented to scale (§6.22). These properties (as we've partly proved, partly will prove) hold also for B. So B's co-ordinate system must be related to his world line and scale exactly as A's are related to his. And we are now in a position to compare the two.

The next few articles consist largely of a description of how a rather complicated diagram is built up step by step. To get a full grip of it, you *must* redraw it for yourself bit by bit as I describe it. This is another occasion on which it will pay you to draw carefully and accurately. *And keep your drawing – it will be very useful in future.*

8.26 We start Figure 8.26 with A's and B's world lines (appropriately labelled). These meet at the origin – the point O representing the event
of A and B passing. Both their clocks read zero at O. Observer A's
world line is also his time axis (§8.10) and is marked *T*-axis, because
we are using *T* to stand for the time of a typical event, Q, according to
A (§8.18). If we use *t* for the time of Q according to B, then we can
similarly use the label '*t*-axis' for B's time axis (world line).

Similarly, if *x* is used to stand for 'the distance of event Q from B
according to B', then B's distance axis will be labelled *x*-axis, just as
A's is labelled *X*-axis (§§8.7, 8.20). The two signal lines through O
are also named. And each observer's distance axis is obtained by
measuring the angle between his time axis and a signal line, and then

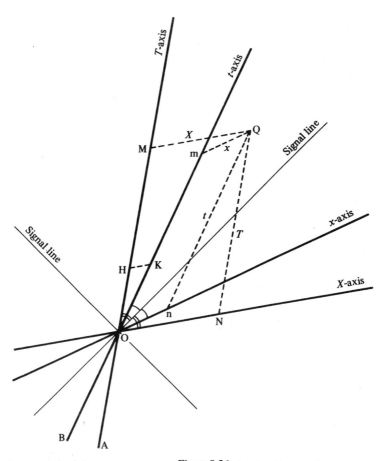

Figure 8.26

drawing a line that makes the same angle with this signal line but on the other side (§8.7). Put otherwise, his distance axis makes the same angle with the horizontal as his time axis makes with the vertical (§8.16).

8.27 We can now use the improved slot technique to produce a moving picture of happenings on the 1D universe as they appear to whichever observer we choose – merely substituting B for A in the instructions of §8.13 when we want B's viewpoint. **Use the slot now to discover whether A and B agree about which events are simultaneous with O.**

8.28 Now let's choose *any* event Q and mark the corresponding point in the diagram. And then we put in four lines: MQ parallel to A's distance axis, meeting his time axis at M; NQ parallel to his time axis, meeting his distance axis at N; mQ parallel to B's distance axis, meeting his time axis at point m; and nQ parallel to his time axis, meeting his distance axis at n. (Note the systematic use of capital and small versions of the same letter in relation to A and B, respectively.)

Each observer has his own scale for representing his on-the-spot times along his world line (§§6.22, 8.24), and he uses this same scale for over-there times and for distance measurements expressed in natural units (§§8.4, 8.8–9, 8.11). **Now work out which of the various lengths in the diagram represent (on the appropriate scale) the time/distance of Q according to A/B.**

The time of Q according to A is represented on A's scale by the length NQ (or OM) (§8.9). And with the agreed abbreviation of §8.18 we can label NQ as T. Similarly, the distance of Q from A according to A is represented on A's scale by the length MQ (or ON) (§8.4) – which, with the abbreviation of §8.20, is labelled X.

To get B's co-ordinates we have only to repeat this, substituting his scale and his axes and lines parallel to them for A's (changing M to m, etc.). **Please do so.**

The time of Q according to B is represented on B's scale by length nQ (or Om). And the distance of Q by length mQ (or On). And since we've adopted abbreviations t and x for these quantities, we can label nQ with t and mQ with x. (*Please* bear in mind the advice about reading these shorthand symbols in full.) We can summarise those prescriptions:

> An observer's version of the time of event Q is represented by the length of the line drawn parallel to his time axis from his distance axis to point Q;
> His version of the distance of event Q from himself is re-

presented by the length of the line drawn parallel to his distance axis from his time axis to point Q
   – the observer's own scale being used in both cases.

   If you can remember those indented formulations, it will save you a
lot of trouble in future.

8.29 **Now I'd like you to spend some time studying Figure 8.26 carefully,
making full use of the slot in the manner of §8.27.**

   I'm sure you'll be able to see, without me rubbing in every detail,
how this diagram contrasts with the old-fashioned one of Figure 6.2,
where (for instance) lengths representing times are always measured
vertically and those representing distances horizontally – no matter
which observer is supposed to be measuring. If the new diagram is a
good representation of space–time, then it begins to be rather obvious
that observers will disagree about many things on which we used to
think they would agree. Let's exemplify by comparing opinions about
(*a*) which events happen at the same distance (from the observer) and
(*b*) which events happen at the same time.

8.30 Events which A regards as happening at the same distance are represented by points of a line parallel to his time axis (§8.12) – for example,
points of line NQ. Similarly, events which B regards as happening at
the same distance are represented by points of a line parallel to *his*
time axis – for example, the line nQ. So it's clear that A and B *disagree
about what events happen at the same distance.*

   Check this with the slot as in §8.27. **Does it seem odd?**

   If I'm in a moving train and you're on the station platform, I regard
all events that occur in the locomotive as happening at the same distance from me, but you regard all events at the end of the platform as
at the same distance from you. We disagree! This deduction corresponds to everyday experience.

8.31 Events which A regards as happening at the same time are represented
by points of a line parallel to his distance axis (§8.12) – for example,
points of the line MQ. Similarly, events which B regards as happening
at the same time are represented by points of a line parallel to *his*
distance axis – for example, the line mQ. So it's clear that A and B
disagree about what events happen at the same time.

   Again please check up with the slot. We've come back to the Relativity of Simultaneity (§2.19).

8.32 So according to Figure 8.26 the disagreement about at-the-same-time is
on a par with the disagreement about at-the-same-distance. The one
corresponds to the differing slopes of MQ and mQ, the other to the

differing slopes of NQ and nQ. It's just a question of interchanging the time and distance axes and the things that go with them. I've emphasised this point by writing the opening paragraphs of §§8.30 and 8.31 in such a way that you can turn either into the other by interchanging 'time' with 'distance', N with M and n with m.

The diagram seems to say that the disagreement about at-the-same-time is very much the same thing as the disagreement about at-the-same-distance. **Then why do we take the latter as an obvious everyday affair, whereas we've never noticed the former and find it hard to imagine?**

8.33 'We don't move fast enough', you say. True, we made that point in §2.11. But it's not the whole truth. Small relative speed merely means that the angle between the two world lines is small; and so the angle between the distance axes will also be small. Then MQ and mQ will nearly coincide, and the disagreement about simultaneity will be too small to notice. But equally, the diagram says, NQ and nQ will nearly coincide, and the disagreement about distance will be too small to notice. Whatever the relative speed, the one effect should apparently be as noticeable as the other. Yet that's not our experience. **Why?**

8.34 The answer lies in the contrast between the sort of distance that we feel to be normal in ordinary life and the 'natural' distance unit (§7.11) on which the diagram is based. One second seems to us to be a modest, rather short time. The corresponding natural unit of distance – equivalent to 300 000 kilometres – seems enormous. Yet these are represented by *equal* lengths on the time and distance axes of Figure 8.26.

Suppose the relative speed is such that when two events are equidistant according to A, B says their distances differ by one-millionth of a natural unit. That's 300 metres, and of course we notice it. If the events are simultaneous for A, but separated by a millionth of a second for B, that's the *same* difference as before – a millionth of a time unit instead of a millionth of a distance unit. But with our ordinary notions of magnitude, it's very tiny.

If we were used to travelling at very high speeds, then presumably we should regard a millionth of a second as just as important as 300 metres. And then the Relativity of Simultaneity would be just as real for us as the familiar disagreement about distance. That's how 'don't move fast enough' (§8.33) comes into it.

8.35 I promised in §4.13 that we should eventually construct 'a new coherent theory of space and time and motion into which things like the three-clocks affair and the Relativity of Simultaneity will fit naturally'. Much of that new theory is implied in Figure 8.26 (though we have quite a way to go before we can make it all explicit), just as Figure 6.2 embodies many of our old beliefs (§6.4).

Figure 6.2 denies the possibility of disagreement about simultaneity. But Figure 8.26 says that this disagreement is very closely related to the familiar at-the-same-distance disagreement – brother and sister in the new theory that is beginning to emerge. So am I making some progress towards fulfilling my promise? Is the Relativity of Simultaneity growing a little easier to live with?

8.36 Here's a way of putting the Relativity of Simultaneity which seems to display absolute absurdity. As A and B pass (event O), A says 'Event N is happening now' – to which B replies 'Rot! It has happened already' (slot!). Surely that's ridiculous. **Criticise, please.**

This ignores the warning of §8.17. You can't know about what's happening *now* – information takes time to reach you. Neither observer can express any opinion about the time of event N until considerably later when they receive signals from N. Then A radios 'I calculate that N happened as you and I were passing'; and B answers '*My* calculations show it had already happened by then.' And that past-tense interpretation is a good deal easier to stomach. Add the present article to the note that reminds you to revisit §§7.14 and 8.17 occasionally.

8.37 The line HK in Figure 8.26 is drawn parallel to A's distance axis – just as it was in §8.23, where we proved that in A's opinion B travels distance represented by length HK in time represented by length OH. So (speed being distance-travelled divided by time-taken) we deduce that

$$\text{speed of B relative to A} = \frac{\text{length HK}}{\text{length OH}}. \qquad (8.1)$$

The (8.1) on the right – like others in similar form that will follow – serves to identify the equation when we want to refer to it again.

8.38M The thing that looks like a fraction on the right-hand side of (8.1) means 'length HK *divided by* length OH' – just as $\frac{3}{4}$ is what you get by dividing 3 by 4. We shall normally write division in this 'fraction' form.

8.39 Equation (8.1) refers to speed measured in natural units, in which the speed of light is 1 (§7.12). So it expresses speed *as a fraction of the speed of light.*

In natural units, speed is a time divided by a time – in other words a *pure number* that does not involve any units of measurement (cf. §7.12).

8.40 In Figure 8.23 the line HK has been extended to meet the signal line at L. Then, with L playing the role of K, (8.1) gives

$$\text{speed of light relative to A} = \frac{\text{length HL}}{\text{length OH}} \, . \qquad (8.2)$$

But line HKL (being parallel to the X-axis) makes the same angle with a signal line as does the T-axis; and so angles HOL and HLO are equal. Hence (by a well known bit of geometry) the sides HL and OH of the triangle HOL are equal in length, and so (8.2) says that the speed of light relative to A is 1. But A could be any observer, and so the speed of light is always 1.

Of course we decided that the speed of light was going to be 1 way back in §§7.10–12 when we chose natural units. So here we are merely showing how the diagram presents the consequences of this choice. You must have felt that there was something odd about the equal-angles-with-signal-line rule for the axes. But now you see that this is how the diagram arranges to exhibit the invariance of the speed of light. So in this relation between the slopes of the axes we have the very essence of a *relativistic* space–time diagram.

8.41 In Figure 8.23, the length HK is clearly less than length OH. So (8.1) says that the speed of B relative to A is less than 1 – less than the speed of light. On the other hand, if you extend HL further to the right, and then swing B's world line round to a slope less than 45° (taking K with it), you'll see that HK will become greater in length than OH. Then by (8.1) B's speed relative to A will be greater than 1. So (check this with the slot)

> An observer's world line is steeper or less steep than 45° according as he is moving, relative to A, slower or faster than light.

8.42 If we give A a vertical world line, then his distance axis will be horizontal (§8.26) – as in Figure 8.42. So the two axes are at right angles, and the co-ordinate system is said to have *rectangular axes* (as contrasted with the 'oblique axes' of other cases). The lengths MQ and NQ, representing the distance and time of Q according to A (§8.28), are now horizontal and vertical.

8.43 Of course there's nothing special about an observer with vertical world line. All inertial observers are equivalent (§5.11). *Every* straight line steeper than 45° is potentially the world line of an inertial observer. And some of these world lines happen to be vertical. That's all.

By way of analogy, every map has a centre, but there's nothing special about the place it represents. On the other hand, if you are manager of London airport, you may find it convenient to *draw* your

map with London at the centre. And similarly, if there is some observer that *we* are specially interested in, we are entitled to give him
rectangular axes – which we find easier to draw and think about.

8.44 We have now finished the dull labour of building up this space–time
diagram, and are ready for the more exciting business of using it. So
now is the time to sort out any remaining difficulties. For the most part
you must wrestle with them yourself. The solutions are certainly hidden in what we've done, and you must search them out. I have space
for a few hints only.

The biggest difficulty, conceptually, is perhaps to convince yourself
that the slope of a world line has no significance – just as speed, as an
absolute, has no significance. The *relation* between the slopes of two
world lines corresponds to the *relative* speed of two observers (§8.37).
We can put in *one* observer's world line at any slope we like (steeper
than 45°), and then the slopes of all other observers' world lines will be
fixed by their speeds relative to him.

We can even draw several diagrams telling the same story, but as-
signing the world line of one particular observer a different slope in
each diagram. These will be like perspective views of the same scene
from different angles.

One's first response to a world line just a little steeper than 45° may
be that it must represent an observer 'going at almost the speed of
light'. Ah, but relative to whom? Relative to an observer with almost
vertical world line. But then the second is going at almost the speed of
light relative to the first. It's not the slope that counts, but the relation
between slopes.

8.45 No matter what the slope of A's world line is, §8.37 gives a recipe for
placing B's world line so that the two shall be moving at any relative
speed you care to name. You draw a line HK parallel to the X-axis, as
in Figures 8.23 and 8.26; and if, for example, the speed is to be $\frac{3}{4}$, then
you measure out length HK as $\frac{3}{4}$ of length OH, and draw B's world
line through O and K.

You can use this recipe to cope with what I call 'squeezing in' problems. If A's world line is tilted almost to the 45° slope, that leaves little
room between it and the signal line. **Then how can we squeeze in as
many world lines of other observers as we could if A's were vertical?**

Take any world line in the latter diagram – call the observer B – and
use (8.1) to find the relative speed of A and B. Then use the recipe
with which this article opened to place B's world line on the other

diagram. If a world line can be got onto one diagram, this gives a method of squeezing it into the other.

Finally, keep remembering that Figure 8.26 (for instance) is a *diagram* of something concerning space and time and motion – *not a picture*. You must not interpret it naively – any more than you would take a diagrammatic map of London's Underground as instructing you to turn through a right angle west of Earls Court in order to get to Wimbledon.

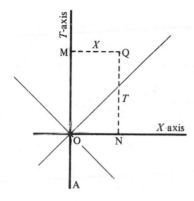

Figure 8.42

9

Combining speeds

9.1 I am standing by a railway line. You are sitting in a train that passes at 100 kilometres per hour, and you notice the ticket collector walking forward along the corridor at 5 kilometres per hour. How fast is he going relative to me? The ordinary answer is 105 kilometres per hour – got by *adding* his speed relative to you and your speed relative to me. But I want to question whether that answer is completely accurate.

9.2 Let's put the problem in more general form – in terms of three observers on the 1D universe (§5.19). We'll use symbolic abbreviations for quantities – as we formerly did for T, X, etc. (and do take note of §8.19).

We'll use the symbol u to stand for the phrase 'the speed of observer B relative to observer A'; v to stand for 'the speed of C relative to B'; and w for 'the speed of C relative to A' – all speeds in natural units (§§7.11, 8.39). Then we are asking whether it is true that

$$w = u + v . \tag{9.1}$$

9.3M If you have been taking due notice of §8.19, you will have no difficulty in understanding (9.1). This is a complete sentence in which the verb is '='–'is equal to'. And '+' is a conjunction we usually read as 'plus' or 'added to'. **Now read (9.1) in full.**

You could put it as 'The speed of C relative to A *is equal to* the speed of B relative to A *plus* the speed of C relative to B'; or a little more briefly as 'The speed of C relative to A is equal to the sum (§3.15) of the speeds of B relative to A and C relative to B.'

The clumsiness of these formulations illustrates the virtue of symbolic abbreviations. But it might also tempt you to ignore my read-it-in-full advice. So let's compromise. Let's refer to the speed of B relative to A and the speed of C relative to B as the 'given speeds', and to the speed of C relative to A as the 'resultant speed'. **Now read (9.1).**

'The resultant speed is equal to the sum of the given speeds' – tolerably brief.

9.4 **Why have we come to believe that (9.1) is true? Experience?**

Certainly. But limited experience which does not exclude the possibility that (a) even at ordinary speeds w might differ slightly from $u + v$, and (b) at higher speeds they might be very different.

9.5 To calculate w, you have to combine u and v in some way or other. But it's not necessarily true that when you combine quantities you must do so by adding. If the cost of living rises by 20 per cent one year and 30 per cent the next, the resultant rise is not 50 per cent, but 56 (£100 rises to £120 in one year; and 30 per cent of that is £36). So maybe combining speeds could also be more complicated than merely adding.

9.6 People often feel they can *prove* that it's only a matter of adding – by a simple process of calculating how far the train and the ticket collector advance in one hour. **Why do such arguments fail?**

   Some of the times and distances are measured by you on the train, others by me at the trackside. And we don't agree about times and distances (§§3.10, 3.38). **Think this out more carefully; consider the train as a measuring rod.**

9.7 Suppose C were a light signal, instead of an observer. **Would (9.1) be true?**

   In this case we should have $v = 1$ and $w = 1$ (§7.12). And (9.1) would say that $1 = u + 1$ – which is obviously false. So (9.1) won't work for light. (If you can't see how I got to the statement that $1 = u + 1$, you're ignoring my read-it-in-full advice.)

9.8 Let's appeal to the space–time diagram for help. We may as well draw A's world line vertical (Figure 9.8), so that he has rectangular axes (§8.42). His axes and B's are labelled as usual (§8.26). And C's world line is merely marked with his name.

   With the aid of §8.41 we reach a shocking result straightaway. If B is moving slower than light relative to A, that article tells us that B's world line *must* be steeper than 45° (as, indeed, I've drawn it). And if C is moving slower than light relative to B, a further application of the same theorem (with A changed to B – what's in a name?) shows that C's world line must also be steeper than 45°. Finally, using the §8.41 proposition the other way round, we deduce that the speed of C relative to A is less than the speed of light. So this ridiculously simple bit of reasoning has led to the striking conclusion that: If the given speeds are both less than the speed of light – no matter how slightly – then the resultant speed is also less than the speed of light.

   And this means that (9.1) can't be true (consider, e.g., the case where u and v are both $0 \cdot 9$). Surely (9.1) is taking a hammering.

9.9 To find the true relation, let's develop Figure 9.8 further. Take the point H on the *T*-axis, such that length OH = 1. Draw HK parallel to the *X*-axis (horizontal) meeting the *t*-axis at K. Then draw KL parallel to the *x*-axis, meeting C's world line at L. From there draw LN parallel to the *X*-axis, meeting the *T*-axis at N. And finally draw KM parallel to the *T*-axis, meeting LN at M. Use the slot to make sure you understand the diagram, considering it from both A's and B's viewpoints, and paying attention to questions of simultaneity.

We are about to embark on what may seem, if you're inexperienced, a long and involved piece of mathematics. Don't be discouraged. First re-study §§3.13–22 to put you in the mood. Then work through §§9.10–19 as well as you can, keeping faith with the read-it-in-full attitude to symbols. Give serious thought to any point that presents difficulties; but if necessary, pass on, leaving the trouble to be sorted out later. I'll offer further advice in §§9.20–1.

9.10 Referring to §8.37 if help is needed, see if you can work out **what, in Figure 9.8, represents the speed of** B **relative to** A.

The line HK plays exactly the same role here as it did in §8.37 (running from the *T*-axis to the *t*-axis parallel to the *X*-axis). So equation (8.1) provides the answer. But we've made length OH = 1, and we're writing *u* for 'the speed of B relative to A'. So we have

$$u = \text{length HK} . \tag{9.2}$$

What represents the speed of C **relative to** B?

This time the lengths OK (along B's world line) and KL (parallel to his distance axis) take the place of OH and HK. And with these alterations, (8.1) gives

$$v = \frac{\text{length KL}}{\text{length OK}} . \tag{9.3}$$

And the speed of C **relative to** A?

Here ON and NL replace OH and HK. So

$$w = \frac{\text{length NL}}{\text{length ON}} . \tag{9.4}$$

And from (9.3) we deduce (see §9.11) that

$$\text{length KL} = \text{length OK} \times v . \tag{9.5}$$

9.11M We get (9.5) from (9.3) by using one of the standard tricks for manipulating equations (cf. §3.19). If you multiply two equal quantities by the same amount, you end up with two equal quantities. In other words, if you multiply both sides of a true equation by the same amount, the result is another true equation. Here we multiply both sides of (9.3) by length OK. So the new left-hand side becomes length OK × *v*. And the new right-hand side is

$$\frac{\text{length KL}}{\text{length OK}} \times \text{length OK}$$

But dividing by length OK and then multiplying by length OK brings you back where you started; so the right-hand side is just length KL. Putting these bits together, we get

length OK $\times v$ = length KL .

And turning that back to front (§3.17) gives (9.5).

9.12M The 'multiply both sides' process in §9.11 is an example of what we call an *allowable operation* – an operation which, when it is applied to a true equation, yields another true equation. We've already met allowable operations in §§3.17 and 3.19. Further obvious examples are: adding the same quantity to both sides; and dividing both sides by the same quantity.

9.13 We've expressed u, v and w in terms of certain lengths. In order to find the relationship between the speeds, we need to discover a bit of geometry connecting the relevant lengths together.

Line KL is parallel to B's distance axis, which makes the same angle with the horizontal as his time axis does with the vertical (§8.26). So the angles MLK and HOK are equal. Also angles LMK and OHK are both right angles. And so the triangles LMK and OHK are the *same shape*. You could produce triangle LMK by making a scale drawing of

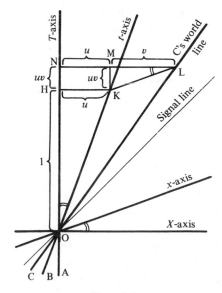

Figure 9.8

OHK (you'd also have to reverse it – by drawing from the back of the paper or using a mirror). So the sides of the smaller triangle are got by reducing those of the larger, all in the same proportion. In other words, there is a certain fraction, which we can call the 'scale factor', such that

> length of each side of triangle LMK
> = length of corresponding side of triangle OHK × scale factor.

There is an equation like that for *each* pair of corresponding sides, and it's the *same* scale factor in all three. **What is this scale factor?**

One pair of corresponding sides is KL and OK. And (9.5) says that

> length of side KL
> = length of corresponding side OK × v.

Comparing this with the previous equation shows that the scale factor is v. Then we can put v instead of 'scale factor' in the equations for the other pairs, getting

$$\text{length ML} = \text{length OH} \times v\,, \tag{9.6}$$

and

$$\text{length KM} = \text{length HK} \times v\,. \tag{9.7}$$

But length OH = 1 (§9.9) and so (9.6) reduces to

$$\text{length ML} = v\,. \tag{9.8}$$

And by (9.2), length HK = u, so that (9.7) becomes

$$\text{length KM} = u \times v\,, \tag{9.9}$$

which is conventionally written as

$$\text{length KM} = uv\,. \tag{9.10}$$

9.14M Mathematicians in their thirst for brevity resent even having to write the '×' in $u \times v$. So they agree to the convention that

> When two symbols standing for numbers or quantities come one after the other (with nothing in between), then they are to be multiplied – exactly as if there were a '×' between them.

So (9.10) means the same as (9.9). Read them both in full.

By the way, the result of multiplying two numbers or quantities is called their *product*.

9.15 Mark OH, HK, ML and KM on your own diagram with 1, u, v and uv, which we have shown to be their lengths. And then **try to work out the correct formula for** w.

Length NM is also equal to u (opposite sides of a rectangle), and so

$$\text{length NL} = u + v \tag{9.11}$$

–the sum of the given speeds. Similarly, length HN is equal to length KM, which is *uv;* and so

$$\text{length ON} = 1 + uv .\tag{9.12}$$

You can conveniently read the right-hand side there as 'one plus the product (§9.14) of the given speeds'. **Can you do it now?**

The two lengths we've just calculated come into equation (9.4). And so we have

$$w = \frac{u + v}{1 + uv} .\tag{9.13}$$

(See below if that step's not clear.) And *that*'s the equation we're looking for to replace (9.1).

9.16M The process of getting from (9.4) to (9.13) is one that we take for granted in everyday thinking. If you know that (1) John's father is Charles, and you're told that (2) John's father is ill, then you deduce that (3) Charles is ill. The thought process (normally unconscious) is that (1) allows you to replace the words 'John's father' by 'Charles'; and that if you make this *substitution* (as it's called) in (2), you end up with (3).

Now take (9.4) and substitute for the lengths NL and ON the expressions which (9.11) and (9.12) say you may. You'll end up with (9.13).

9.17M **Now can you read** (9.13) (perhaps using the semi-abbreviated forms suggested in §9.15)?

The horizontal line indicates division (§8.38). And you already know how to translate $u + v$ and $1 + uv$ into longhand. So you can read it as

> The resultant speed is equal to the sum of the given speeds divided by one plus the product of the given speeds.

Actually the form given in the next article may appeal to you more. The main point is to think of (9.13) as a shorthand way of writing whatever longhand statement you find easiest to work with.

9.18 So the answer to the question of §9.2 is that you don't just add the given speeds. You *start* by adding them, *but then* you divide by a correcting factor – which is got by adding the product of the given speeds to 1.

9.19 Before we come to some help towards understanding §§9.9–17, let's have a numerical example, to see what this new formula implies. If B's speed relative to A is 0·9 (nine-tenths of the speed of light, §8.39) and C's speed relative to B is 0·8, what is the speed of C relative to A?

In shorthand symbolism, the statements about the given speeds are

$u = 0.9$ and $v = 0.8$. We can take these as permission to substitute (§9.16) 0·9 for u and 0·8 for v in (9.13) which gives

$$w = \frac{0.9 + 0.8}{1 + 0.9 \times 0.8} = \frac{1.7}{1.72} = 0.988$$

– instead of the 1·7 that you would previously have expected. (Second and subsequent occurrences of '=' within one statement should be read as 'which is equal to'.)

If you think further examples would be good for you, find w when (i) $u = 0.6$ and $v = 0.8$; (ii) $u = 0.7$ and $v = 0.9$; (iii) $u = v = 0.9$. (Answers at end of chapter.)

These examples illustrate the point we proved in §9.8 – that you can't produce a speed faster than light by combining two speeds slower than light. (Mathematical adepts should prove that if u and v both lie between -1 and 1, then w will do likewise.)

9.20M Were you dismayed by the process of reasoning and calculation that began in §9.9 and eventually led to equation (9.13)? I can assure you that after a little practice this sort of thing will seem simple. Meanwhile, let's wrestle with present difficulties.

First let's be sure that you're in command of the mathematical techniques we use. So check up on §§9.3, 9.11, 9.14 and 9.16–17.

9.21M You probably found that you could understand each separate step. (If not, go back now to any point that gave trouble. It's all explained in the text, and if temporary mental blockage prevented understanding first time, a little further study will clear it up. *Do remember* about reading things in full!) Your real difficulty (I'm pretty sure) is in appreciating the general strategy – you can't see the wood for the trees.

The total argument is built out of four main blocks:

(1) Constructing the diagram – Figure 9.8.

(2) Finding pairs of lengths in this diagram whose ratios give the speeds we're interested in. (The *ratio* of two quantities is what you get by dividing one by the other.)

(3) Working out a piece of geometry that reveals relations between these various lengths.

(4) The final knitting together of the previous three.

The key to it all is in the cunning construction of Figure 9.8. This incorporates the essence of relativistic space–time (§8.40). And its bits are arranged to fit together in a way that facilitates the geometrical task of block (3). Revise §9.9 as necessary.

Block (2) is §9.10. And this is just a case of applying (8.1) three times, but changing the lettering appropriately.

Block (3) is §9.13. Here we prove that triangles LMK and OHK are the same shape. Hence we infer that each side of one is equal in length to the corresponding side of the other multiplied by the scale factor. From a known relation between one pair of corresponding sides, we find that the scale factor is v. And applying this knowledge to the other pairs leads to (9.6) and (9.7) – which, after a bit of tidying up, become (9.8) and (9.10).

In block (4) – which is §9.15 – we bring these bits together. Equation (9.4) tells us that to find w we shall need to know lengths NL and ON. And there's no difficulty in calculating them from those we already know, with the help of the rectangle HKMN. Finally we do the substitution for these lengths in (9.4) to reach the end-product (9.13).

I hope the main strategy is now clear. Each of the main blocks consists of a series of manoeuvres that I have just been briefly describing. And each manoeuvre can itself be analysed into the details that you will find in the appropriate earlier articles. If you will concentrate on each of these levels in turn – perhaps working downwards from general strategy through the four blocks to the details, and then back up again to the general strategy – you will learn to grasp the process as a whole.

When in future you come to some other rather complex train of reasoning, will you please try to analyse it and study it on similar lines.

9.22 **Why has experience led us to believe that (9.1) is true, rather than (9.13)?**

At the speeds we're used to, u and v (in natural units) are very small fractions. So their product (§9.14) is an even smaller fraction, and the correcting factor $1 + uv$ is only slightly different from 1. So (9.1) is nearly true.

Suppose that jet planes are approaching you from opposite directions at 1080 kilometres per hour. They are A and C, while you are B. Then in natural units u and v are both a millionth (§1.8). So their product is a millionth of a millionth, and the correcting factor differs from 1 by that tiny amount. Thus the speed of either plane, as measured by the other, will differ from the expected 2160 kilometres per hour by only one part in a million million – less than 2 centimetres per year. No wonder we don't notice!

9.23 The proof of (9.13) depends essentially on the triangles LMK and OHK having the same shape. This in turn is derived from the relation between the slopes of the time and distance axes, which itself is how the space–time diagram incorporates the invariance of the speed of light (§8.40). And so this proof shows particularly clearly that the formula for combining speeds takes the form that it does precisely *because* the speed of light is invariant.

9.24 Now this is another case in which we can do an experimental test to see if Relativity's predictions fit the facts. Light travels through a material medium more slowly than in vacuum – only three-quarters as fast in water, for example. And this speed is *not* invariant. The speed of light in water is $\frac{3}{4}$ *relative to* an observer who is stationary with respect to the water. And so the speed of this light as measured by some other observer, past whom the water is moving, should be that predicted by (9.13) – taking as given speeds the speed of light in stationary water and the speed of the water relative to the observer. The water is the train of §9.1, and the light is 'walking along the corridor'.

9.25 An experiment testing this prediction was done (actually with a different purpose) as early as 1859 by H.L. Fizeau (and repetitions by Michelson and others confirmed his results). The water flows through a bent tube (Figure 9.25) from A to B. Using prisms, lenses and mirrors, a beam of light coming from point L is split in two, one half travelling as shown in the diagram, while the other (not shown) goes the opposite way round the tube, the two coming together again at M (close to L). The moving water should speed the light up in one direction and slow it down in the other – both in accordance with (9.13). And the experimenter must measure the difference between the two speeds to see if it fits the prediction.

The results of all such experiments have agreed very closely with the relativistic prediction and disagreed decisively with traditional (9.1). So

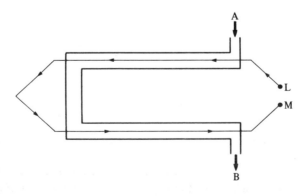

Figure 9.25

this is a further verification of Relativity Theory, a further piece of evidence of its validity (cf.§§3.32–3).

9.26 Consider now a whole series of observers, A, B, C, D, . . . (on the 1D universe). Observer B is moving at speed 0·3 (natural units) relative to A; C is moving at speed 0·3 relative to B; D at the same speed relative to C; and so on. What are the speeds of B, C, D, . . . relative to A? Applying (9.13) we get

$$\text{speed of C relative to A} = \frac{0\cdot3 + 0\cdot3}{1 + 0\cdot3 \times 0\cdot3} = \frac{0\cdot6}{1\cdot09} = 0\cdot5505 \ .$$

Combining this with the speed of D relative to C gives

$$\text{speed of D relative to A} = \frac{0\cdot5505 + 0\cdot3}{1 + 0\cdot5505 \times 0\cdot3} = \frac{0\cdot8505}{1\cdot1652} = 0\cdot7299$$

– and so on. Rounding off to three decimal places, we get the speeds listed in Table 9.26.

Table 9.26

Speeds relative to A (as fractions of speed of light)							
A	0·000	D	0·730	G	0·952	J	0·992
B	0·300	E	0·845	H	0·974	K	0·996
C	0·550	F	0·913	I	0·986	L	0·998

9.27 So the successive speeds, as observed by A, go on increasing, but more and more slowly. And no matter how far we extend the series, we shall never reach an observer moving faster than light (§9.8). One of my students used a computer to show that the 31st in the series has speed 0·999 999 98 relative to A.

9.28 Now I want to apply those figures to a piece of 'super-rocketry'. Suppose we have a rocket capable of an acceleration that will take its speed up to 0·3 in one year (quite a modest acceleration, actually; but we want it to work for years!). At the start the rocket is stationary relative to A. At the end of one year, with a speed of 0·3 relative to A, it is travelling *with* B – stationary relative to B.

Let's assume that it has *constant* acceleration – always the same acceleration. It begins the second year stationary relative to B – related to B now in exactly the same way as it was related to A at the start. So with the same acceleration as before, it will end the second year with a speed of 0·3 relative to B. It will be stationary relative to C in fact. Similarly at the end of yet another year it will have the same speed as D.

And so we could go on. Relative to A, the speeds attained by this rocket at yearly intervals will be as listed for the successive observers in Table 9.26.

9.29 Now this rocket is clearly putting up the same performance at all stages of its career. Any one of observers A, B, C, . . . , provided he starts reckoning from the moment the rocket is stationary relative to him, will get the same answer to the question 'What speed does it reach by such-and-such a time?' So it is certainly going with constant acceleration. And yet, as one observer views it, its speed goes on increasing more and more slowly.

In ordinary slow-speed conditions, when we say that something has constant acceleration, we mean that its speed increases by equal amounts in equal intervals of time. But our example shows that when high speeds are involved, we have to modify that definition – we must add the important qualification that the measurement for each time interval is to be made by an observer relative to whom the rocket was stationary at the beginning of that interval. So A will make measurements showing that the speed increases by 0·3 during the first year; B will check that it also increases by 0·3 during the second year; and so on. Let's make a formal definition:

> A body has *constant acceleration* if its speed increases by
> equal amounts in equal intervals of time, where the measure-
> ment for each time-interval is made by an inertial observer
> relative to whom the body is stationary at the beginning of
> that interval.

9.30 I've been guilty of one deceptive over-simplification. The successive years (in the numerical example) are measured by different observers – the first by A, the second by B, and so on. Because of the dilation of time (§3.36), the second interval according to A will be longer than a year, the third longer still . . . So the speed of this rocket, relative to A, increases even more slowly than we thought at first.

Naturally, similar results would emerge from calculations about a constant acceleration of any size. The easiest way to see this is merely to change the time-interval – by considering an acceleration that gives speed 0·3 in 38 days, or in 0·01 second (!!), or what you will, we could cover all possible cases by this one calculation.

9.31 It takes force to accelerate a body. And as we cannot have indefinitely increasing forces, there is a limit also to the size of acceleration we can produce. The biggest possible effect, then, would be what you would get by having a constant acceleration of this maximum possible size.

And still you could not accelerate the body to or beyond the speed of light.

This does not prove that nothing can go faster than light. There remains the possibility that there are some things that always move slower than light and others that always move faster than light, but nothing that can be accelerated through the speed of light. All the same, I think we've got to the point where the idea that nothing can move faster than light would be worth investigating.

Answers to §9.19. (i) 0·946; (ii) 0·982; (iii) 0·994.

10

Causality and the speed of light

10.1 Let's try another approach. Take any two events happening on the 1D universe – call them O and Q. Under what conditions would it be possible for a cause operating at O to have an effect at Q?

The signal lines through O divide the diagram into quadrants, which we can name as in Figure 10.1. We can now rephrase our question: If a cause at O has an effect at Q, where could point Q lie on the diagram?

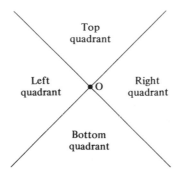

Figure 10.1

10.2 **Could Q lie in the bottom quadrant?**

A straight line through Q and O (Figure 10.2) could be the world line of an inertial observer, B. **Try again.**

He finds that effect Q happens before its cause O, which is contrary to experience. So the answer is No.

10.3 Actually that argument depends on an assumption which I shall call the *Causality Principle* – namely that *an effect cannot happen before its cause.* This comes from outside Relativity Theory. And of course it's not a certainty. When people claim to foretell the future, they are saying that future events can affect present knowledge. But I think they're wrong, and you probably agree. So we'll assume the validity of the Causality Principle.

To forestall a possible objection to §10.2, time must run *upwards* on any world line. Otherwise, in a case like Figure 7.3 the signals could bring A at event R news of future event P. (Of course the diagram is constructed on the basis that signal lines *do* run upwards, §6.17).

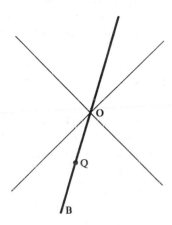

Figure 10.2

10.4 **Could Q lie in the right quadrant,** as in Figure 10.4?

   A much harder question! Draw a line through O, passing between Q and the signal line above Q. Draw another line making the same angle as this one with the signal line, but on the other side. The latter, being steeper than 45°, could be the time axis (world line) of an inertial observer, C (§§6.18–20); and then the first line we drew would be C's distance axis (§8.26). **Can you answer now?**

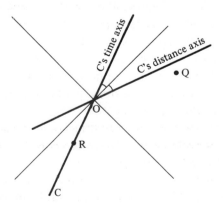

Figure 10.4

In C's opinion Q happened *before* O (check with slot) – effect before cause again. **So can we say that this case also is impossible?**

10.5 It's not so simple. Draw another line between Q and the signal line below Q (Figure 10.5). By the same argument as before, this could be the distance axis of another observer, D, whose time axis would be given by the equal-angles-with-a-signal-line rule. In his opinion (slot, please!) effect Q happens *after* its cause O. **Are we to believe C or D?**

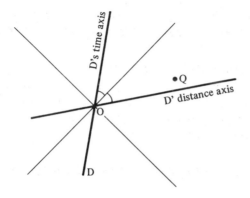

Figure 10.5

On such an important question we daren't rely on anything that's a mere matter of opinion. Better run through Chapter 7 again, especially §§7.2–3,7.9,7.14–15) to remind yourself that all we can *know* is the times at which an observer sends a signal to the event and gets one from it, and that we merely *decided* to define the time of an over-there event in a certain way in terms of these. Superimpose the set-up of Figure 7.14 on your own copies of Figures 10.4 and 10.5, and **think about the above 'before' and 'after' statements in terms of event M.**

We may *say* (§10.4) that Q happens before O in C's opinion. But all we *mean* is that if you take the tick of C's clock which is halfway between sending a signal to Q and getting one back from Q, then this tick happens before cause O. Surely it's not obvious that *that* conflicts with the Causality Principle.

10.6 To feel confident, we must restate the Causality Principle in a way that avoids questions of opinion: *for any one observer, an on-the-spot effect cannot occur before its on-the-spot cause.* And now we can carry the argument of §10.4 a bit further.

If a cause at O could have an effect at Q, which is earlier than O in C's opinion, then a cause at Q could equally well have an effect at on-the-spot event R (Figure 10.4), which is earlier still according to C. So now an on-the-spot cause at O has an on-the-spot effect at R (through the intermediary of Q), and *this* very clearly contradicts the Causality Principle. Thus Q can *not* lie in the right quadrant – nor the left. So

> A cause at O can have an effect at Q *only if* Q lies in the top quadrant.

10.7 If you object that from O to Q is from on-the-spot cause to over-there effect, whereas from Q to R is from over-there cause to on-the-spot effect, just bring in another observer who is stationary relative to C and present at Q.

If the unsymmetrical appearance of Figure 10.4 makes you feel that the relation between Q and R is different from that between O and Q, use the slot (*à la* C). And if that fails to convince you that there's no real difference, draw another diagram (as you are entitled to, §8.44) with B's world line vertical, and consider how Q and R would appear on this.

10.8 **Where on the diagram must Q lie to make it possible for a cause at Q to have an effect at O?**

In the *bottom* quadrant. (Draw the signal lines through Q in each of its possible positions, and apply the §10.6 conclusion to the quadrants into which these divide the diagram.)

Figure 10.8 summarises our results. Since a light signal can transmit a causal influence, the signal lines are included in the top and bottom quadrants.

How does all this work out on the old-fashioned space–time diagram?

The horizontal through O represents events that happen at the same time as O, and so cannot be causally connected with it. You can do the

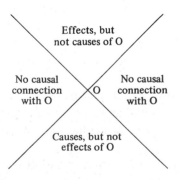

Figure 10.8

rest yourself. So the horizontal line, representing 'the present' on the old-fashioned diagram, is spread out in the relativistic diagram into the left and right quadrants. **Think about that.**

10.9 Now we connect this up with our question about travelling faster than light. Consider our old friend A, the inertial observer that we first put in the diagram (§6.20), giving him a world line steeper than 45° because that corresponds to the experience of observers we know can exist – like ourselves.

Suppose B is an inertial observer who is present at both events, O and Q. If Q lies in the top quadrant (Figure 10.9), then clearly B's world line is steeper than 45°. And so, by §8.41, B is moving slower than light relative to A (slot!). On the other hand, if Q were in the left or right quadrant (draw this for yourself), B's world line would be less steep than 45°, and B would be travelling, relative to A, faster than light. **Tie that up with the conclusion of §10.6.**

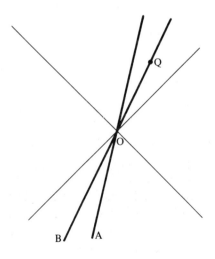

Figure 10.9

What comes out is that a cause at O can have an effect at Q *only if* it would be possible for an inertial observer to get from one event to the other without going faster than light – relative to A.

10.10 To sum up, if the Causality Principle is valid, then

No causal influence can travel faster than light.

That's only a shorthand way of saying that cause and effect cannot be so related that to get from one to the other it would be necessary to travel faster than light – relative to A, as representative of slower-than-light observers like ourselves. Another way of putting this is that

Information cannot be transmitted faster than light

(since a causal influence transmits the information that the cause has happened; and information is a causal influence, affecting the recipient).

10.11 It follows pretty obviously that no observer – inertial or not – can travel faster than light (nor as fast as light, §2.1). And then, using §8.41, we see that

The world line of *any* inertial observer must be steeper than 45°.

This clears up a doubt that has been with us since §8.23, or indeed since §§6.18–20. And by §8.26, his distance axis must be less steep than 45°.

10.12 **Prove that no material object can travel faster than light.**

Attach a time-bomb to it – the act of priming the fuse is a cause that later produces the effect of an explosion.

You have probably come across the stronger statement that, according to Relativity '*Nothing* can move faster than light.' **Is that statement justified by what we've been doing?**

10.13 A laser beam can be used to throw a quite small spot of light, searchlight fashion, onto the Moon's surface. If we swivel the laser at a mere 45° per second, the spot will travel across the Moon faster than light. Again, consider two rulers crossed at an acute angle, as in Figure 10.13. If you move the slanting one bodily downwards, the point P where they cross will move to the right. By making the angle between the rulers small enough, you could make P move faster than light.

So apparently some kinds of things *can* move faster than light. However, these examples do not involve the transmission of cause and

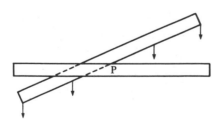

Figure 10.13

effect. Nothing you can do *to the light spot* at one time will have any effect on it at a later time. Similarly for the rulers.

But if the rulers fitted very closely and a speck of dust got jammed between them at P, so that the motion of the sloping ruler would force it forward, then this operation would suddenly become impossible (§10.12). And if you saw that Moon spot going faster than light, you could be sure that it wasn't a lamp carried by a hyperactive lunar astronaut.

10.14 Maybe you're thinking 'In these faster-than-light cases there's only an *appearance* of motion. But when it's a case of something *really moving*, then *that* can't go faster than light'. **But what do you mean by 'something really moving'? Is it a case of whether there is an actual transfer of matter from one place to another?**

Even that won't do. Consider wave motion. If you lay a rope on the ground and waggle one end of it up and down, you can see the waves moving along it; but when you stop waggling, all the *material* of the rope is where it was. And that's true of all wave motion: all that travels is the *state* or *condition* of being raised above normal, or compressed beyond normal, or something like that. No matter is transferred. **So can waves move faster than light?**

Waves transmit cause and effect, and so the answer is No.

10.15 So the theorem of §10.10 denies the possibility of moving faster than light in some cases where no transfer of matter is involved. And conversely it has to allow the possibility that material particles *might* be able to go faster than light, provided they do not transmit causal influences. Such hypothetical particles have been called *tachyons*.

It might be that we can affect tachyons. Or it might be that they can have effect on us. But not both. One could not, for example, hope to deflect a tachyon at some point in its career and observe where the deflection takes it at some later time – for that observation would be an effect caused by the deflection. But we cannot deny the possibility of observing the same tachyon several times, provided the act of observation does not affect it in a way that could be detected by later observations. That would put it on a par with the spot-on-the-Moon of §10.13. However the experimental search for tachyons has had no success so far.

10.16 The proposition of §10.10 is a very simple one. And the proof of it was fairly simple too. Yet the consequences that flow from it are very far reaching indeed. I can only spare space for one.

The planets and satellites of the Solar System rather obviously do not move in straight lines at constant speed. They follow complicated paths – the Moon, for instance, orbiting round the Earth as the latter orbits round the Sun.

Newton's Theory of Gravitation asserts that the *cause* which makes them do so is the gravitational attraction of the other bodies around them. According to §10.10 this gravitational influence must be transmitted at the speed of light or slower. And so the Sun's pull on the Moon at any particular moment will not be towards where the Sun actually lies. It will be in a direction depending on the relative positions of Sun and Moon at the earlier time when the Sun transmitted the gravitational influence that reaches the Moon at the moment in question.

Yet if we put these considerations into the calculations, Newton's theory gives the wrong results. Laplace showed that to give the right answer for the Moon's motion, the gravitational influence must be transmitted at least 30 million times faster than light – which our discoveries in this chapter forbid. Newton had assumed *instantaneous* transmission.

Clearly, then, in spite of the fact that it gives right answers time after time, there must be something radically wrong with Newton's theory. This was one of the considerations that led to Einstein's *General* Theory of Relativity.

11

The nature of spacetime

11.1 The 'old-fashioned' space–time diagram (Figure 6.2) gives graphical expression to our traditional ideas about space and time (§6.4). The new space–time diagram that we've been developing since Chapter 6 should lead to sounder conceptions on these matters.

Let's start by comparing old and new ideas about past, present and future. In the pre-Relativity view all observers agree about which events, at any one moment, belong to the past, which belong to the present, and which to the future. Now take up again the line of thought suggested in §10.8. Think about Figures 10.4 and 10.5. According to the Relativistic space–time diagram, **do observers always agree about these past, present and future divisions?**

11.2 To discuss the past, present and future of event O, consider inertial observers with world line through point O. Their distance axes are all less steep than 45° (§10.11). If Q lies in the left or right quadrant, then we can find an observer who says that, at the time of event O, event Q belongs to the past – he would be like C of §10.4. Similarly, there is an observer – like D of §10.5 – who says that Q belongs to the future. And if you took an in-between observer whose distance axis passes through point Q, he would say that event Q (at the time of O) is in the present. The question of past, present and future is relative to the observer.

On the other hand, if Q is in the top quadrant, it lies above *all* the distance axes, and so all observers agree that it is in the future (compared with O). And similarly they all agree that an event represented by a point in the bottom quadrant is in the past when O happens.

For these reasons, the regions of space–time represented by the top and bottom quadrants are known as the *absolute* future and past of O (and the signal lines are included in these – **check that**), while those corresponding to the left and right quadrants are the *relative* present of O. See Figure 11.2 and **ponder for a moment on how conveniently these past, present, future divisions fit in with the causality categories of Figure 10.8.**

11.3 Given any two events, one can ask whether they are separated by space, or time, or a combination of the two. In the old-fashioned diagram of Figure 6.2, S and T represent events that happen at the same time and differ only in position, so that they are separated by *pure space*. Events, T and U, on the other hand, happen in the same place; they are separated by *pure time*. Finally, S and U are separated by a *mixture of time and space*. And according to the old beliefs embodied in this diagram, all observers agree about these three statements.

Time is represented by the vertical, space by the horizontal. And the clear distinction between vertical and horizontal enshrines the notion that time and space are distinct things that cannot be interchanged. Furthermore, it's the same time and the same space for all.

11.4 The new-style space–time diagram of Figure 8.26 (your own copy of which should be at your elbow for the next few pages) explodes these beliefs. Since each observer has his own axes, each also has his own time and his own space. (If you have difficulty with what follows, you probably need to reread parts of Chapter 8, particularly §§8.7, 8.10, 8.12 and 8.26–31. And I can't keep reminding you to use the slot.)

In A's opinion events M and Q (Figure 8.26) happen at the same time – they are separated by *pure space*. But B thinks they happen at different times and places – and so are separated by a *mixture of time and space*. Again, N and Q happen in the same place according to A, and are therefore separated by *pure time*. But once more B says that what separates them is a *mixture of time and space*. **Work out similar examples with the roles of A and B reversed.**

If merely changing your observer can mix up time and space like that, then they can't be as distinct as we used to think.

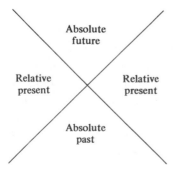

Figure 11.2

11.5 **Would it be possible to find a case in which one observer's pure space is another's pure time?**

If any observer considers two events to be separated by pure time, the line joining the corresponding points on the diagram must be parallel to his time axis (§8.12), and so (§10.11) its slope must be steeper than 45°. And for some other observer to think they are separated by space only, this slope must be less steep than 45°. It can't be both. So the answer is No.

Thus changing the observer can change pure time or pure space into a mixture of the two, but it *can't* change pure time into pure space or vice versa.

11.6 I'm sure you're beginning to see that space and time, as separate things, exist only in relation to this or that observer. The deeper reality, transcending the parochialism of particular observers, is a homogeneous and undifferentiated *spacetime* – a thing in its own right, not just what you get by talking about space and time in the same breath. It's in this integral spacetime that our observers live – that *we* live – not just in mere space plus time.

11.7 However, you and I do not think naturally in spacetime terms. We insist on splitting it up into space *and* time. So naturally we make our observers do the same. And we find, contrary to previous belief, that different observers do the splitting in different ways.

Space for observer A is represented by the slope of his distance axis (X-axis, Figure 8.26). Any line with that slope represents the state of space at some particular moment according to A. But B's notion of space is represented by lines parallel to the x-axis – he does not agree with A.

You can think of any line on the diagram as representing a sort of cross-section of spacetime. So what A calls space is a certain set of cross-sections – represented by lines parallel to the X-axis. And B's idea of space is the *different* set of cross-sections represented by lines parallel to the x-axis. Similarly, both are convinced that there is a thing called time, clearly distinguishable from space. But when you ask them what it is, A picks out cross-sections represented by lines parallel to the T-axis, while B chooses cross-sections represented by lines parallel to the t-axis.

Every observer divides the totality of spacetime into what he calls space and what he calls time. But observers in relative motion do this dividing up differently.

Since §11.6 I've been writing 'spacetime' instead of 'space–time'.

Dropping the hyphen serves to emphasise the radically new outlook we've now reached – space and time as a rather artificial splitting up of the natural spacetime, rather than space–time as what you get by putting space and time together. The mere absence of a hyphen does not sufficiently stress this fundamental conceptual revolution. But do try to use it as a mnemonic.

11.8 However, even within this monolithic spacetime there still remains *some* distinction between space and time – see §§11.2, 11.5. Let's look at that from another angle.

Consider any two events and the corresponding points of the space-time diagram. The line joining these points will not be parallel to either the time axis or the distance axis of an observer picked at random. So a typical observer will say that the events happen at different times and places. But we can ask whether there is a *particular* group of observers who consider them to be separated by time only or by space only. We use §§8.12, 10.11 and 11.4; and Figure 8.26 will help you think out details that I omit.

(1) If the line joining the points is steeper than 45°, there are inertial observers whose time axes are parallel to it. According to them (but *only* according to them) the events are separated by *pure time*.

(2) If this line is less steep than 45°, there are observers whose distance axes are parallel to it. They say the events are separated by *pure space*.

(3) If this line has 45° slope, no observer can have either time or distance axis parallel to it (§6.18). Every observer says the events are separated by a *mixture of time and space*.

This classification applies to the pairs of events as such, not just some observer's opinion, and so it preserves, within the unified space-time, some vestige of the former distinction between space and time. We say that the separation of the two events is *time-like* in case (1), *space-like* in case (2) and *light-like* in case (3).

11.9 What distinguishes our new concept of unified spacetime from the old one of separate space and time is essentially this: *observers disagree about the very quantities* – the time and space co-ordinates (§8.15) – *that they use to specify an event*. Furthermore, the cause of their disagreement is simply this: though they may use the same word 'time' (or 'distance'), they are really talking of different things (§11.7).

Odd though this may seem to you, analogous situations are so familiar that you don't notice them! I'm going to spend the next few pages discussing one of them.

11.10 For the purpose of this analogy, we're back in the old familiar slow-speed world, in which space behaves as we used to think it does. A surveyor, called A, is required to specify how various positions in a flat field are related to a fixed point O. He might do so as follows, for a typical point Q (Figure 11.10).

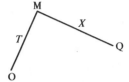

Figure 11.10

Starting from O, he walks forward, counting paces, until – at point M – he sees that the line from him to Q is perpendicular to the direction he's walking in. Then he turns to face Q and counts how many paces are needed to get there. (An impenetrable wood prevents him from going direct to Q.) The length OM is called 'the thatway of Q according to A' (because on his very first job, the foreman pointed and told him to 'walk that way'). And we'll use *T* as shorthand for this clumsy phrase (cf. §8.18). Similarly, we'll call length MQ 'the cross of Q according to A', abbreviating that to *X*. Then we can mark *T* and *X* against lengths OM and MQ in the figure.

11.11 Adding the lines marked *T*-axis and *X*-axis in Figure 11.11, we have a co-ordinate system analogous to that of Figure 8.16. Each co-ordinate in the plane (the field) is the length of a line drawn parallel to one axis from the other axis to Q – just as in the spacetime case (§8.28).

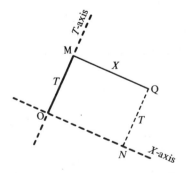

Figure 11.11

11.12 Now that's a perfectly good method for fixing the position of a point in relation to O. Yet if another surveyor, B, sets out to specify the position of Q by the same method, he won't necessarily get the same answer – for he may be facing a different way at the start.

In fact, B has his own co-ordinate system. Figure 11.12 shows the two superimposed (cf. Figure 8.26). The thatway of Q according to B (abbreviated to *t*) is the length Om. And his version of the cross (*x* for short) is length mQ. Obviously A and B will disagree about the thatway and cross measurements. Or, in symbolic language,

$$t \neq T \text{ and } x \neq X. \tag{11.1}$$

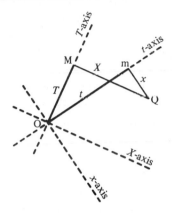

Figure 11.12

13M Just as '=' is short for 'is equal to', so '≠' is an abbreviation for 'is not equal to'. Keep reading sentences like (11.1) in full.

1.14 So the surveyors disagree about the very quantities – the thatway and the cross co-ordinates – that they use to specify the position of a point. That is just like what we said of the spacetime observers in §11.9. Yet in the present instance we are not worried. We simply say that the thatway and cross do not themselves measure the reality (in the sense of §2.20) of the relation between O and Q. Reality lies deeper. The measurements certainly carry information about reality, but this is diluted with something that relates to the surveyor himself.

1.15 The experiences of these surveyors in the plane (their field) and of the observers in spacetime (the 1D universe) are so alike that it's worth tabulating a comparison. ('Observer' means 'surveyor' when referring to the plane.)

To specify the relation between226

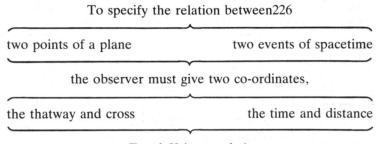

two points of a plane   two events of spacetime

the observer must give two co-ordinates,

the thatway and cross   the time and distance

T and X (or t and x).

Different observers disagree about the co-ordinates,
i.e., they find that t ≠ T and x ≠ X.

We conclude that the

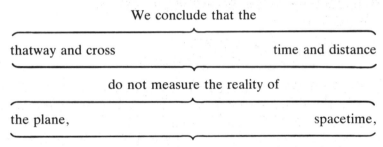

thatway and cross   time and distance

do not measure the reality of

the plane,   spacetime,

but represent only the viewpoint of the particular observer.

Notice that the bottom of the right-hand column has led to a conclusion that we partially anticipated in §§11.6–7. Using the equal-angles-with-a-signal-line rule (§8.26), we can add:

To get from one observer's co-ordinate system to another's
we rotate the two axes through the same angle in

the same direction.   ·   opposite directions.

11.16 The only difference of substance between the two columns lies in the contrast of 'same' and 'opposite' at the very end. Other differences are merely verbal – 'points' against 'events', 'thatway' and 'cross' against 'time' and 'distance'. But what's in a name? The two systems resemble each other so closely as to suggest that the analogy might be worth following further. **And I rather think you may have a suggestion to make.**

11.17 There *is* something the surveyors agree about. If each takes *his own* version of the thatway and *his own* version of the cross, calculates the square (§3.14) of each, and adds these two squares together, then they will both get the same answer. Or putting it symbolically,

$$t^2 + x^2 = T^2 + X^2 . \qquad (11.2)$$

18M Are you clear what (11.2) means? It's only a question of following earlier advice about reading symbolism in full (§§8.19, 9.3, 9.17). In the present context, t stands for 'the thatway of Q according to B'. And so, by an obvious extension of §3.14, t^2 means what you get when you multiply the thatway of Q according to B by itself.

This threatens to grow too unwieldy. So let's agree on the compromise 'B's thatway' for t. Then you could translate t^2 as (B's thatway)^2 – where the brackets remind us that it is the quantity described by the whole phrase inside them that is to be multiplied by itself (cf. §3.14). **Write out (11.2) in this semi-shorthand language.**

It comes to

$$(\text{B's thatway})^2 + (\text{B's cross})^2$$
$$= (\text{A's thatway})^2 + (\text{A's cross})^2. \qquad (11.3)$$

(Compare a similar statement in §3.15.) By this time I'm sure you can look at (11.3) and read it mentally as: B's thatway multiplied by itself *plus* B's cross multiplied by itself *is equal to*. . . **You finish it.** Or maybe you've got to the stage where you can safely dispense with 'multiplied by itself' and read: The square of B's thatway *plus*. . . **Finish it, please.** And you'll notice that a freer translation is the second sentence of §11.17.

Of course 'the thatway of Q according to B' is just the name we've given to length Om (Figure 11.12). So you could read the t of (11.2) as 'length Om', and similarly for x, T and X. **Using this language, express (11.2) in a form like (11.3).**

You get

$$(\text{length Om})^2 + (\text{length mQ})^2$$
$$= (\text{length OM})^2 + (\text{length MQ})^2. \qquad (11.4)$$

And now think out the meaning of this in the same way as we thought out (11.3).

If you find any of this sticky, the only thing to do is to have a rest and then go over it again – till you've mastered it.

19M Now we know what (11.2) means. But is it true? Thinking in terms of the lengths Om, mQ, etc., of Figure 11.12, **see if you can deduce its truth from what we already know.**

If you don't succeed in a few minutes, draw the line joining O and Q, and **try again.**

The way in which the equation refers to (a length)^2 plus (another length)^2 **should have reminded you of something.**

Triangle OMQ has a right angle at M. So by Pythagoras' Theorem (§3.15),

$$\text{(length OM)}^2 + \text{(length MQ)}^2 = \text{(length OQ)}^2.$$

But length OM is the thatway of Q according to A, which we abbreviate to T (as marked in the figure); and similarly length MQ can be written as X. Then the last equation becomes

$$T^2 + X^2 = \text{(length OQ)}^2. \tag{11.5}$$

Now use triangle OmQ to prove that

$$t^2 + x^2 = \text{(length OQ)}^2. \tag{11.6}$$

Can you finish the job?

The last two equations state that $T^2 + X^2$ and $t^2 + x^2$ are both equal to the same thing. So they must be equal to each other – which is just what (11.2) says. Now reread §11.17, and then carry on from here.

11.20 The outcome of §11.17 is that, although the surveyors disagree about the separate co-ordinates of Q, yet if each uses his own values of the co-ordinates to calculate the quantity

$$\text{(thatway)}^2 + \text{(cross)}^2,$$

they will *agree* about the result.

This is an example of an *invariant* – a quantity that does not vary when you change from one surveyor's (or observer's) point of view to another's, a quantity that is the same for all surveyors (observers). And I'm sure it's obvious to you that an invariant is something very important – it gives information about reality, as contrasted with the differing viewpoints of those who are observing reality.

We've already met this notion of invariance in relation to the speed of light – which also stays the same when you change the observer (§1.22). But let's agree that when the word 'invariant' occurs as a noun – *an* invariant, or *the* invariant – it will mean the new type of invariant we've just met, which is a *combination of the co-ordinates*. This type of invariant will play a very special role in future developments.

11.21 You're probably eager to point out that these surveyors have made no wonderful discovery. They've merely found that equations (11.5) and (11.6) give recipes allowing each of them to calculate the *distance* from O to Q, which they would agree about in any case if they measured it directly.

True. But suppose it really was impossible to make the direct measurement. Can you not imagine the joy of these chaps when, after having to admit that they couldn't agree about the quantities they actually measure, they discover at last that they can calculate a quantity that they do agree about? They would want a name for their new-found invariant. So they would *define* the quantity 'distance' to be such that

(distance from O to Q)^2
= (thatway from O to Q)^2 + (cross from O to Q)^2.

In symbols, using (11.5) and (11.6) (but with 'length' changed to 'distance'), they would define

(distance from O to Q)$^2 = T^2 + X^2 = t^2 + x^2$.

And they would now be able to study the geometry of their plane more profoundly, in terms of this invariant, even if they never learnt how to find distance by direct measurement.

Surely the analogy of §§11.15–16 suggests that we should look for a spacetime invariant. Don't miss the next exciting instalment!

12

Interval

12.1 To help our search for an invariant we need to learn one more trick with spacetime diagrams. An observer's time and distance measurements are represented to scale in the diagram (§§6.22, 8.4, 8.8–9, 8.11, 8.28 – please revise). We don't yet know how the scales of different observers are related to each other (§8.24). But when we have only two observers to deal with, it would be convenient if they both had the same scale. **So could we choose their world lines in such a way as to arrange this?**

The scale used for any observer must depend on the slope of his world line. Equal slopes should give equal scales. In fact the symmetry of the situation (Figure 12.1) makes it obvious that

> When the world lines of two observers A and B have the
> same slope (same angle on either side of the vertical) then

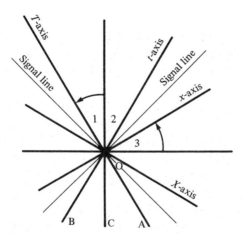

Figure 12.1

the scales used for representing time and distance measurements on the diagram are the same for both observers.

Please check that I've drawn the distance axes and labelled all the axes in conformity with §8.26.

12.2 There's nothing special about this pair of observers. If we pick *any* two observers, calling them A and B, we can construct a diagram of this type for them. **Can you prove that?**

Consider another observer C whose motion is such that A and B are moving, relative to him, at the same speed in opposite directions. Give C a vertical world line (cf. §§8.42–3). Then it becomes obvious that A's and B's world lines are equally inclined to the vertical and so they have the same scales.

12.3 In Figure 12.1 it *looks* as if

Each observer's distance axis is at right angles to the other's time axis.

Can you prove that it must be so?

Figure 12.1 also includes the vertical and horizontal through O. Angles 1 and 2 are equal, since the time axes have equal slopes. And angles 2 and 3 are equal, since B's distance axis makes the same angle with the horizontal as his time axis makes with the vertical (§8.26).

It follows that angles 1 and 3 are equal. So to get the x- and T-axes, we have only to take the horizontal and vertical lines – which *are* at right angles – and rotate them through the *same* angle (arrows in figure). And of course they end up at right angles. Similarly for the X- and t-axes.

12.4 Back to our search for an invariant! Given any two events, O and Q, observers will usually disagree about the time and distance between them. But is there *any* measurement connecting O and Q that observers will *agree* about?

First consider observers A and B who are both present at event O. If we give equal slopes to their world lines and add the distance axes (as in Figure 12.1), we have a diagram to which the propositions of §§12.1 and 12.3 will apply. To complete Figure 12.4 we put in line mQ parallel to B's distance axis meeting B's time axis at m, and NQ parallel to A's time axis meeting his distance axis at N. So mQ and NQ play the same role as the similarly named lines in Figure 8.26. Join mN.

12.5 With A's clock set to zero at O, the time from O to Q according to A is simply the time *of* Q according to A. And since O happens at zero distance from him, the distance from O to Q according to A is the same as the distance *of* Q *from* A according to A. We abbreviate these, as usual, to T and X. And t and x stand for the similar quantities according to B.

Now use the descriptions of §8.28, aided by Figure 8.26, to **check for yourself** that T is represented by length NQ, X by length ON, t by length Om, and x by mQ. And all these representations use the same scale (§12.1). (In Figure 8.26 it is MQ that is marked X; but ON is the same length).

12.6 **Can you find an equation involving T, X, t and x?**

The surveyor analogy suggests using **what geometrical proposition?**

Pythagoras' Theorem (§3.15); and §11.19 may suggest how to **apply it to Figure 12.4.**

Lines NQ and mQ are parallel, respectively, to the T- and x-axes, which are perpendicular (§12.3). So NQm is a right angle, and applying Pythagoras' Theorem to triangle NQm gives

$$(\text{length NQ})^2 + (\text{length mQ})^2 = (\text{length mN})^2.$$

Translating this into symbols, we get

$$T^2 + x^2 = (\text{length mN})^2.$$

Find a similar equation containing t and X.

Angle NOm is a right angle (§12.3). So

$$t^2 + X^2 = (\text{length mN})^2.$$

We've proved that $t^2 + X^2$ and $T^2 + x^2$ are both equal to (length mN)^2; hence they must be equal to each other. That is

$$t^2 + X^2 = T^2 + x^2.$$

12.7 To find a quantity on which the observers agree, we need an equation that contains only A's measurements on one side, and on the other side only the corresponding measurements of B. **Can you see how to get that?**

Subtract X^2 from both sides of the last equation (§3.19), which gives $t^2 = T^2 + x^2 - X^2$. And then subtracting x^2 from both sides leads to

$$t^2 - x^2 = T^2 - X^2. \tag{12.1}$$

Make sure you're abreast of the reasoning from §12.4 on. Then, before reading further, compare (12.1) with (11.2) of §11.17 and **think about the significance of what we've done.**

12.8 Observers A and B *disagree* about the time between O and Q and about the distance between them. But if each uses *his own* version of time and distance to calculate (time)^2 − (distance)^2, then they *agree* about this quantity.

So far we've only established the agreement for two observers. But it

does look as if we're on the track of a spacetime invariant analogous to
the one the surveyors found in their plane (§11.20–21).

12.9 **Can you prove that all observers whose world lines pass through O will
agree about this quantity?**

The same argument could be used to show that *any two* of them
agree (we have only to draw a new diagram – §8.44 for justification – in
which the world lines of *these two* have equal and opposite slopes).
And if any two agree, they all agree.

12.10 **Now what about an observer, D, whose world line does not pass
through O?**

Consider another observer E whose world line *does* go through O,
and who is stationary relative to D. Relatively stationary observers
agree about time and distance separately, and so also about (time)^2 −
(distance)^2. So D agrees with E, and therefore with all observers having world lines through O. We have now covered all cases, and so have
proved that

All inertial observers agree about the quantity

(time from O to Q)^2 − (distance from O to Q)^2.   (12.2)

Stressing that word 'all' and remembering that O and Q can be *any*
two events, we shout in triumph that

WE HAVE FOUND AN INVARIANT!

In ordinary slow-speed life much of our understanding of the world
and much of our command over nature depends on knowing that time
and distance are invariants – that all observers, no matter how they do

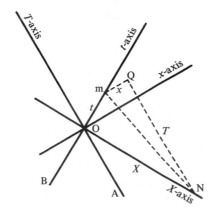

Figure 12.4

the measuring, will get the same time and distance measurements. In high-speed conditions time and distance let us down. But we have found a new spacetime invariant to replace them. Surely it will play as big a part in our understanding of the high-speed world as time and distance did for the slow-speed one. (If you are at home with negative quantities, you can skip to §12.14).

12.11M To develop this subject further we need to learn some more properties of negative quantities (§§6.23–5). To prove them properly would mean working through a lot of things that we shall not need for a long time yet. So I shall just state them now – with only a weak attempt to make them plausible – leaving strict proof till later.

We all know how to subtract 3 from 5. But what do you get when you subtract 5 from 3? Well '3 seconds − 5 seconds' means 5 seconds earlier than time 3 seconds, which is 2 seconds earlier than zero – or (as we put it in §6.23) 2 seconds measured backwards. And we write that as '−2 seconds'. So

$$3 - 5 = -2 .$$

Note that we have −2 in place of the 2 that would result from 5 − 3. More general thinking on these lines leads to the conclusion that

> The result of subtracting a larger quantity from a smaller one is the *negative* quantity that you would get by subtracting the smaller from the larger and then making the result negative.

12.12M We shall also need to use the point that

> The square of a negative quantity is positive (§6.24) and is equal to the square of the corresponding positive quantity (the one you would get by suppressing the '−').

For example, $(-3)^2 = 9$, which is the same as 3^2 (the −3 is put in brackets to indicate that it is negative quantity −3 that is to be multiplied by itself; -3^2 would mean 'the negative of what you get from multiplying 3 by itself'). And in general if $-a$ is any negative quantity, then

$$(-a)^2 = a^2. \tag{12.3}$$

Since we already know that the square of a positive quantity is positive, we can add that

> The square of *any* (non-zero) quantity is positive.

12.13M To make that at least plausible, consider first the multiplication of a negative quantity by a positive one. Think of it in terms of pacing on a line, as in §6.25. For instance $3 \times (-2)$ would be an instruction to take 2 paces backwards 3 times over, 6 paces backwards in all. So $3 \times (-2) = -6$. Thinking more generally on the same lines, you'd conclude that the product of a positive quantity and a negative quantity is negative.

The effect of the '$-$' is to reverse the direction of pacing. So it's plausible to suggest that the *two* negative signs of $(-3) \times (-2)$ would tell us to reverse direction twice, bringing us back to the positive direction again. Thus it's a good guess that $(-3) \times (-2) = 6$; and similarly that the product of two negative quantities is always positive. Then if we take the case of two *equal* quantities, we have the statement of §12.12 about squares.

12.14 We have proved that the quantity (12.2) is invariant. However, when the surveyors reached a similar situation (§11.21), they found it better to choose as their invariant the quantity whose *square* is equal to $(\text{thatway})^2 + (\text{cross})^2$. This suggests that in spacetime the invariant that matters might similarly be the quantity whose *square* is equal to (12.2).

But in view of §12.12, we must be careful.

The time from O to Q may be a positive quantity or a negative one (§8.14). So may the distance from O to Q. But §12.12 tells us that $(\text{time from O to Q})^2$ and $(\text{distance from O to Q})^2$ are both positive. If the latter is the greater of the two, then (12.2) is negative (§12.11). And there is no such thing as a quantity whose square is negative (§12.12) – which seems to defeat our purpose.

However, if (12.2) is negative, then the quantity

$$(\text{distance from O to Q})^2 - (\text{time from O to Q})^2 \qquad (12.4)$$

(got by subtracting the other way round) is positive; and we avoid the difficulty by taking the quantity whose square is (12.4) as the invariant. On the other hand, if the time is greater than the distance, then (12.2) is positive, and there is no problem in taking the invariant to be the quantity whose square is equal to (12.2).

12.15 The invariant obtained in this way is called *interval*. Let's have a formal definition:

> The *interval* between two events O and Q is defined to be the quantity whose square is equal to
>
> *either* $(\text{time from O to Q})^2 - (\text{distance from O to Q})^2$ (12.5)
>
> *or* $(\text{distance from O to Q})^2 - (\text{time from O to Q})^2$,   (12.6)
>
> whichever of these is positive (the time and distance being measured by the same inertial observer).

If the time and distance are equal, either alternative makes the interval zero.

From §12.10 it follows that

INTERVAL, THUS DEFINED, IS AN INVARIANT.

12.16 For example, if time from O to Q is 5 seconds and distance from O to Q is 13 seconds, we use (12.6); and (interval between O and Q)^2 = $13^2 - 5^2 = 144$, so that the interval between O and Q is 12.

This example reminds us that we are using *natural units* (§7.11), with distance measured in the same units as time. Thus (12.5) and (12.6) both take the form:

$$(\text{a time})^2 - (\text{another time})^2;$$

and so (interval)^2 also has the status of (a time)^2. And interval itself has the status of a time, and is measured in time units.

12.17 Using s to stand for 'the interval between O and Q' and with the notation of §12.5, **write the definition of interval in shorthand symbols.**

If you use A's measurements, you would abbreviate 'time from O to Q' to T and 'distance from O to Q' to X. And the definition of §12.15 becomes:

either $s^2 = T^2 - X^2$
or $s^2 = X^2 - T^2,$

whichever of these is positive. Using B's measurements, you would get the same equations with T and X changed to t and x. We can put these together as:

either $s^2 = T^2 - X^2 = t^2 - x^2$ (12.7)
or $s^2 = X^2 - T^2 = x^2 - t^2,$ (12.8)
           whichever of these is positive.

12.18 Now this is the point at which – I hope – we are going to start dragging ourselves out of the Slough of Despond. Our explorations so far have taken us ever deeper into a quicksand where one thing after another that we thought of as hard reality turned out to be merely a matter of this or that observer's point of view. Now at last we have found a piece of firm ground – something they all agree about.

The time and distance between two events are only relative quantities – depending on the observer. But this particular combination of time and distance that we call interval turns out to be independent of the observer. So it is an *absolute characteristic* of the relation between two events. Time and distance do not refer to reality as such. They represent nothing more fundamental than how things *appear* to a group of observers who are stationary relative to each other (e.g., all humanity so far, in effect). But interval is the same for *all* inertial observers, and so it does refer to reality (in the sense of §2.20).

And interval is a mixture of time and distance. Here is a reinforcement of the conclusion we reached in Chapter 11 that the reality of the world we live in is not space and time, but an undifferentiated spacetime.

12.19 Please glance through §11.15 again. The analogy can now be completed by adding:

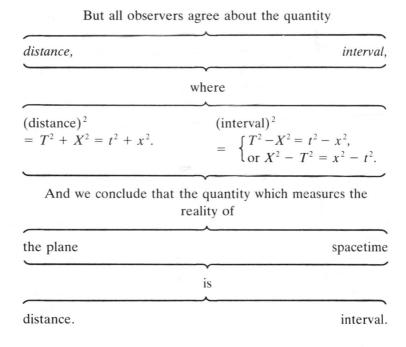

But all observers agree about the quantity

distance,   interval,

where

$$(distance)^2 = T^2 + X^2 = t^2 + x^2.$$

$$(interval)^2 = \begin{cases} T^2 - X^2 = t^2 - x^2, \\ \text{or } X^2 - T^2 = x^2 - t^2. \end{cases}$$

And we conclude that the quantity which measures the reality of

the plane   spacetime

is

distance.   interval.

12.20 The analogy has become doubly impressive. There are two real – nonverbal – differences between the columns: the different ways in which the axes move in changing from one observer to another (§11.15); and the fact that the invariant involves $T^2 + X^2$ in one case and $T^2 - X^2$ in the other. All the other differences are merely verbal (§11.16). So the analogy is really very close.

And very encouraging! When the surveyors learned of the invariance of distance, they were well on the way to a complete understanding of the plane they lived in. When we fully appreciate the consequences of the invariance of interval, perhaps we shall have just as complete an understanding of spacetime.

12.21 The surveyors and their co-ordinate systems are concerned with the geometry of the plane. Don't think of there being just one thing called Geometry. There are many different geometries – the geometry of the plane, the geometry of the surface of a sphere, and so on. And they have different rules. In the geometry of the plane, for example, the angles of a triangle always add up to two right angles. But in the geometry of the surface of a sphere they always add up to *more* than two right angles (think of a triangle with two corners on the equator and one at a pole).

Many geometries, then, each with its own set of rules. We can define the geometry of this, that or the other as 'the rules for working with measurements in' this, that or the other. And then we can speak also of the *geometry of spacetime* – meaning the set of rules for working with measurements in spacetime.

12.22 Now a geometry is completely characterised by its invariant. By that I mean that if you know how the co-ordinates are measured and if you know the formula for calculating the invariant from the co-ordinates, you can deduce *everything* about that geometry. (Sorry, the proof is difficult; I must ask you to take it on trust.) You can, for example, deduce every detail of the geometry of the plane, once you've been told how the surveyors measure their co-ordinates (§§11.10–12) and also the formula for calculating the invariant (distance) from these (§11.21). Similarly, to characterise the geometry of spacetime, all you need to know is (1) how time and distance are measured, and (2) the rule for calculating interval in terms of these co-ordinates.

12.23 Of course we've been assuming that there is no gravity. In fact the Special Theory of Relativity is just the set of rules for working with measurements made by inertial observers in gravity-free spacetime (§5.5). In the language of §12.21 it is the *geometry* of gravity-free spacetime as interpreted by inertial observers. So we ought to be able to deduce the *whole* of this Special Theory from our knowledge of (1) the methods by which inertial observers measure time and distance, and (2) the rule for calculating interval from these measurements.

As I said before, I can't prove that point for you. But at least I can give an impressive example. We'll start the next chapter with that.

13

Old friends revisited

13.1 Starting from the formula for interval, I am going to deduce something that we've had before. I'll not tell you what it is, so that you can have the joy of meeting an old friend unexpectedly. **See how soon you can recognise him.** There will be no diagram so that you can be sure that this deduction is made from the invariance of interval and nothing else.

We'll consider two inertial observers, A and B, who pass at event O with relative speed v (expressed as a fraction of the speed of light – §8.39). Let Q be the tick of B's clock which happens, in A's opinion, 1 time unit after O; and let t stand for the time from O to Q by B's clock.

13.2 **What is the distance between O and Q according to B?**

Zero – both happen where he is. **And what is the interval, s, between O and Q?**

We use (12.7), but with x changed to 0 to fit the present case. So $s^2 = t^2 - 0^2$, or simply

$$s^2 = t^2. \tag{13.1}$$

13.3 **What is the distance from O to Q according to A?**

Between events O and Q, as A sees it, B has travelled away at speed v for 1 unit of time – thus covering distance v. **Work out the interval from O to Q in A's terms.**

This time we put $T = 1$ and $X = v$ in (12.7), getting

$$s^2 = 1 - v^2. \tag{13.2}$$

(If any of this algebra seems difficult, it's because you're forgetting about reading things in full.) **Can you now spot anything interesting?**

From the invariance of interval, these two versions of s must be the same. And so

$$t^2 = 1 - v^2. \tag{13.3}$$

We've found a connection between A's and B's times.

13.4M To say that one number or quantity is the *square root* of another means that the second is the square of the first. Thus 3 is the square root of 9, because 9 is the square of 3. **What is the square root of t^2?**

It is t – since the square of this is t^2.

The symbol $\sqrt{}$ stands for 'the square root of'. So we can write the second sentence of this article as: Thus $3 = \sqrt{9}$, because $9 = 3^2$. And the definition of the first sentence could be put symbolically:

To say that $b = \sqrt{a}$ means that $a = b^2$.

The square root of a complex expression like $1 - v^2$ has to be written as $\sqrt{(1 - v^2)}$, where the brackets indicate that we mean what you get by first calculating $1 - v^2$ and then taking its square root. (You may be used to another form involving a bar across the top; but that's too expensive to print.)

13.5 From (13.3) it follows that

$$t = \sqrt{(1 - v^2)}. \tag{13.4}$$

(If you don't find that obvious, change a to $1 - v^2$ and b to t in the symbolic form of the definition in §13.4.) When we said that the time between O and Q according to A is to be 1 unit, we didn't specify what unit. So in fact it could be any size we like (cf. §3.13, 3.21). Thus (13.4) tells us that, in all cases,

> time from O to Q by B's clock
> = time from O to Q according to A $\times \sqrt{(1 - v^2)}$.

Using T to stand for the time from O to Q according to A, this can be written as

$$t = T \sqrt{(1 - v^2)}, \tag{13.5}$$

where, by the convention of §9.14, $T\sqrt{(1 - v^2)}$ means $T \times \sqrt{(1 - v^2)}$. This is the equation we've been aiming at.

13.6 **Have you recognised an old friend?**

If not, study again §§3.7–8 (and §§3.1–2 if needed). In the present case, **which time is proper and which improper?**

Observer B is present at O and Q. So t is the proper time between them. But A is not present at Q, and so T is improper. Thus (13.5) states a relation between proper and improper times. **Now do you recognise it?**

13.7 We're back to the dilation of time (§3.10). Equation (13.5) says that the proper time is equal to A's improper time multiplied by the factor $\sqrt{(1 - v^2)}$. **Is this greater or less than 1?**

By §12.12, v^2 is a positive quantity. Thus $1 - v^2$ is less than 1, and so is $\sqrt{(1 - v^2)}$. So (13.5) states that the proper time is less than the improper. This, of course, was the conclusion we reached in §3.10. But now we've done better: we've found *how much* less.

13.8 Now the only information we used to make this deduction (apart from the meanings of the symbols) was the fact that observers agree about interval. It does look as if the invariance of interval is a very important discovery – since time dilation follows so simply from it. This illustrates the point of §12.23 about it being possible to deduce the complete Special Theory of Relativity from the invariance of interval.

In the process of living we use a vast body of geometrical knowledge. All of it can be deduced from the proposition that distance, calculated from the co-ordinates as in §11.21, is invariant. Even those who know no formal geometry survive and prosper by their intuitive appreciation of this fact.

The invariance of interval would play a similar role in high-speed life. Aware of the relativity of time and distance, people would depend on their knowledge that interval is ('obviously' to them) invariant. And time dilation would appear as just a simple deduction, the equivalent of some elementary geometrical theorem that every child should know. Interval can't have such a vivid significance for us, but it will help to deepen our understanding of our spacetime environment.

13.9 Let's check that (13.5) gives the same answer as the calculation of §§3.13–22. With $v = 0.995$, we get $v^2 = 0.990\ 025$, $1 - v^2 = 0.009\ 975$, and $\sqrt{(1 - v^2)} = 0.099\ 87$ (check by squaring) – or near enough 0.1. If you need practice, work problems (1) and (2) of §3.23 by this method.

13.10 Equation (13.5) shows why we don't notice the dilation of time. Our speeds are tiny compared with light's. So v is a small fraction and v^2 even smaller. Thus $1 - v^2$ is very nearly 1, as is $\sqrt{(1 - v^2)}$, and so t is very nearly equal to T.

You'd like some actual figures? But working them out for slow speeds brings you to an arithmetical obstacle (try it and see!) – which we circumvent by using an approximate formula.

.11M The sign $\approx$ is used as an abbreviation for 'is approximately equal to'.

.12M In Figure 13.12 the large square has sides of length 1. So length FC is a fraction, which we'll call $\frac{1}{2}x$ (this gives a neater calculation than calling it x). We use EBCG, for example, to stand for 'the area of the rectangle whose corners are E,B,C and G'. Then

$$\text{ABCD} - \text{EBCG} - \text{FCDH} \approx \text{AEKH.} \qquad (13.6)$$

This is only approximate, because FCGK has been subtracted twice. But if $\frac{1}{2}x$ is a *small* fraction, the error will be negligible.

Using the ordinary *length × breadth* formula for each area, (13.6) gives

$$1 - \tfrac{1}{2}x - \tfrac{1}{2}x \approx (1 - \tfrac{1}{2}x)^2.$$

But subtracting $\frac{1}{2}x$ twice is the same as subtracting x, and so

$$1 - x \approx (1 - \tfrac{1}{2}x)^2, \tag{13.7}$$

whence

$$\sqrt{(1 - x)} \approx 1 - \tfrac{1}{2}x \tag{13.8}$$

(cf. opening of §13.5). This gives a handy method of calculating square roots of numbers just less than 1. **Try one or two (like 0·994) and check your answers by squaring.** When x is changed to v^2, (13.8) gives

$$\sqrt{(1 - v^2)} \approx 1 - \tfrac{1}{2}v^2. \tag{13.9}$$

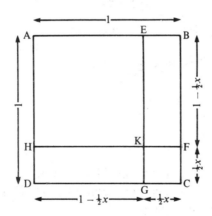

Figure 13.12

13.13 For a jet plane, $v = 0{\cdot}000\ 001$ (§1.8). **How large would the dilation effect be for somebody on the ground?**

Here $v^2 = 0{\cdot}000\ 000\ 000\ 001$. Then by (13.9),

$$\sqrt{(1 - v^2)} \approx 1 - 0{\cdot}000\ 000\ 000\ 000\ 5 = 0{\cdot}999\ 999\ 999\ 999\ 5.$$

You'd hardly expect to notice a discrepancy of 1 part in 2 million million! **Work a few more examples for yourself** – e.g., comparing opinions of observers on Earth and Sun (§1.8); and investigating what speed would be needed to produce an effect that we'd actually notice.

13.14 There's something new to be learnt from yet another approach to the dilation of time – starting from the spacetime diagram in its §12.1 form, where A's and B's world lines have equal slopes. **Check that Figure 13.14 represents the story told in** §13.1.

Line MQ is parallel to the X-axis, and therefore perpendicular to the t-axis (§12.3). So length OQ is less than length OM. But these represent, respectively, B's proper time and A's improper time from O to Q, both on the same scale (§12.1) – showing again that proper time is less than improper. **As an exercise, consider the case where Q happens before** O. Equation (13.5) can easily be deduced from Figure 13.14.

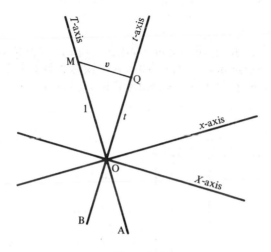

Figure 13.14

.3.15 In §§3.34–6 we met another way of stating the dilation of time. And in §3.37 I drew your attention to an apparent contradiction. Did you manage to sort that out? **If not, try again.**

Using Figure 13.14, we conclude that A thinks B's clock is running slow because the length OQ representing a time actually measured by B's clock is less than length OM representing what A thinks it should be. To decide what B thinks of A's clock, you would have to take a point q on A's world line, draw mq parallel to the x-axis, and go through an exactly parallel argument – reaching a conclusion that merely interchanges A and B.

The key point is that MQ and mq, being parallel to the two distance axes, are *not* parallel to each other – which is how the diagram presents the relativity of simultaneity. And if you now refer to *(a)* and *(b)* of §3.37, you'll see there's no contradiction.

Mark equal time scales along the two time axes. Then use the slot to show A's point of view (§§8.13, 8.27) and notice how, between the appearances of O and Q in the slot, more time passes on A's scale than on B's. Then do the same for O and q from B's point of view.

13.16 The dilation of time prediction has been checked by many experiments besides that of §§3.28–32. Further tests have been done with muons, at different speeds in different conditions. Other tests used pions (§1.9), whose shorter life allows the whole experiment to be done inside a laboratory; the results agreed with prediction to within 0·4 per cent. Subtle experiments using laser beams to measure how much the inbuilt clocks of moving atoms appear to run slow are described in *New Scientist,* 22 April 1976, p.184. They gave about the same accuracy.

The very short-lived particles studied by high energy physicists must travel a few metres from where they are produced to where they are to be subjected to experiment. Will enough survive to make the experiment worthwhile? To answer that question correctly, the experimenters must allow for time dilation. So every experiment that works is incidentally a further verification of this prediction – and therefore (§3.33) of the Special Theory of Relativity.

13.17 **What is the proper time between two events at which a light signal is present?**

Your answer was presumably 'Zero'. And presumably you got it by putting $v = 1$, giving $\sqrt{(1 - v^2)} = 0$, and therefore $t = 0$ by (13.5). Fine! But a cause for some doubt is that (13.5) was established for inertial *observers,* and they can't go at the speed of light (§2.1). But if you assume (plausibly) that (13.5) holds good for light also, then the zero answer is correct. *Time does not pass for a light signal.* (Was I being too cautious in §6.11, as you probably thought at the time?)

13.18 The spacetime diagram in the simple equal-slopes version of §12.1 can also help us to understand the Fitzgerald Contraction (§3.38; also §§2.23–6 if needed). We'll temporarily suspend the convention of §8.5, to allow us to talk about the length of a rod. Observer A (Figure 13.18) is carrying a rod, holding it by one end. So the other end is always at the same distance from him and its world line is a line parallel to the *T*-axis (§8.12) – the broken line in the figure.

Though A is carrying the rod, it still has a spacetime existence, independently of what he thinks about it. **What represents the rod as a spacetime reality?**

The whole area between A's world line and the broken line. Every point in that area represents the event of a bit of the rod being 'instantaneously there'. Now remind yourself from §11.7 of how each observer splits spacetime, in his own way, into space and time. **So what represents the rod at one instant according to A?**

Any line parallel to his distance axis (§8.12) – for example, PQ or LN. **And what represents the rod at one instant according to B?**

A line parallel to *his* distance axis, like RS or KN. Other sample lines are drawn showing what constitutes the rod at this or that instant according to one or other observer. Check up on all this with the slot.

The length of a rod is the distance between its ends at the same time (§2.23) – length LN for A, KN for B (on the same scale, §12.1). And now we see that the disagreement about length is a simple consequence of the way observers split spacetime differently into space and time (§11.7).

Since KN is perpendicular to the *T*-axis (§12.3), clearly it is shorter than LN. And so B thinks A's rod has contracted – by a factor of $\sqrt{(1 - v^2)}$, as you can prove by the method of §3.38 or by compar-

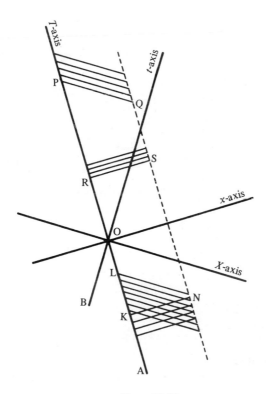

Figure 13.18

ing the geometries of Figures 13.14 and 13.18. And of course the situation is symmetrical, so that A thinks exactly the same about B's rod. So we can improve the conclusion stated in §3.38: ' . . . *contracted* by a factor of $\surd(1 - v^2)$'.

I hope that this has helped your understanding of spacetime, and the way observers split it into space and time. If not, I've been wasting *your* time. For unlike time dilation, the Fitzgerald Contraction has no practical significance – situations in which fast-moving objects can be treated as measuring rods (analogously to treating muons as clocks) do not occur in practice.

13.19M Just as = means 'is equal to', so the sign > is used as an abbreviation for 'is greater than', and < for 'is less than'. Their shapes provide obvious mnemonics.

13.20M The *absolute value* of a quantity or number is what we get by ignoring a negative sign (§§6.23–4) if there is one. So the absolute value of both 3 and −3 is 3. By §12.12, the square of any number (positive or negative) is equal to the square of its absolute value.

13.21 The rest of this chapter is used to clear up a few points about interval – dull, but necessary. Consider any two events O and Q. Look at them from the point of view of observer A who is present at O (since interval is invariant, it doesn't matter what observer we use). Give him rectangular axes (§8.42) with T and X, as usual, standing for his co-ordinates. The signal lines through O divide the spacetime diagram into quadrants, named as in Figure 10.1.

13.22 If Q is in the top or bottom quadrant (Figure 13.22), it is obviously farther from the distance axis than the time axis. Hence (whether T and X are positive or negative)

absolute value of T > absolute value of X,

and so $T^2 > X^2$. Thus $T^2 - X^2$ is positive; and the interval between O and Q is given by (12.7) of §12.17.

In this case, the line joining O and Q is steeper than 45°, and so the events have a time-like separation (§11.8). Correspondingly, we say that the interval between them is *time-like*.

The straight line through O and Q, being steeper than 45°, could be the world line of an inertial observer, B, who is present at both events. **What is the relation between the time from O to Q by his clock and the interval between O and Q?**

Exactly as in §13.2 we can prove that $s^2 = t^2$, and so $s = t$. So we've proved that

> If the interval between two events is time-like, it is equal to the time between them as measured by the inertial observer who is present at both.

13.23 You should have no difficulty in **proving for yourself** the contents of this article and the next. If Q is in the left or right quadrant, the separation of O and Q is *space-like,* and so we use the same adjective for the interval between them, which is given by (12.8). There is an observer who says the events happen at the same time; and

> If the interval between two events is space-like, it is equal to the distance between them as measured by an observer according to whom they are simultaneous.

13.24 Finally, if Q lies on a signal line through O, either (12.7) or (12.8) gives $s = 0$. We call such an interval *light-like.* There is no observer according to whom the separation between O and Q is purely spatial or purely temporal.

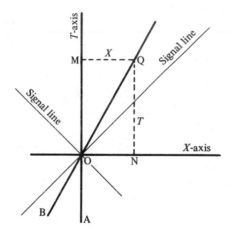

Figure 13.22

14

The scales of the spacetime diagram

14.1 It's about time we settled the question of how one observer's time scale is related to another's (§8.24). We can do so, as you may have guessed, by using the invariance of interval (§12.15).

Consider all the events in the 1D universe which are separated from event O by some fixed interval – all events Q such that the interval between O and Q has the same constant size. These will be represented in the spacetime diagram by the points of a curve, which we shall call an *isoval* (shortened from 'isointerval', on the analogy of 'isobar', etc.) – the word is my personal coinage, not part of the official language. We can obviously distinguish between *time-like* and *space-like isovals,* according to whether the interval concerned is time- or space-like (§§13.22–3).

An isoval is the spacetime analogue of a circle. The latter consists of all points at constant distance from a fixed point; the former represents all events at constant interval from a fixed event.

14.2 Clearly isovals are independent of the observer. So we can study them from the point of view of one observer A, whose world line goes through O and to whom, for convenience, we assign rectangular axes (§8.42). Let's look at time-like isovals first.

If the co-ordinates, T and X, of some point Q are such that

$$T^2 - X^2 = s^2, \tag{14.1}$$

then (12.7) tells us that the interval between O and Q is time-like (§13.22) and equal to s. So this Q is a point of the *time-like isoval for interval s*. And the complete isoval will consist of all points whose co-ordinates are such as to make (14.1) true.

14.3 Now to find what an isoval looks like. In Figure 14.3, OE has been given length s. We take *any* point N on the X-axis and draw NQ vertically, equal in length to NE. **Can you prove that the point Q obtained in this way is a point of the isoval?**

Applying Pythagoras' Theorem to triangle OEN, we get

(length NE)^2 = (length OE)^2 + (length ON)^2,

or (since length NE = length NQ = T) $T^2 = s^2 + X^2$. And subtracting X^2 from both sides gives equation (14.1) – showing that Q *is* a point of the isoval.

14.4 Now imagine the point N moving along the X-axis. As it goes over towards the right, point Q moves to the right with it, but also rises higher up the paper (to keep length NQ = length NE). The curve that Q will thus trace out is the isoval. **Try to visualise what it will look like. Better still,** use this process to mark several positions of Q, and draw a smooth curve through them – rough sketch only.

14.5 The result of your efforts should look like Figure 14.5. Are you surprised to see the curve going off to the left as well as the right? But we said that N could be *any* point on the X-axis. And are you surprised to see a similar curve below the X-axis? But NQ could be drawn vertically down, just as well as vertically up. (If you're at home with negative quantities, you'll see another reason why the curve is symmetrical about both axes).

You'd probably think of the pieces above and below the axis as two distinct curves. But mathematicians prefer to consider them as two 'branches' of *one* curve – because they both result from applying the procedure of §14.3, or from equation (14.1).

14.6 Of course the curve could be continued as far as you like at each end of each branch. The figure suggests that the farther it goes out towards right or left, the nearer does it get to the signal lines through O, yet without actually reaching them. And you could easily prove this by noticing that the farther N is from O (Figure 14.3), the nearer does NE, and therefore NQ, come to being equal in length to ON.

14.7 If $s = 1$, we have what we can call the *unit time-like isoval* – the curve

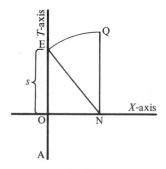

Figure 14.3

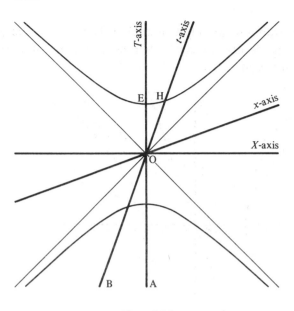

Figure 14.5

representing all events separated from O by time-like interval 1. Figure 14.5 still serves perfectly well, provided we make length OE = 1 – i.e., take OE to be one unit of A's time scale. We'll use this unit isoval to solve the scales problem.

14.8 Let B be another inertial observer whose world line (through O) meets the upper branch of the unit isoval at H (Figure 14.5). Then the time from O to H by B's clock is equal to the interval between O and H (§13.22) – which is 1, according to the definition of the unit isoval. In other words, the length OH is one unit of B's time scale, just as length OE is a unit of A's (§14.7). **Think about how this scale unit varies in size as B's world line tilts farther away from the vertical.**

So here we have a recipe for relating the scales of all inertial observers whose world lines go through O. And for those whose world lines don't pass through O, we simply use the scale on a parallel world line through O (since relatively stationary observers agree about time and about simultaneity).

14.9 The co-ordinates of any point of a *space-like isoval for interval s* will be connected by the equation

$$X^2 - T^2 = s^2 \qquad (14.2)$$

– obtained from (12.8) (cf. §14.2). We could draw this curve by a method like that of §§14.3–4, but with everything swivelled through a right angle. The result would work out to be what you get by taking the time-like isoval and turning it through a right angle to occupy the left and right quadrants. Figure 14.9 repeats Figure 14.5 with the space-like isoval added.

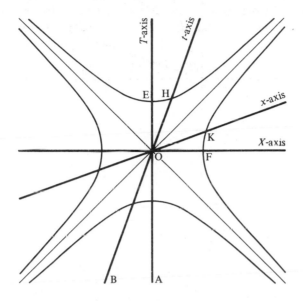

Figure 14.9

14.10 To get the *unit space-like isoval* (cf. §14.7) we take $s = 1$ – make length OF = 1 in Figure 14.9. And then an argument like that of §14.8 (**work out the details**) shows that length OK represents unit distance according to B (K being where B's distance axis meets the isoval).

14.11 If we take the spacetime diagram that we worked out in Chapter 8 – particularly as embodied in Figure 8.26 – and add to it the two unit isovals (as in Figure 14.9) to define the scales, then we have what is known as the *Minkowski diagram* (after Hermann Minkowski who first interpreted Einstein's theory in this geometrical fashion). The whole of the Special Theory of Relativity can be deduced from this diagram.

So the Minkowski diagram is, as it were, an accurate *map* of spacetime – of the gravity-free spacetime of Special Relativity.

14.12 Please note: I said a *map* of spacetime – not a picture or a model. A map often distorts. If it maps something on something else of a different nature – maps the spherical Earth on a flat page, for instance – then it *necessarily* distorts. Treated as a picture, it tells silly lies. Interpreted according to the correct rules, it gives accurate information.

Consider, for example, the following way of mapping the Earth. Start with an undistorted map of the Earth drawn on a transparent globe, G (Figure 14.12). Wrap round this a sheet of paper to form a cylinder, M, which touches the globe all round the equator, E. A point source of light at the centre, C, of the globe will cast a shadow which forms a map on the cylinder. Recording this photographically, unwrap the cylinder – and you have a *Mercator projection* map of the Earth.

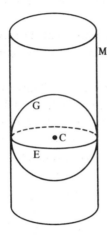

Figure 14.12

If you think of a fly crawling on the globe and how its shadow would move on the cylinder, you will realise that such a map produces gross distortions in its representation of distances. The distance scale depends on latitude; the higher the latitude, the larger the scale.

14.13 Thinking of the spacetime diagram as a map should help with points that may be troubling you. The way the scale varies with slope is no more mysterious than the way a Mercator projection scale varies with latitude.

Or you may be worried about the way the isovals run off to infinity in spite of being the analogues of circles (§14.1). But that's also like the Mercator map – which puts the North Pole an infinite distance up the paper, so that any circle passing through the Pole is mapped by an open curve running off to infinity.

Suppose you and I were incapable of imagining a sphere, but had to work out the Earth's geometry by studying flat maps. We should have to deal with these open curves and work out by pure reasoning that, despite appearances, they must represent curves, on this unimaginable sphere, consisting of points at the same distance from a fixed point.

Now – since I can't spoonfeed you all the time – **work out the analogy for interval.**

4.14 I haven't been reminding you recently to use the slot. And you may have fallen into the trap of thinking of the diagram too geometrically – forgetting that it represents change and motion. I suggest that you spend some time now studying Figure 14.9 with the slot – making sure you understand it in terms of time and space and motion, and from either observer's point of view.

   If you play around enough, you may hit on an important discovery before I get to it. And even if you don't, you'll be better prepared when the time comes.

4.15 The Great Plum Pudding Eating Race takes place on the 1D universe. The competitors are provided with standard plum puddings, which they start to eat as they all pass each other at event O; and the object is to be first to finish one's pudding. Although they don't know it, they all actually eat at the same rate. **What impression does each get about the result?**

   If it takes time s to eat a pudding, then each swallows his last mouth-ful at the event represented by the point where his world line crosses the isoval (upper branch) of Figure 14.5 – point E for competitor A; H for B. But in A's view of things a horizontal line through E represents the events that happen elsewhere just as he is finishing (§8.12). The whole upper branch of the isoval lies above this line. So he thinks all the others finished after him (Slot!). But A is no different from any of the others – even if *we* have given him a special representation on the diagram. So *every* competitor thinks himself the winner.

   This, you'll notice, is the dilation of time approached from yet another angle. **Think it out in detail.** (And if you're mathematically expert, work it out from B's point of view. What does it imply about the tangent at H in relation to the x-axis?)

4.16 In Figure 14.16 we still have rectangular axes, but the observer whose axes they are has been called A_1, because I want to use A temporarily for something else. Both branches of the unit time-like isoval (§14.7) are shown. The diagram also shows the world lines of three other inertial observers A, B and C. Use the slot if you need it to **work out what story the diagram tells about these three.**

   It's the story we first met in §§4.1–2 – the 'three-clock problem'. Reread §§4.1–5, and then see if you can **use Figure 14.16 to give another demonstration of the conclusion we reached there.**

14.17 By §14.8 (with Q playing the role of H), length OQ is the unit of C's
time scale, and so the time from O to Q by C's clock is 1. Similarly the
time from P to O by B's clock is 1. **What about the time from P to Q by
A's clock?**

   Length OE represents unit time by A's clock as well as by A_1's
(§14.8). And since length PQ is clearly greater than twice length OE, it
follows that the time from P to Q by A's clock is greater than 2 –
greater than the time P to Q by B's and C's clocks jointly, just as we
concluded in §4.5.

14.18 Maybe you suspect that it only worked out that way because we gave
A a vertical world line. But there's nothing special about an observer
with vertical world line – only about our representation of him.

   Still you'd better convince yourself by giving A a sloping world line
and seeing what happens. Make a new diagram by tracing the origin O
and the unit isoval from Figure 14.16, but omitting the rest. For A's
world line take *any* line steeper than 45°. Complete the three-clock
diagram, as exemplified in Figure 14.18; but use a *different* world line
for A. Check that the diagram still tells the three-clock story. Next
draw A_1's world line through O parallel to A's, meeting the isoval at E.
Measure PQ and OE, and work out the times again.

   Every word of §14.17 applies to the new figure – with the same con-

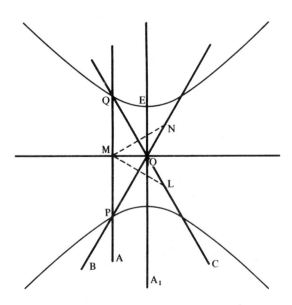

Figure 14.16

clusion emerging. You ought at least to do one or two more similar examples. And if your mathematics is up to it, you might prove that we *must always* get the result that A's time is greater than that of the B–C team.

14.19 Now please go through §§4.6–10 again, but referring constantly to Figure 14.16 and taking note of the following comments.

The problem raised in §4.7 vanishes when you remember that 'the time from P to O according to A' really means the time from P to M by A's clock (§7.14).

It becomes obvious why attempted back-to-front arguments, as in §4.8, always break down. For example, by applying reasoning like that of §4.4 to the times from P to M and M to Q, you can easily prove that 'Time from P to Q by A's clock is *less* than time from P to M according to B plus time from M to Q according to C.' **But can you produce anything meaningful by adding these last two?**

Imagine B's distance axis drawn in by the usual rule (§8.26) and MN drawn parallel to this. Then in B's opinion M and N happen at the same time (§8.12). So when we speak of the time from P to M according to B, we really mean the time from P to N by B's clock (compare the thinking of §7.14). Thus (with LM similarly parallel to C's distance axis) the sentence in quotation marks in the previous

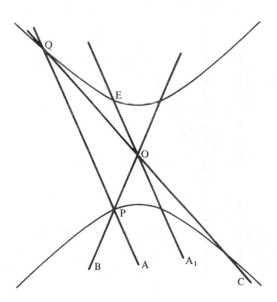

Figure 14.18

paragraph merely says that time from P to Q by A's clock is less than time from P to N by B's clock plus time from L to Q by C's clock. These last two *don't* add up to make the time from P to Q by B's and C's clocks jointly – and so there's no conflict with the three-clock result of §4.5.

14.20   In relation to §4.9, temporarily think of Figure 14.16 as a geographical map of a flat piece of territory. Surveyors A, B and C are engaged to measure three straight roads joining town centres P, O and Q. You'd just laugh if somebody pointed out that A and the B–C team start at the same point and finish at the same point, and argued that they are therefore measuring the same thing. Considering how well this sort of analogy worked in Chapter 11, ought you not to agree that – with Figure 14.16 back in its spacetime role – the clocks of A and the B–C team are measuring different things between events P and Q?

In the geographical case, we say that surveyors A and B–C are following different routes. So surely we can reasonably interpret the spacetime diagram as showing that A and B–C are taking *different routes through spacetime* from event P to event Q. Furthermore we take it for granted that surveyors measuring distances along different routes will get different results. Ought we not (in regard to §4.9) to expect observers measuring times along different spacetime routes to get different results?

14.21   We used to think that the time between one event and another didn't depend on the route followed. Now we know it does – just as distance between points depends on route. In common parlance, surveyors B and C have 'gone the long way round' from town P to town Q. And similarly in spacetime observers B and C have 'gone the *short* way round' between event P and event Q. (It's the scale recipe of §14.8 that allows us to represent spacetime's 'short way round' by means of the 'long way round' POQ of the diagram – we're back to the oddities of maps again, §§14.12–13.)

In space the indirect route is *longer* than the direct one. In spacetime the indirect route is *shorter*. This is linked mathematically with other differences we've met before – see §12.20. But that's a connection I'll have to leave unproved.

Now reread §4.12. You will see that the Public Time idea is equivalent to the idea that time does not depend on route – and both beliefs are untenable. Saying that each observer has his own private time is equivalent to saying that the amount of time that passes for him depends on his route through spacetime.

Am I getting nearer to fulfilling my promise of §4.13?

14.22 Figure 14.22 is derived from Figure 14.16 by suppressing the isoval and lines LM and MN, reducing B's and C's world lines to broken lines, and adding a new world line – that of D, which coincides with A's at bottom and top but departs from it in the middle. **What's the story of A and D?**

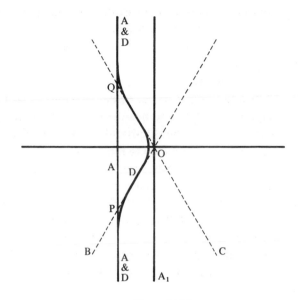

Figure 14.22

It's the space-travel tale of §4.14. Check up with slot. And then consider: **What answer would you now give to the question at the end of §4.14?**

I suspect you will be more sympathetic to the suggestion that between parting and reunion D's clock records less time than A's.

14.23 Now please read §§4.15–18 again, reinterpreting them in terms of Figure 14.22, and taking note of the comments that follow.

The argument of §4.15 (first paragraph) can now be put more neatly by saying that A and D follow different routes through spacetime, and so we'd tend to expect their time measurements to be different.

In §4.17 we are saying that A's and D's experiences are different because A is always inertial, whereas D has three periods of accelerated (non-inertial) motion – represented by the curved portions of his world line in Figure 14.22. And the 'geographical' analogy again proves helpful. If A and D are surveyors, measuring the distance between two towns, and if we are told that A always follows a straight line, whereas D sometimes goes round curves, then *from that fact alone* we could deduce that D's distance will be greater than A's. Similarly, if we are told that observer A's motion is always inertial, while D's is sometimes accelerated, it seems at least plausible to suggest that *from that fact alone* we could infer that D's time measurement between parting and reunion will be different from A's – presumably less.

14.24 Finally consider the counterargument described in §4.18. This amounts to saying that: in the case of Figure 14.16 the time from P to Q is shorter by the route POQ than by the direct route PMQ; *but* that in Figure 14.22 the curved portions of D's world line near P, O and Q (representing accelerations) in some way *increase* the time that D's clock measures. **How does this argument appeal to you now?**

In §4.18 I suggested that the fair answer would be 'I don't know'. **Would you still feel that's as much as we can say?**

When you think of Figure 14.22 as representing alternative roads between towns P and Q, you wouldn't seriously suggest that the curves at P, O and Q in some way decrease the distance measured by D, so as to compensate for the increased distance he covers in going 'the long way round'. Considering how good the analogy has proved to be, it's surely rather unlikely that observer D's periods of acceleration – curves on the diagram – would have the alleged compensating effect.

So on every point the analogy suggests rather strongly that less time will elapse for the space traveller than for those she left at home. But analogies can be dangerously misleading. So let's agree to keep an open mind till we're ready to tackle the question again.

15

The radar point of view

15.1 Our spacetime diagram has been a very useful aid to both logic and imagination. Yet it is also unpleasantly complex. The rules that relate the co-ordinates and scales of different observers are too complicated. Now I want to show that this complication arises because, when we thought we were being revolutionary, we were actually being pigheadedly conservative. Our perversity consisted in constructing the diagram in terms of the old familiar time-over-there and distance – even though we knew that these were only relics of slow-speed life, which prove to be nearly useless in high-speed conditions.

Shall we try the effect of working instead with the quantities that are actually measured – the times of sending a signal to an event and receiving one from it (§§7.2–3)? We can call these the *radar co-ordinates* of the event (cf. §§5.20, 7.1). Now please revise §§6.13–22. We're starting afresh from there.

15.2 We need shorthand symbols for these radar co-ordinates. But we're running short of convenient letters of our ordinary alphabet, and so we'll use two Greek letters:

> *theta* – printed as Θ for the capital and θ for the small letter;
> and *phi* – Φ for capital and ϕ for small letter.

We'll use the capitals for A's radar co-ordinates and small letters for B's. And when we want to talk about these radar co-ordinates in general terms, without specifying an observer, we can speak of 'the theta' or 'the phi' (just like 'the time' and 'the distance') of this or that event.

An observer sends one signal to an event and receives one from it. One of these is positive-going and the other negative-going; but which is which depends on whether the event happens on the positive or negative side of the observer. We'll say that

> The *theta* is the time of sending or receiving the *positive-going* signal which is received at or sent from the event; and

The *phi* is the time of sending or receiving the *negative-going* signal which is received at or sent from the event.

The distinction is illustrated by Figure 15.2 (where A's Θ and Φ are marked, starting from the zero of his clock). In (i) with Q on the positive side of A, Θ is the time of sending and Φ that of receiving. In (ii) it is the other way round.

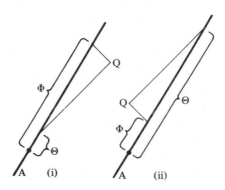

Figure 15.2

15.3M To say that one number or quantity is *proportional* to another means that they vary together in such a way that one of them is always equal to the other multiplied by a constant (fixed) number or quantity – often called *a constant factor*. So the number of human fingers in a house is proportional (barring accidents) to the number of people there. Again, the time between two ticks of A's clock is proportional to the length between the corresponding points of his world line (that's what we mean by representation 'to scale', §6.22). And if $t = kT$, then t is proportional to T. As an alternative way of putting it, we can say that the two quantities vary *proportionally* or *in proportion* to each other.

15.4 In Figure 15.4 observers A and B have been given world lines of equal and opposite slopes. And A is shown sending signals (at events marked with capital letters) which B receives (at events labelled with small letters). Slot please!

It's geometrically obvious that length pq is proportional to length PQ. And since the time scales are equal (§12.1), that means that time between events p and q is proportional to time between P and Q. In other words,

time between p and q
= (time between P and Q) × (constant factor).   (15.1)

Length pq is greater than length PQ, and so the constant factor is
greater than 1. (Each observer's events are timed by his own clock, of
course.) Similarly you can **prove that**

> time between r and s
> = (time between R and S) × (constant factor),   (15.2)

where this constant factor is *less* than 1.

So if the observers are *receding* from each other, the constant factor
is *greater* than 1. If they are *approaching,* it is *less* than 1. And if they
are relatively stationary, the constant factor *is* 1.

Using equal slopes for the world lines made it easier to prove (15.1)
and (15.2). But their truth cannot depend on how *we choose* to draw
the diagram (§8.44). So they will continue to hold good whatever
slopes the world lines may have.

Of course all this is what commonsense would suggest anyway. If the
sender is receding from you, the second signal has farther to travel and
so will take a little longer . . . **You finish it.**

15.5 It's clear from Figure 15.4 that the size of the constant factor depends
on the angle between the world lines, which is in turn decided by the
relative speed of the observers (§8.37). And indeed we've been describing a standard radar method of measuring speed, which is used,
for example, in radar speed traps.

So in our new system we can use this constant factor, instead of
speed, to specify relative motion – just as we use the theta and phi

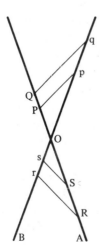

Figure 15.4

co-ordinates, instead of time and distance, to specify events. Then all our measurements will refer directly to on-the-spot times of sending and receiving signals.

15.6 Now suppose that A sends to B a *continuous* beam of light. But let's consider the successive waves (§1.4) of this continuous light as successive signals. Then, in the case where A and B are moving apart (I'll stick to that in future; **you work out the approaching case**), the time between any two successive waves as received by B is equal to the time between them as sent by A increased (multiplied) by a constant factor (§15.4). And so the frequency (§1.5) of the light received by B will be its frequency as sent by A *de*creased (divided) by this constant factor.

This is called the Doppler effect (after its discoverer, 1842). With sound waves, where frequency corresponds to pitch, you can easily arrange to observe it. For light waves it has been verified by astronomical observations and laboratory experiments for speeds up to about a three-hundredth of that of light. So (15.1) and (15.2) are not mere deductions from theory. They are found to correspond to physical reality.

15.7 And that fact allows us to prove the point about scales that we had to assume in §8.24. In Figure 15.7, A is the observer whose world line and scale we specified in §§6.20 and 6.22, while B is any other inertial observer.

From §15.4 the time between events p and q is proportional to the

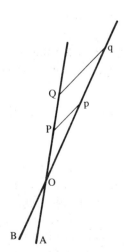

Figure 15.7

time between P and Q, which is in turn proportional to the length PQ (§15.3). And the geometry of Figure 15.7 shows that length PQ is proportional to length pq. Putting these three proportionalities together, we deduce that time from p to q is proportional to length pq – in other words, B's on-the-spot times *are* represented to scale by lengths on his world line.

15.8 The constant factor of §15.4 has no agreed name, but I propose to call it the *Doppler factor,* and we shall usually abbreviate it to *k*.

In the first place it is the Doppler factor *from* A (sender) *to* B (receiver). But from §5.11, this must be the same as the Doppler factor from B to A – otherwise A and B would *not* be equivalent. So we'll call it the Doppler factor *between* A and B. As we said in §15.5, *k* takes over the role previously played by *v*.

15.9 From the cases discussed in §15.4, we see that there are actually two distinct Doppler factors between A and B. One applies to signals sent and received while they are approaching (before event O), the other to signals sent and received while receding (after O).

We exclude the case of one signal being sent before O and the other after. But we can allow a signal to be sent *at* event O, the moment of passing. The other observer will, of course, receive it at O. And we can use it in either the receding or the approaching context.

15.10 Bearing these points in mind, we can now state the conclusions of §15.4 more neatly:

> If either of two observers sends two light signals to the other, then
> time between events of receiving these signals
> = $k \times$ (time between events of sending them),   (15.3)
> where the Doppler factor k is greater than 1 if the observers are receding, less than 1 if they are approaching, and equal to 1 if they are relatively stationary.

(The times of sending and receiving are, of course, times by the clock of sender and receiver, respectively.)

.11M When writing a division in 'fraction' form (§8.38), it is not always convenient to place one symbol above another, as in $\frac{3}{8}$. We can use instead the form analogous to 3/8. Thus Φ/k means 'Φ divided by k'.

15.12 Figure 15.12 represents two inertial observers and their radar co-ordinates for event Q (as defined in §15.2). Slot! If k stands for the Doppler factor between A and B, **find an equation connecting** θ, Θ **and** k.

The signals that A sends at events O and P reach B at O and p, respectively. So Θ and θ are the times between the signals at sending and receiving, and (15.3) gives

$$\theta = k\Theta . \tag{15.4}$$

And an equation for the phi co-ordinates?

Similar reasoning applied to signals sent by B at O and r and received by A at O and R gives $\Phi = k\phi$. And on dividing both sides by k (§9.12) we have

$$\phi = \Phi/k . \tag{15.5}$$

This proof of (15.4) and (15.5) assumes that Q is on the positive side of both observers with P and R happening after O. It can, however, be generalised to cover all cases. Maybe you'll take my word for that, in which case you should skip to §15.15. But if you're at ease with the mathematics, you can do the general proof sketched in §§15.13–14.

15.13 If k stands for the Doppler factor between A and B after they have passed and κ (the Greek k, called 'kappa') for the Doppler factor between them before they pass, **prove that**

$$\kappa = 1/k . \tag{15.6}$$

Hint: Apply (15.3) to the signals which A sends to B at O and time T after O; and to those which he receives from B at time T before O and at O.

15.14 In the situation of §15.12, define k to be the Doppler factor when B is on the positive side of A (whether that is before or after O). Consider the two signal lines separately. Then (15.3) and (15.6) can be used to show that (15.4) and (15.5) are true in all cases.

15.15 **Think for a few moments about the significance of equations (15.4) and (15.5).**

Although observers disagree about the radar co-ordinates of an event (just as they do about the time and distance co-ordinates), at least the rule for changing from one to the other is simple – multiply the theta, and divide the phi, by the Doppler factor. The use of radar co-ordinates really does simplify things.

15.16 **Can you find an invariant** (§§11.20, 12.10, 12.15)?

If we multiply corresponding sides of (15.4) and (15.5), we get

$$\theta\phi = \Theta\Phi.$$

So the quantity (theta) $\times$ (phi) is the required invariant.

15.17M As an obvious extension of §9.11, multiplying corresponding sides of two equations is an allowable operation (§9.12). In the previous article we multiplied the left-hand sides of (15.4) and (15.5) to get $\theta\phi$ on the

left of that last equation. And for the right-hand side we had to multiply $k\Theta$ by Φ/k. But the 'multiply by k' in one of these is cancelled by
the 'divide by k' of the other, giving $\Theta\Phi$.

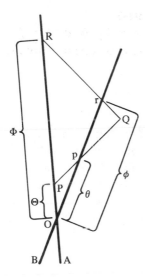

Figure 15.12

15.18 Articles 15.18–20 take us down an interesting side street to which we
shall never return. So I'll treat them concisely, and it doesn't matter if
you don't completely follow.

Figure 15.18 is got from Figure 15.12 by adding the signal lines
through O (and suppressing unwanted clutter). As Q takes up different
positions, lengths OP and OU will change, and they will obviously vary
proportionally (§15.3). Then time Θ, being represented by length OP
on A's time scale, could be represented equally well by length OU on
another time scale marked along the signal line. The 1 second mark on
the new scale will be the position of U when P is at the 1 second mark
of A's scale. Similarly, co-ordinate Φ can be represented by length OV
on a suitable scale marked on the other signal line.

Labelling these lengths on the figure, we have produced a new co-
ordinate system whose axes are the signal lines. And – very remarkable – these axes are *independent of the observer*. We can call this the
'radar spacetime diagram'.

15.19 We have not, however, discovered some new sort of absolute measurement. For a light signal doesn't provide any physical means (e.g., a
clock) for marking the new scales. The scales for representing A's

radar measurements on the new axes have to be derived from A's own time scale.

The radar co-ordinates of Q according to *any* observer (with world line through O) will be represented by the *same lengths* OU and OV, but on *different scales*. And (15.4) and (15.5) show how these scales are related: length OU represents k times as much θ as it does Θ; and length OV represents $1/k$ times as much ϕ as Φ.

So the radar diagram uses the *same axes* for all observers who have origin O, and the *same lengths, OU and OV to represent the co-ordinates*. And the only change in passing from one observer to another is in the scales. This is a great simplification compared with the spacetime diagram based on time-over-there, distance and speed (axes, etc., as in §§8.26 and 8.28; scales fixed by the roundabout procedure of Chapter 14). You can guess which system the high-speed navigators of the 25th century will prefer for their spacetime charts.

15.20 This future generation will find the nature of spacetime far easier to understand – because they will split it up differently. Instead of talking about an event happening at a certain time and a certain distance, they will speak of it happening at a certain theta and a certain phi. So they would think of the fundamental divisions of spacetime as those which correspond to a constant value of one or other co-ordinate – those cross-sections (cf. §11.7) that are marked out by signal lines. They would thus avoid our disagreement as to what constitutes space (all events at the same time) and time (all events in the same place). They would all agree about which events happened at the same theta (i.e., at which a certain positive-going signal is present). Presumably they would group these in some named category taking the place of our 'space' – a category that actually would be as clear and unambiguous as we used to think 'space' was. Similarly for the phi co-ordinate. And their only (rather minor) disagreement would be about which *numbers* to use in specifying these cross-sections.

If spacetime seems cussedly complicated to us, we have only ourselves to blame!

15.21M The rest of this chapter is for those who still need to learn how to add and subtract negative quantities – including those who can only do it by rule, without understanding why. I shan't have space to set practice problems to work out, but you'll learn much better if you try to keep a step or two ahead of me. Please revise §§6.23–5 and 12.11.

15.22M We start from the basic idea that if q is *any* positive quantity (or number), then

$$q + (-q) = 0 .$$ (15.7)

In the language of §§6.23–5, $-q$ is just q measured backwards. The brackets tell us it is the negative quantity, $-q$, that is to be added to q. And it seems fairly obvious that adding a quantity and the same quantity measured backwards gives zero – think, for example, of pacing on a line (§6.25).

A strict mathematician would reject the 'measured backwards' idea as loose talk. He'd prefer to say that the negative quantity $-q$ is *defined* by (15.7). In other words, he'd say that $-q$ *means* the quantity that gives zero when you add it to q. Whichever way you take it, (15.7) is now our starting point.

5.23M What is the result of adding a negative quantity $-b$ to a positive quantity a? Thoughts about pacing on a line (§6.25) would suggest that

Adding $-b$ has the same effect as subtracting b.

or in symbols

$$a + (-b) = a - b .$$ (15.8)

Proving (15.8) is so simple that it looks like a conjuring trick. Since (15.7) is true for *any* quantity, we have $b + (-b) = 0$. Adding a to both sides (§9.12) gives

$$a + b + (-b) = a .$$ (15.9)

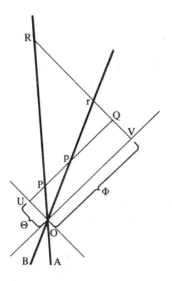

Figure 15.18

And subtracting b from both sides (§3.19) leads to (15.8).

15.24M Again, subtracting $(-b)$ from both sides of (15.9) and then interchanging the sides (§3.17) gives

$$a - (-b) = a + b .\qquad(15.10)$$

So

Subtracting $-b$ has the same effect as adding b.

15.25M Let's translate into symbols the statement at the end of §12.11. If b stands for the larger of these quantities and a for the smaller, the result of subtracting the smaller from the larger is $b - a$. To make this negative, as §12.11 instructs, we put '−' in front of it, giving $-(b - a)$. (We need the brackets to indicate that it is the whole quantity inside them that is to be made negative.) You can now complete the translation of the §12.11 statement as

$$a - b = -(b - a) .\qquad(15.11)$$

However, we can't be content with the §12.11 procedure of working a numerical example and stating that the same must hold good in all cases. To Prove (15.11) properly we start with the equation

$$b - a + a - b = 0 ,$$

which is obviously true since we've both added and subtracted a, and ditto for b. Grouping the first two terms as a single quantity $(b - a)$ and then adding $-(b - a)$ to both sides (§9.12), we get

$$-(b - a) + (b - a) + a - b = -(b - a) .$$

But (15.7), with q changed to $b - a$, tells us that $-(b - a)$ and $(b - a)$ add up to zero; and what remains is (15.11) – which we have thus proved. If b is greater than a in (15.8), then (15.11) shows how to finish the job.

15.26M Using the rules we have just worked out, every case of adding or subtracting numbers or quantities (which may be positive or negative) can be reduced to one of the cases $a + b$, $a - b$ and $-a - b$ (a and b both being positive). We already know how to deal with the first two of these. And for the third you can see that

$$-a - b = -(a + b) .\qquad(15.12)$$

Prove this for yourself by starting from the truism that $a + b - a - b = 0$, and adding $-(a + b)$ to both sides.

15.27M Adding corresponding sides of two equations (left side to left and right to right) is obviously an allowable operation (§9.12) – as is subtracting corresponding sides. Thus if we're given that u, v, x and y are connected by the equations $u = 2x + 3y$ and $v = x + y$ (read as 'Quantity

u is equal to. . .'), we can deduce (adding) that $u + v = 3x + 4y$, and (subtracting) that $u - v = x + 2y$.

5.28M But suppose the given equations are

$$u = x + y \tag{15.13}$$

and

$$v = x - y . \tag{15.14}$$

How do we deal with the '−' sign?

We use (15.8) (with a and b changed to x and y) to rewrite (15.14) as

$$v = x + (-y). \tag{15.15}$$

Adding corresponding sides of (15.13) and (15.15) gives

$$u + v = x + y + x + (-y)$$
$$= 2x + y + (-y) .$$

So, using (15.7),

$$u + v = 2x . \tag{15.16}$$

And subtracting corresponding sides of (15.13) and (15.15) gives

$$u - v = x + y - x - (-y)$$
$$= y - (-y)$$
$$= y + y,$$

by (15.10). That is,

$$u - v = 2y. \tag{15.17}$$

As an exercise, add and subtract corresponding sides of (15.16) and (15.17) and show that, after dividing both sides by 2, you get back to (15.13) and (15.14). Don't worry if the last few pages have been a strain. If you've partly understood, you can hobble along for quite a while, coming back from time to time till you've really mastered them.

16

Relations between the radar and time–distance systems

16.1 When we talk in radar terms, we are only using a different language to describe the same reality that we formerly described in terms of time-over-there and distance. There must be a way of translating one language into the other. **Can you see how to express T in terms of Θ and Φ?**

We have only to put the conclusion of §7.9 into symbols. This says that T is halfway between Θ and Φ – in other words, T is the average of these two, which is half their sum. So

$$T = \tfrac{1}{2}(\Phi + \Theta) . \tag{16.1}$$

16.2 According to §7.13 the distance of an event is half of the time between sending a signal to it and getting one from it. But §15.2 suggests a slight alteration:

> The distance of an event is equal to half of the time from the theta to the phi of that event (all quantities according to the observer concerned)

– whether theta is sending and phi receiving or the other way round. But the time from the theta to the phi is $\Phi - \Theta$, and so

$$X = \tfrac{1}{2}(\Phi - \Theta) . \tag{16.2}$$

It follows (§§12.11, 15.25–6) that X is a positive or negative quantity according to whether Φ is greater or less than Θ. And (**work this out on Figure 15.2**) this means that the distance of an event from A according to A is a positive or negative quantity according as it happens in the positive or negative direction from him as defined in §6.14 – a convenient arrangement. We already know (§8.14) that the time of an event according to A is positive or negative according as it happens, in his opinion, after or before O.

16.3 The equations for translating in the opposite direction are

$$\Phi = T + X, \tag{16.3}$$

and

$$\Theta = T - X. \tag{16.4}$$

You can easily **deduce these** by writing (16.1) and (16.2) as $T = \frac{1}{2}\Phi + \frac{1}{2}\Theta$ and $T = \frac{1}{2}\Phi - \frac{1}{2}\Theta$, and then using §15.28 with u, v, x and y changed to T, X, $\frac{1}{2}\Phi$ and $\frac{1}{2}\Theta$. And you get the corresponding equations for B by merely changing capitals to small letters.

If you are familiar with the identity $(T - X)(T + X) = T^2 - X^2$, you can skip to §16.17.

16.4M You have had to learn a lot of basic mathematics lately. And another chunk follows now – at the end of which you'll have enough (barring details) to keep you going for a long time. If you don't manage to absorb it fully at one go, be content to understand the general trend and return at intervals till you've coped.

What do you think of this sentence: 'Sir Phillip Anders kissed his wife and the maid and the butler helped him into his coat'?

It's ambiguous – the meaning depends critically on where you put a comma. A similar difficulty arises with an expression like $a + bc$. Does it mean 'What you get by multiplying b by c (§9.14) and adding the result to a' or 'What you get by adding a and b and multiplying the result by c'? We need *punctuation conventions* to avoid such ambiguities.

16.5M The first of these conventions says that

> Multiplication and division take precedence over addition and subtraction.

So in the previous article the first answer was correct.

16.6M And if we want to say that a and b are to be added and the result multiplied by c, then – as you've doubtless guessed from earlier examples – we use brackets. We write $(a + b)c$. The convention is that

> Everything inside the brackets is to be regarded as a single symbol when you are considering its relations to things outside. So operations inside the brackets take precedence over operations linking the bracket to anything else.

And by §9.14, $(a + b)c$ will then mean $(a + b)$ multiplied by c.

16.7M Now consider this statement:

> If a, b and c stand for *any* three numbers or quantities, then $a(b + c) = ab + ac$. (16.5)

What does it mean?

A correct answer could take many different forms. If we refer to a, b and c as the first, second and third (numbers or quantities) respectively, then one possibility is:

What you get by adding the second and third and multiplying the result by the first is equal to what you get by multiplying the first by the second and the first by the third and adding these two products.

Or sticking closer to the symbols, on the left you have orders to add b and c and multiply the result by a. On the right you are told to multiply a by b, then multiply a by c, and add the results of these two operations. And the statement is that the two procedures will always give the same answer.

16.8M Is the statement true?

Yes. This is a well-known piece of arithmetic that we constantly use. If you buy 5 articles at 7 pence and another 5 at 4 pence, you can calculate the total cost as $5(7 + 4)$ or $5 \times 7 + 5 \times 4$ pence.

Actually this is a basic property of numbers that we are not required to prove. But Figure 16.8 may make it easier to grasp. In terms of area, the large rectangle is equal to the sum of the two small ones. Using the length $\times$ breadth formula for area, **show that this gives (16.5).**

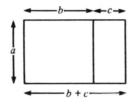

Figure 16.8

16.9M An equation like (9.13) of §9.15 is only true if u, v and w are the quantities defined in §9.2. But (16.5) is true no matter what numbers or quantities a, b, and c may stand for. An equation with this property is called an *identity*. Identities are useful tools, especially in helping us to dig out hidden relationships.

16.10M Now study the statement that

If a, b and c stand for *any* three numbers or quantities, then
$$a(b - c) = ab - ac \tag{16.6}$$

in the same way as we studied (16.5). Work out its meaning in forms like those of §16.7 – or any others you can think up. Consider whether you treat it as true in your arithmetical practice. You do! And see how it is illustrated by Figure 16.10.

This is another identity. Actually it can be deduced from (16.5), but we'll not bother. Just take it as obviously true.

5.11M The *difference* of two numbers or quantities is what you get by subtracting the second from the first. With the help of that word we can express both identities (16.5) and (16.6) in the following verbal statement:

5.12M

Multiplying a number or quantity by the $\left\{\begin{array}{c}\text{sum}\\\text{difference}\end{array}\right\}$ of two others gives the same result as multiplying it by each of the others in turn and $\left\{\begin{array}{c}\text{adding}\\\text{subtracting}\end{array}\right\}$ the results.

(Read both upper or both lower alternatives.)

5.13M When I talk of 'multiplying' *a* by *b*, I can write *either ab or ba* – the order of the symbols is not specified. Similarly for 'sum' and 'adding'. (But with 'difference' and 'subtracting' the order *is* specified.) Thus the statement of §16.12 conveniently covers several forms of the identities which, if you use the symbolic notation, would require you to do things like rewriting $a(b + c)$ as $(c + b)a$, etc.

5.14M If, for instance, I take the three quantities to be p, p and q in that order, and use the lower alternative in each bracket of §16.12, I get

$$(p - q)p = p^2 - pq, \tag{16.7}$$

where you'll notice I've written $(p - q)p$, rather than $p(p - q)$, as well as contracting $p \times p$ to p^2 (§3.14). By choosing the three quantities appropriately in each case, **show that the identities**

$$(p - q)q = pq - q^2, \tag{16.8}$$
$$(p - q)(p + q) = (p - q)p + (p - q)q. \tag{16.9}$$

also follow from §16.12.

The three quantities have to be q, p and q in the former and $(p - q)$, p and q in the latter. Lower alternative in the former, upper in the latter.

.15M Fitting those last three equations together leads to an interesting out-

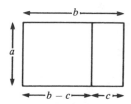

Figure 16.10

come. Equation (16.7) says that wherever $(p - q)p$ occurs, we are permitted to substitute $p^2 - pq$ in its place (§9.16). **What substitution does (16.8) permit?** Making these substitutions in (16.9) we get

$$(p - q)(p + q) = p^2 - pq + pq - q^2.$$

Are you ahead of me?

The two middle terms come to zero and we're left with

$$(p - q)(p + q) = p^2 - q^2. \tag{16.10}$$

Explaining the details may have made all this look complicated. But coming back to it after another chapter or so, you'll be surprised at its simplicity.

Since we used nothing but identities in proving it, *equation (16.10) is also an identity.*

16.16M **Make sure you understand what (16.10) says by expressing it in words – preferably in several different ways.**

Here's a neat formulation:

> The difference of any two numbers or quantities multiplied by their sum is equal to the difference of their squares.

If you still feel more comfortable speaking of 'number multiplied by itself' rather than its square (§3.14), you'll need a slightly clumsier version. **Do a few examples with particular numbers substituted for p and q.**

16.17 **Use equations (16.3) and (16.4) and the corresponding ones in t, x, θ and ϕ to translate the invariant of §15.16 into the language of time-over-there and distance.**

We get $\Theta\Phi = (T - X)(T + X)$; and then (16.10) gives

$$\Theta\Phi = T^2 - X^2. \tag{16.11}$$

Similarly

$$\theta\phi = t^2 - x^2. \tag{16.12}$$

So we're back to the same invariant that we found in Chapter 12. We've not discovered anything new – merely approached interval (§§12.15–17) from a different direction. Now you can see what I meant in §16.9 about the usefulness of identities in digging out hidden relationships.

16.18 In the radar system, the Doppler factor takes the place of speed (§§15.5, 15.8). So there must be an equation connecting these two. Although there is basically only one such equation, this can be put in several forms, all of them useful.

In Figure 16.18 (slot!) observer A has been given rectangular axes. At event P he sends a signal which reaches B at event Q; and a signal which B returns from Q reaches A at R. With M halfway between P and R, the line MQ is obviously parallel to the X-axis.

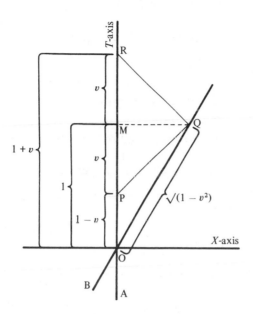

Figure 16.18

16.19 Take the time from O to Q according to A as the time unit. The distance of Q from A according to A is the distance that B travels in time 1 (in A's opinion)–i.e., it is distance v, where v is the relative speed of the two observers. So by §7.14 the times from P to M and M to R are both equal to v (cf. §8.4 if in difficulties). **Check the other times marked along A's world line in Figure 16.18.**

What is the time from O to Q by B's clock?

This is the proper time from O to Q, while the corresponding improper time according to A is 1. So by the dilation of time (§§3.10, 13.5–7) the answer is $\sqrt{(1 - v^2)}$–obtained by putting $T = 1$ in (13.5).

16.20 Next let's do a similar set of calculations in terms of the Doppler factor, k, between A and B. It's now convenient to take the unit of time to be the time from O to Q by B's clock (Figure 16.20). Signals sent by B at O and Q are received by A at O and R. **Use §15.10 to find the time from O to R by A's clock.**

Applying (15.3) to the appropriate events, it is $k \times 1$, which is just k. **What about the time from O to P?**

Applying (15.3) to signals sent by A at O and P and reaching B at O and Q, we find that

$$1 = k \times \text{(time from O to P by A's clock)},$$

so that (§9.12)

time from O to P by A's clock $= 1/k$.

Since M is halfway in time between P and R, the time from O to M is $\frac{1}{2}(k + 1/k)$–cf. §16.1. And the time from P to M or M to R (being half the time from P to R) is $\frac{1}{2}(k - 1/k)$.

16.21 The times marked in Figures 16.18 and 16.20 are exactly the same, of course, but measured in different units. Changing units changes all the measurements in the same proportion. And so the ratio (§9.21) of any two times marked in Figure 16.18 is equal to the ratio of the same two times in Figure 16.20. For example, the ratio OR/OQ (time O to R divided by time O to Q) is $k/1 = k$ in one figure and

$$\frac{1 + v}{\sqrt{(1 - v^2)}}$$

in the other. And so

$$k = \frac{1 + v}{\sqrt{(1 - v^2)}} . \tag{16.13}$$

Check for yourself that each of the following equations is obtained by stating that the ratio on the left is the same in the two figures.

OP/OQ
$$\frac{1}{k} = \frac{1 - v}{\sqrt{(1 - v^2)}} , \tag{16.14}$$

OM/OQ
$$\frac{1}{2}\left(k + \frac{1}{k}\right) = \frac{1}{\sqrt{(1 - v^2)}} , \tag{16.15}$$

PM/OQ
$$\frac{1}{2}\left(k - \frac{1}{k}\right) = \frac{v}{\sqrt{(1 - v^2)}} , \tag{16.16}$$

OR/OP
$$k^2 = \frac{1 + v}{1 - v} , \tag{16.17}$$

PM/OM
$$v = \frac{k - \dfrac{1}{k}}{k + \dfrac{1}{k}} . \tag{16.18}$$

And the last can also be written as

$$v = \frac{k^2 - 1}{k^2 + 1} . \tag{16.19}$$

These are merely seven different ways of saying the same thing – any one can be deduced from any other.

.22M To get the last three we made use of the point that the size of a fraction is not altered if we multiply the numerator and denominator (the quantities above and below the fraction line) by the same amount – since this is just a matter of multiplying by the amount in question and then dividing by it again. In Figure 16.20 the ratio OR/OP is $k/\frac{1}{k}$, but multiplying numerator and denominator by k brings it to $k^2/1 = k^2$. And the same process is involved in getting the right-hand side of (16.18) and in the transition from this to (16.19), the multipliers being 2 and k respectively.

16.23 You've noticed that I'm now making life tougher for you by pruning explanations severely. I have to do so – otherwise the book will grow intolerably long. But I'm not leaving anything essential out – just compressing. And you have to adjust. All the explanations you need are there, but to extract and master them you will have to give more thought per sentence than hitherto – and continue all the time to improve the ratio of your thought to my sentences.

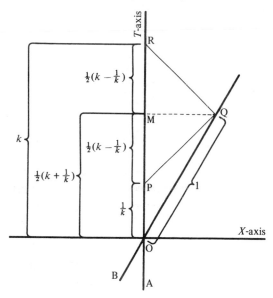

Figure 16.20

16.24 The rule for combining speeds – (9.13) in §9.15 – turned out to be unexpectedly complicated. What if we had used Doppler factors instead of speeds to specify relative motion?

Figure 16.24 shows the world lines of three inertial observers, A, B and C; and also two signal lines, PRU and QSV (so chosen that none of the world lines cross each other between them). Slot!

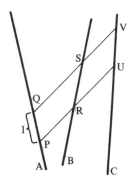

Figure 16.24

Let k_{AB} stand for the Doppler factor between A and B, and k_{BC} and k_{AC} for the Doppler factors between B and C and between A and C, respectively. Take the time from P to Q by A's clock to be 1. **Now use (15.3) to establish the following three equations:**

time from R to S $= k_{AB} \times 1 = k_{AB}$;
time from U to V $= k_{BC} \times$ (time from R to S)
$= k_{BC}k_{AB}$;
time from U to V $= k_{AC} \times 1 = k_{AC}$.

The last two give different versions of the same time. And so

$$k_{AC} = k_{AB}k_{BC}. \qquad (16.20)$$

In the radar scheme this replaces (9.13). As usual, the radar is simpler. We do *not* combine speeds by adding, as we used to believe – it's more complicated. But we *do* combine Doppler factors by simple multiplication. Furthermore, (16.20) is much easier to prove than (9.13).

16.25   To illustrate these various relationships, let's do exercise (i) of §9.19 by converting speeds to Doppler factors, using (16.20), and converting back. Applying (16.17) gives

$$k_{AB}^2 = \frac{1 + 0.6}{1 - 0.6} = 4, \quad \text{and} \quad k_{BC}^2 = 9.$$

A very obvious extension of (16.20) is

$$k_{AC}^2 = k_{AB}^2 k_{BC}^2 \qquad (16.21)$$
$$= 4 \times 9 = 36 \quad \text{in the present case.}$$

Then by (16.19), $w = \frac{35}{37} = 0.946$ – which checks.

Do one or two more numerical examples by this method, checking your answers by (9.13). If you're good at algebra, you could apply (16.17) to each k^2 in (16.21) and hence deduce (9.13).

16.26 If the relative motion of two observers is specified by their Doppler factor, and if we know one observer's radar co-ordinates of an event, then (15.4) and (15.5) enable us to calculate the other's radar co-ordinates of it. We could translate these equations into time–distance–speed language by using (16.3), (16.4) (and their small letter equivalents), (16.15) and (16.16), with help from §§16.12 and 15.28. If you're fairly good at algebra, work it out. Otherwise trust me that the result is

$$t = \frac{T - vX}{\sqrt{(1 - v^2)}},$$ (16.22)

and

$$x = \frac{X - vT}{\sqrt{(1 - v^2)}}.$$ (16.23)

(If these seem difficult to understand, be content to appreciate their general significance, and get what you can out of the rest of the chapter – which is a dead end anyway.) These equations do a job like that of (15.4) and (15.5): if the relative motion is specified by speed, and if we know one observer's time and distance co-ordinates, they enable us to calculate the other's. Once again the time–distance–speed equations are much more complicated than the radar ones.

16.27 Equations (16.22) and (16.23) taken together are known as the *Lorentz Transformation* – 'transformation' because it enables us to transform one observer's viewpoint into another; and 'Lorentz' after H. A. Lorentz, who actually arrived at the equations before Einstein did his work (though their true significance could not be seen till after Einstein).

These equations are the algebraic equivalent of the spacetime diagram in the time–distance–speed form, as we developed it in Chapters 6–8 and 14. They can be derived directly from §1.22 or §5.11. And just as we unfolded our theory by making deductions geometrically from the diagram, so a more orthodox approach to Relativity evolves by making deductions algebraically from (16.22) and (16.23).

The Lorentz Transformation approach is far quicker and shorter than the one we've been following, and requires much less mental effort (from those who possess the mathematical technique). If the objective is efficient, practical mastery of the theory, particularly with a view to using it in other branches of science or in engineering, then without doubt that is the best method. But I still think *our* approach gives us, in return for our hard work, a deeper understanding (cf. last few paragraphs of the Introduction).

17

Constant acceleration

17.1 Did your researches with the slot at §14.14 yield the 'important discovery' I hinted at. **Perhaps you should try again.**

Revise as necessary on isovals – a minimum of §§14.1, 14.5 and 14.9. If you didn't ask yourself whether a branch of an isoval could be a world line – representing something moving on the 1D universe – **consider the question now.**

A time-like isoval couldn't be a world line. For its slope is always shallower than 45°, and this would imply a causal influence travelling faster than light, which is impossible (§§10.10–12).

17.2 But a branch of a space-like isoval, being always steeper than 45°, *could* be a world line. Check with the slot that it represents something always moving slower than light. **What sort of motion will it represent?**

World lines of inertial observers are straight. So this curve must represent the motion of something that is *non-inertial*. **What does 'non-inertial' mean?**

The meaning is in the definition of §5.2 – which please reread. The test particle moves away.

But when there is no gravity – as we are assuming (§5.5) – an inertial observer has constant speed relative to any other inertial observer (§5.6). Therefore a non-inertial observer moves with changing speed – he is *accelerated* in the ordinary sense of the word. So this space-like isoval represents accelerated motion.

17.3 Note that acceleration has a quite different status from speed (§§1.1–3, 5.1–2). Speed is purely relative. But everybody can see whether an observer and his test particle stay together or move apart; and so all must agree whether he is accelerated or not. Acceleration is an absolute. **Work out how this important distinction is represented on the diagram.**

Drawing different diagrams (§8.44) can give an inertial observer's world line *any* slope steeper than 45°. But a curve always stays curved.

17.4 Let's fix on the right-hand branch of a space-like isoval (Figure 17.4).

This, we've seen, could be the world line of somebody (or something) with accelerated motion. We don't know enough about him yet to give him the status of an observer (who interprets information from signals sent and received). So let's call him an accelerated *traveller* – who can be observed, but does not do any observing. Call him W (with 'wanderer' for mnemonic).

Figure 17.4 also shows the rectangular axes of the observer A that we used to study isovals in Chapter 14, and another inertial observer, B, whose axes also pass through O. And P and Q are the points where W's world line crosses their respective distance axes.

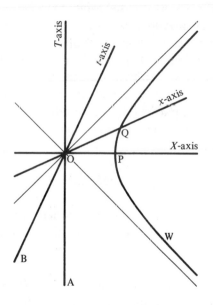

Figure 17.4

17.5 Let's get a general idea of W's motion – from A's point of view to begin with. At first – bottom right of the figure – W is coming in towards A. Since his world line is nearly parallel to the signal line, he is moving at nearly the speed of light relative to A. But his rocket engine is giving him an acceleration in the positive direction – so that his motion relative to A gets slower and slower.

At event P, W's speed is down to zero – he is momentarily stationary relative to A. Then, as his rocket engines continue working, he accelerates away again. And as the world line disappears off the edge of the diagram, his speed (relative to A) is again approaching that of light. Use the slot (A's point of view, §8.13) to check up on that description.

With the isoval defined in terms of interval (§14.1), about which all observers agree, it is clear that W's motion will appear exactly the same to B as to A. Slot! (If you gave B the rectangular axes, the isoval would be unchanged.) **At what event will W be stationary relative to** B?

In A's opinion the event P, at which W is stationary, is simultaneous with O (§8.12). Similarly, W will be stationary relative to B at the event which B regards as simultaneous with O, that is at event Q. Or, as somebody else would see it, Q is the event at which W's speed is the same as B's.

17.6 Did you feel that I should have called W's motion before event P 'deceleration', not 'acceleration' – since he is slowing down? But this is another 'according to whom?' question. Between P and Q, for example, W is starting from zero speed and speeding up, according to A, but slowing down to zero according to B. So deceleration is just acceleration from another point of view. And we stick to the latter word.

However, the *direction* of acceleration is real. At all times and according to all observers, W's test particle will move away from him in the negative direction. And so he is always accelerating in the positive direction.

17.7 Is W's acceleration constant – always the same size – or varying? **No hasty answers, please.**

The hasty answer would have been: 'His speed is changing rapidly near P, but much more slowly at the bottom or top of his world line. So he has varying acceleration – bigger at P than before or after'. **Criticism?**

In §§9.28–30 we did a preliminary analysis of constant acceleration and found that, according to any one observer, the speed changes at a varying rate. The results we got there are consistent with W's behaviour. So his acceleration *could* be constant.

And that opens up exciting prospects. For constant acceleration is the key to dynamics – force, mass, energy, and the famous mass–energy relation. Again, if space travel is ever to take us much beyond our tiny solar system, it will have to be done in constant acceleration rockets – technologically inconceivable though these may seem at present. And the travellers' experiences would be very strange indeed. Finally, when we learn how to use the constant acceleration traveller as an *observer*, we shall be on the threshold of the General Theory of Relativity (page 00).

17.8 This traveller's motion will appear exactly the same to any observer (§17.5). And that fits with what we worked out in our first crude approach – see the second sentence of §9.29, for example. The constant acceleration idea looks good!

With the revised definition of constant acceleration (§9.29) the size of W's acceleration at event P must be calculated from A's measurements, and its size at Q must be calculated in B's terms. So to clinch the argument we have only to calculate these two sizes and see if they come out the same. **Can you anticipate what the answer is going to be?**

Since W's motion looks the same to B as to A, with Q instead of P as the event when W is momentarily stationary, it follows that this calculation must yield the same answer in both cases – in *all* cases. So W's acceleration *is* constant. (But you'll probably feel happier when we've done the actual calculation.)

17.9 Let's specify W's world line as the right-hand branch of a space-like isoval for interval *s*. So it represents events which are all separated from O by interval *s* (§14.1).

According to A, O and P are simultaneous. And so (§13.23)

distance OP according to A = *s* .   (17.1)

17.10 *All the quantities and calculations that follow are 'relative to A' or 'according to A' unless otherwise stated.* To find the size of W's acceleration, we have to be able to calculate his speed – say at Q. And this is (momentarily) the same as B's (constant) speed – see end of §17.5. So we want a recipe for calculating the speed of B relative to A.

17.11 Figure 17.11 is an obvious development of Figure 17.4. Lines MQ and NR are vertical and horizontal, and lengths ON and OM are equal. Since B's axes make equal angles with horizontal and vertical (§8.26), it follows that lengths NR and MQ are also equal. If *v* stands for the speed of B relative to A, then (§8.37)

$$v = \frac{\text{length NR}}{\text{length ON}} = \frac{\text{length MQ}}{\text{length OM}} ;$$

or in symbols

$$v = \frac{T}{X} ,$$ (17.2)

where *T* and *X* as usual stand for the time and distance of Q according to A (cf. §8.42). And (§17.10) this is also the instantaneous speed of W at event Q (relative to A).

(A speed equal to a time divided by a distance! Don't worry: we're not saying anything silly like W's speed is equal to *his* time of travel divided by the distance *he* travels. And remember our distances are times – §§7.10–13.)

17.12 Acceleration is concerned with the rate at which speed changes – 'rate of speed change' for brevity. (In such a context 'acceleration' means the *size* of the acceleration, of course.) Between P and Q, W's speed increases, according to (17.2) by amount T/X in time T. So *taken over the whole time-interval between* P *and* Q, the rate of speed change is T/X divided by T, which is $1/X$.

17.13 We are concerned with W's acceleration *at* event P – which is clearly some sort of measure of the rate at which his speed is changing *at* that event. But you can't measure a rate of speed change *at* an event – you need a time-interval over which to calculate the change of speed and then divide by the time (as in the previous article).

On the other hand, the acceleration *at* P is obviously not the same as the rate of speed change over the time-interval between P and Q. But we could reasonably regard the last as an *approximation* to the acceleration at P. And the nearer Q is to P, the better this approximation will be.

Thus we can't directly calculate the acceleration *at* P, but we can get as close to it as we wish by calculating the rate of speed change over a *small enough* time-interval, P to Q. In fact the only way we can give a quantitative meaning to the words 'acceleration *at* P' is by taking the idea we've just discussed and turning it into a *definition:*

> The acceleration of W at event P is defined to be the quantity that we can calculate *as accurately as we wish* by taking it to be the rate of speed change over the whole time-interval between P and Q and *making the time from* P *to* Q *small enough.*

17.14 A procedure that only allows us to calculate something 'as accurately as we wish' may give you an unsatisfied feeling. Yet you will find as we go on that it is the gateway to a realm peopled with new mathematical powers. You'll be surprised to discover in the next couple of articles that reasoning which starts from 'as accurately as we wish' can end with 'exactly'.

17.15 Now let's combine the definition in §17.13 with what we learned in §17.12 about rate of speed change. If Q coincides with P, then so does M. So in this case X is the distance OP, which is s by (17.1). Thus the nearer Q is to P, the nearer is X to s, and the nearer is $1/X$ to $1/s$. And then it follows from §17.12 that by taking the time from P to Q small enough we can make the rate of speed increase over the time-interval from P to Q as nearly equal as we wish to $1/s$. And when we put this last into the definition of §17.13, we see that

> The size of the acceleration *at* event P is $1/s$.

17.16 To reach that conclusion we used two as-accurately-as-we-wish statements – the definition of §17.13 and the fifth sentence of §17.15. And yet I claimed to deduce that the acceleration at event P is *exactly* 1/*s*. **Do you feel that this conclusion ought to be modified by some sort of as-accurately-as-we-wish qualification?**

You really can draw an exact conclusion. The two statements say that by taking Q near enough to P, we can make the rate of speed increase as nearly equal as we wish to *both* the acceleration at P *and* the quantity 1/*s*. However the acceleration and 1/*s* are both fixed quantities – they don't change when we bring Q nearer to P. Suppose they were *not* exactly equal. Then the rate of speed change could not be made as nearly equal as we wish to *both* of them. If you keep making it closer and closer to the acceleration, you must eventually drag it away from 1/*s* (or vice versa).

So following up the suggestion that the acceleration is not exactly equal to 1/*s* has led to an impossibility – which forces us to admit that the conclusion of §17.15 is exact. (If this continues to worry you after a bit more study, give it a rest: it will come up again later.)

The ability of as-accurately-as-we-wish reasoning to yield exact conclusions is the main 'state secret' of the new realm that I promised in §17.14.

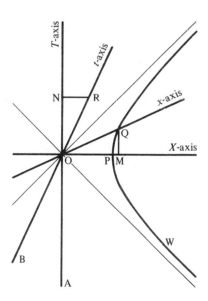

Figure 17.11

17.17 From the invariance of interval, it is clear that if we had calculated the acceleration at any other event of W's life, using measurements of an observer according to whom W is momentarily stationary at that event – for example, at event Q using B's measurements – we should have got the same answer. So W's acceleration is always the same – constant.

We'll use the traditional symbol f to stand for W's acceleration (a is often used nowadays, but we need it for too many other purposes). So we've proved that

$$f = 1/s , \tag{17.3}$$

and

$$s = 1/f . \tag{17.4}$$

17.18 Event Q happens, in A's opinion, at time T, and W is then at distance X from him. To describe W's motion (as seen by A) **we need an equation connecting T and X.**

From the specification of W's world line in §17.9 it follows that T and X are related by (14.2) of §14.9. And using (17.4), this gives

$$X^2 - T^2 = \frac{1}{f^2} . \tag{17.5}$$

Or more neatly (§9.11)

$$f^2(X^2 - T^2) = 1 . \tag{17.6}$$

17.19 Combining (17.1) and (17.4) gives: distance OP $= 1/f$. And $v = 0$ at event P. That is to say, in A's opinion W is momentarily stationary at time zero and is then distance $1/f$ away.

The equations describing W's motion take the simple forms we have found *only* if we use the measurements of an observer (like A or B) who is the right distance ($1/f$) away from W at nearest approach. We shall always arrange it so.

Then in A's view of things, W's distance from him at time T is X, where T and X are connected by (17.5) or (17.6). And at that moment W's speed relative to A is given by (17.2).

However, since W starts at distance $1/f$ from A,

distance travelled by W between zero time and time T (according to A) $= X - 1/f$. (17.7)

Figure 17.19 summarises this information visually.

17.20 Let's do some highly futuristic space travel calculations. Specially interesting is the case in which the vehicle has the same acceleration as a stone falling freely at the Earth's surface – an acceleration of 1*g* as it is called. In such conditions you would feel as if your rocket were standing nose up on Earth, with ordinary gravity pulling you towards the tail. **Can you see why?**

If you released a test particle, it would be inertial. But you, pushed by the cabin floor (and ultimately by the rocket engine) would accelerate away from it. So it would seem to be accelerating towards the tail, with acceleration 1*g*. It would seem to fall, just as on Earth. To keep you accelerating (stop you from falling, as it would seem) the floor would have to exert the same force on you as the Earth's surface usually does. In fact, the effects of the rocket's acceleration would replace in every detail those of the Earth's gravity.

Human beings could not live indefinitely in weightless conditions. So really long journeys would require artificial gravity. And the easiest way to provide it (if the engineers can build the super-rockets!) would be by travelling always with an acceleration of 1*g*.

17.21 In a 1*g* acceleration, the speed increases by 9.81 metres per second every second (provided we stick to short periods, so that relativistic effects don't show up). For calculations about long space journeys we'll need a bigger time unit. Let's choose the *year*. Then the natural unit of distance (§7.11) will be what is commonly called a *light year* (§1.8) –

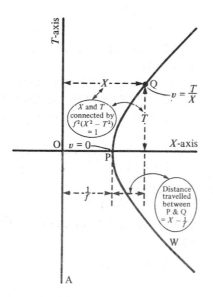

Figure 17.19

and let's use that name, though in our system we ought to speak simply of 'a distance of one year'.

If the speed actually increased by 9·81 metres per second every second, then in a year it would increase by 365·26 × 24 × 3600 × 9·81 metres per second ≈ 310 000 kilometres per second – or about 1·03 in natural units.

Thus reaching a speed greater than that of light in a year? – you wonder. Of course not. Over last Christmas my weight increased *at a rate of* 20 tons per century – but it won't go on doing so for 100 years. Similarly a 1g acceleration increases the speed *at a rate of* 1·03 per year. But only for a short period, after which the speed will increase more and more slowly (§§9.28–9, 17.7).

So near and yet so far! If it had worked out as exactly 1, we could have put $f = 1$ in our equations, making everything much simpler. Well, let's do so anyway. Our calculations will still give a pretty good idea of what 1g space travel would be like.

17.22 If we put $f = 1$ in (17.7) and (17.5), we see that, for this nearly 1g traveller,

$$\text{distance covered in time } T = X - 1 , \tag{17.8}$$

where X is related to T via the equation

$$X^2 - T^2 = 1 . \tag{17.9}$$

And his speed at time T is still given by (17.2). (All measurements according to A, of course.)

17.23 For example, if we put $T = 1$ in (17.9), we get $X^2 - 1 = 1$, so that $X^2 = 2$, and $X = \sqrt{2} = 1·41$. (I'm sure you can see which allowable operations we've used. Square roots are best found from a table or calculator. Otherwise just check that $1·41^2$ is 2, near enough.) So by (17.8), the distance travelled in 1 year is 0·41 light years. And by (17.2) the speed at the end of the year is 1/1·41 = 0·71. **How far would this traveller go in (a) $1\frac{1}{3}$ years, and (b) 9·95 years, and what speed would he reach?**

Of course we can also ask how long it would take the traveller to go a certain distance – say 5 light years. From (17.8), $X = 6$; and (17.9) gives $36 - T^2 = 1$, whence $T^2 = 35$, and $T = 5·92$ years – at which time his speed if 5·92/6 = 0·987. **Work out (c) the time and terminal speed for a distance of $\frac{8}{5}$ light years.** (Answers at end of chapter.)

17.24 If your mathematics is still shaky, skip this article for the present. From (17.5) we have

$$X^2 - \frac{1}{f^2} = T^2 ,$$

and so, by (16.10),

$$\left(X - \frac{1}{f}\right)\left(X + \frac{1}{f}\right) = T^2 . \qquad (17.10)$$

If we stick to small speeds, then Q will be near P (Figure 17.19) and $X \approx 1/f$ (§13.11). Using this approximation in the second bracket of (17.10) (and there's no law compelling us to do the same in the first!), we get

$$\left(X - \frac{1}{f}\right)\frac{2}{f} \approx T^2 ,$$

whence, on multiplying both sides by $\frac{1}{2}f$ and using (17.7),

$$\text{distance travelled} \approx \tfrac{1}{2}fT^2. \qquad (17.11)$$

You may recognise this as a well-known equation of elementary mechanics, which holds good at slow speeds.

.25M **Can you prove the identity $(ab)^2 = a^2b^2$?**

We merely use the definition of a square (§3.14) and do some rearranging: $(ab)^2 = ab \times ab = abab = aabb = a^2b^2$. In the next article we use this to write T^2 as v^2X^2 instead of $(vX)^2$.

17.26 From equations we've already got, we can derive two more which answer the questions: In order to reach a certain speed, how far will W have to travel, and how long will he take? From (17.2) we get

$$T = vX . \qquad (17.12)$$

And substituting for T in (17.6) gives $f^2(X^2 - v^2X^2) = 1$, or (using the identity of §16.12) $f^2X^2(1 - v^2) = 1$. Dividing both sides by $1 - v^2$ yields

$$f^2X^2 = \frac{1}{1 - v^2} ,$$

whence, taking the square root of both sides,

$$fX = \frac{1}{\sqrt{(1 - v^2)}} , \qquad (17.13)$$

which is the equation connecting X and v that we were seeking. (To check that fX is the square root of f^2X^2, we simply square it – §13.4.)

From (17.12) we get $fT = vfX$. And when we use (17.13) to substitute for fX, this gives

$$fT = \frac{v}{\sqrt{(1 - v^2)}} \qquad (17.14)$$

– which is the other equation we wanted.

17.27 Another article to skip for the time being if mathematics is being a worry. You'll notice that if v is small, (17.14) gives

$$v \approx fT \tag{17.15}$$

– the classical, slow-speed, formula. **Can you prove, again as an approximation at slow speeds, that**

$$v^2 \approx 2f \times \text{(distance travelled)}? \tag{17.16}$$

Multiplying both sides of (17.11) by $2f$ gives $2f \times$ (distance travelled) $\approx f^2T^2 \approx v^2$ by (17.15) – which is (17.16) written back to front.

17.28 You've probably noticed, in **playing around with the diagrams and the slot**, that W's motion has some very weird features. I have room for only a few hints. You can have fun following them up.

Suppose we allow W to try his hand at being an observer. Which points in Figure 17.4 represent events to which he could send signals? And from which he could receive signals? Consider each of the quadrants of Figure 10.1. What, if anything, could he know about the part of spacetime which it represents? I think you'll conclude that he can have full knowledge only of the right quadrant; and he'd know of the bottom quadrant's existence, but couldn't investigate it properly (couldn't do the signal-sending). The rest of spacetime would not exist for him. Just as well we decided (§17.4) not to let him join the Observer Corps just yet!

17.29 There are oddities even about inertial observers' observations of W. For example, A can't send signals to W after O, nor receive them from W before O. So if he waits for evidence of W's existence, it's already too late for a proper radar investigation.

Draw for yourself a vertical line through P (Figure 17.4) crossing the lower signal line at K. This could be the world line of an inertial observer A_1. Let's identify A_1 with us on Earth (ignoring the complications of gravity, orbital motion, etc.). Check (with slot or otherwise) that if W switched off his engines at P, he'd make a perfect soft landing. Suppose W is actually a space fiction invading force, using $1g$ acceleration. Prove that the maximum possible warning of the attack would be 1 year (time KP = distance OP = 1).

Of course accelerated motion is possible only for a finite time (you have to expend energy to make it happen). If the invaders began their approach inertially at high speed, we could theoretically observe them; but if they kept strict blackout, we couldn't do so in practice – till they switched on their engines to produce the touch-down acceleration, certainly revealing themselves in the process, but still with less than a year's warning!

17.30 According to A the distance of W at closest approach (event P) is $1/f$ (§17.19). Every inertial observer with world line through O is on a par with A, and so he also says that W's distance at closest approach is $1/f$. Thus, although W is always accelerating away from O, there is a curious sense in which he always stays the same distance from that event. And the bigger the acceleration, the closer (in this odd sense) does he stay to O!

This becomes a little less strange when we remember that an isoval is the spacetime analogue of a circle. The analogue for relative speed is relative direction (recall the surveyors of §§11.10–12) and the analogue for acceleration is therefore curving or swerving. You can think of a circle as a curve that keeps on swerving towards a certain point (the centre) and yet always stays the same distance from it – just as W keeps on accelerating *away* from O and yet always stays the same distance from it. A more vivid if less close analogy (for those who know the relevant bit of mechanics) is something moving at constant speed on a circle – always accelerating towards the centre, never getting any nearer.

You've got plenty to think about before the next chapter! If your thoughts seem to be driving you mad, the escape may lie in remembering that acceleration can't go on for ever.

Answers to §17.23. (a) $T = \frac{4}{3}$, so that $T^2 = \frac{16}{9}$. By (17.9), $X^2 = \frac{25}{9}$ and $X = \frac{5}{3}$. So distance travelled $= \frac{2}{3}$ light year. By (17.2), speed $= \frac{4}{3} \div \frac{5}{3} = \frac{4}{5}$. ($b$) 9 light years; speed 0·995. (c) $\frac{12}{5}$ years; speed $\frac{12}{13}$.

18

Dynamics – mass, momentum, force

18.1 So far we've not asked questions about what causes things to accelerate. When we do so, we enter the subject of *dynamics*. Naturally we shall be concerned with *relativistic dynamics* – the dynamics of things moving at high speed. But curiously enough the main difficulty for most of us will be to get a sound grasp of the fundamentals that belong just as much to Newtonian (slow-speed) dynamics. Making a thorough job of that would fill more pages than you would care to read (or pay for). So I'll make the best compromise I can between full explanation, a bit of fudging and a bit of 'it can be proved that'. (If you're not interested in the dynamics, you can skip to Chapter 20, provided you can do without the mathematical practice that we'll get on the way.)

18.2 We all know that if something is stationary, you need a force to get it moving; and if it's moving, you need a force to stop it. We also know that some things are easier to get moving or to stop than others. A moderate push will get a car going. But try it on a bus! Things vary in the extent to which they resist being speeded up or slowed down. *Mass* is a quantity that gives a measure to this resistance – though we shall have to work out a more precise statement of this vaguely conceived idea.

   (If you have been taught that mass measures the 'quantity of matter', please forget it. Also don't confuse mass with weight. Weight is connected with gravity; mass is not. In an orbiting spaceship, everything is weightless. But it still needs force to get things moving or stop them. Weight has disappeared, but mass remains.)

18.3 The difficulty of stopping a vehicle depends also on its speed (relative to you, of course). A fast-moving cyclist may take as much stopping as a crawling car (even with both vehicles coasting). The total difficulty of stopping the vehicle depends on *both* its mass *and* its speed. Thus we get a first vague idea of something that we might call 'quantity of motion'. For if the fast cyclist and slow car are equally difficult to stop, we can reasonably say that they both possess the same 'quantity of motion' – stopping being essentially the process of robbing the vehicle of what motion it possesses.

'Quantity of motion' was Newton's term. We'll use the briefer, if less descriptive, word *momentum* (plural, *momenta*). But do keep the 'quantity of motion' idea in mind. Obviously momentum, like the motion it measures, is relative to the observer.

18.4 In a collision, motion is transferred from one body to another. So momentum has to be defined in a way that will make sense of collisions. Easiest to deal with are 'stick-and-stop' collisions – in which the colliding bodies stick together to form a single body, which stays motionless.

The result of such a collision is zero momentum. And it seems an obvious inference that the original bodies, before colliding, had equal momenta in opposite directions, which cancelled each other out.

18.5 Obviously the direction of motion matters. On the 1D universe, where we have only two directions to cope with, we take a body's momentum to be a positive or negative quantity according as it is moving (relative to the observer) in the positive or negative direction (§6.14).

This positive/negative momentum arrangement makes good sense of stick-and-stop collisions. Before the collision, one body had momentum (say) p in the positive direction, and the other had the same momentum in the negative direction, that is, $-p$. The sum of these, by (15.7), is zero. So the momenta *have* cancelled out, as we said they should (§18.4).

18.6 For consistency, we need a similar positive/negative distinction for speeds. And there's a useful convention: we use the word *velocity* for a speed in a definite direction. So on the 1D universe the velocity of something is equal to its speed or to the negative of its speed according as it is moving in the positive or negative direction.

18.7 Now you could do a lot of stick-and-stop experiments with the object of discovering how momentum is related to mass and speed. Take, for example, three identical lumps of lead, and use one of them as body A, hammering the others together to make body B. If mass is to measure resistance to change of motion, it seems obvious that B's mass is twice as big as A's. You would then find experimentally that to produce a stick-and-stop collision you must give B half the speed you give to A. *Twice the mass* and *half the speed* gives the same momentum. From many experiments like that you would eventually reach the following conclusions:

(1) You can assign to every body a quantity called its *mass*, which is a characteristic of the body itself and which never

changes unless matter is added to or taken from the body. (At least that's what you find in low-speed experiments. Just what happens at relativistic speeds is an open question.)

(2) With mass assigned in this way, you can define

$$\text{momentum} = \text{mass} \times \text{velocity}. \tag{18.1}$$

Mass is always taken to be a positive quantity. And so (18.1) works out correctly as regards direction – the momentum will be positive or negative according as the velocity is positive or negative (§12.13).

18.8 A multitude of experiments have shown that with the aid of (1) and (2) we can make sense of the arithmetic of collisions. The total momentum, taking account of direction, is the same after the collision as before. As you extend your experiments, you find that this is true for *any* sort of interaction between bodies – collisions with rebound, electric and magnetic force, what you will. The total momentum in every case is the same at the end as it was at the beginning. There is in fact overwhelming evidence that

> If a system of bodies is not acted on by outside forces, then the sum of the momenta of all the bodies remains constant – it is not changed by their mutual interaction.

The simplicity and generality of this *Principle of the Conservation of Momentum* (as it is called) shows that momentum is an extremely important quantity.

18.9 Statement (1) of §18.7 merely asserted that a mass *can* be assigned to any body. But how? The simplest method *in principle* would be to do experiments comparing momenta of bodies, and then use (18.1). We measure, for example, the velocities of two bodies that produce a stick-and-stop collision. We know that their momenta add to zero – i.e., they are equal, apart from one being negative. And so the equation

$$\text{mass} = \frac{\text{momentum}}{\text{velocity}}$$

– derived from (18.1), using §9.12 – enables us to *compare* their masses. (And §§18.7–8 assure us that if we do other experiments involving the momenta of these bodies, directly or indirectly, we shall always get consistent results.)

If in addition we choose the mass of one particular body to be our *unit of mass,* then the above method enables us to measure the mass of any other body in those units (via a chain of comparisons, if necessary). The internationally agreed unit, the kilogram, in terms of which all masses are expressed, is the mass of a certain lump of platinum–

iridium alloy kept at Sèvres, a Paris suburb. It is equivalent to roughly 2·2 pounds (lbs.).

(For *practical* mass measurements we normally use the experimentally discovered fact that, at any one place, mass is proportional to weight. Using a balance to compare weights, masses can be measured far more accurately than is possible with the theoretically preferable method I've described.)

18.10 We can alter the motion of a body by applying a force to it. Force changes its motion – changes its momentum. So what is the relation between the strength of the force and the rate at which momentum changes? The simplest guess would be that

$$\text{strength of force} = \text{rate of change of momentum.} \qquad (18.2)$$

('Rate' in terms of time, of course.)

18.11 This suggestion can often be tested experimentally. A standard spring compressed to a standard length will always exert the same push. And we can check that, no matter what body it is acting on, it always changes the momentum at the same rate. Two such springs acting in parallel will give double the force, and we can check that the rate of change of momentum is also doubled. The results of such experiments have always supported (18.2).

And then we can do the usual business of assuming (18.2) true in more complicated situations and checking that the resulting predictions agree with empirical facts – as indeed they do.

18.12 Using (18.1), (18.2) can be rewritten as

$$\begin{aligned}&\text{strength of force}\\ &= \text{rate of change of (mass} \times \text{velocity)} \qquad (18.3)\end{aligned}$$

– so that it could be the velocity that's changing, or the mass, or both. And it *is* sometimes necessary to study the motion of things whose mass is varying – a rocket losing mass as fuel is ejected, or a raindrop growing by accretion as it falls through a cloud.

In many cases, however, the body neither gains nor loses matter; and so (§18.7) its mass stays constant (at low speeds). And then, as a moment's thought will convince you,

$$\begin{aligned}&\text{rate of change of (mass} \times \text{velocity)}\\ &= \text{mass} \times \text{(rate of change of velocity).} \qquad (18.4)\end{aligned}$$

But in slow-speed conditions, rate of change of velocity is acceleration. And putting this into the right-hand side of (18.4) and using (18.3) to substitute 'strength of force' for the left-hand side, we get

$$\text{strength of force} = \text{mass} \times \text{acceleration.}$$

Or maybe we should put this more explicitly as

> strength of force
> = (mass acted on) × (acceleration produced).   (18.5)

On the other hand, (18.4) can't be true if the mass is changing – it would falsely state that mass change has no effect on momentum. So, although (18.2) is *always* true, (18.5) is true *only if* the mass stays constant.

18.13 Actually the truth of (18.2) and (18.5) depends on choosing our unit of force to be the force that produces unit acceleration in unit mass – which is the natural way to arrange things. A different force unit would bring in a conversion factor, and we should have to substitute 'proportional' for 'equal' in these equations.

18.14 Forces are seldom so tangible as the compressed springs of §18.11. How could we measure the force of air resistance that is slowing down a projectile? Only by observing its acceleration (deceleration). The same is often true of electric and magnetic forces. In fact, in most cases you have no direct means of measuring a force.

In such circumstances we agree to *define* the strength of the force as what is given by (18.2) – or where appropriate, (18.5). And theories using this definition are repeatedly found to give good predictions.

18.15 When the strength of the force is constant (whether the mass varies or not) a useful simplification results. For then (18.2) says that the rate of change of momentum is also constant, and we can measure it over any time-interval we like. The rate of change is simply the total change divided by the time-interval. So (18.2) now becomes

$$\text{strength of force} = \frac{\text{total change of momentum}}{\text{time}},$$

whence (making the wording a bit more explicit)

> (strength of force) × (time during which it acts)
> = total change of momentum in that time.   (18.6)

18.16 That is an outline of the conclusions that emerged from research in *slow-speed* dynamics. To make the transition to *relativistic* dynamics, let us throw away just one of these conclusions. Let us contemplate the possibility that the mass of a body changes with its speed – speed relative to whoever is measuring the mass, of course. However, we do know that at the low speeds of ordinary experience the mass stays constant – or so nearly so that we can't detect any change. This is called the *rest mass* – because it is the mass measured when the body is virtually at rest relative to the measurer.

(Actually you can't measure the mass by the theoretically correct method of comparing momenta – §18.9 – without accelerating the body, and so it will not be strictly at rest all the time. But the speeds involved are so tiny that the point has no practical significance. And we could, if it were worth the trouble, deal with the theoretical difficulty by an as-accurately-as-we-wish definition akin to that used in §17.13.)

18.17 Let us take W, the traveller with constant acceleration from Chapter 17, and turn him into a body ('it' from now on) with rest mass m. Think of it as a body in outer space propelled by rocketry. We can't use the rocket itself, since that loses mass by ejecting fuel. But W could be pushed by a rocket whose controls are adjusted to keep it moving with constant acceleration. And the push can be transmitted through a giant spring balance, which will monitor the strength of the force.

Notice that this force measurement is an on-the-spot affair. Any force is produced by a physical action on the spot – by the fuel flow through the rocket nozzle in our example. And so its strength is measured by the man on the spot – an inertial observer momentarily travelling with W. Relative to him W is momentarily stationary; and so the mass which he measures is the rest mass m – which is constant. So we can use (18.5) to give

$$F = mf, \tag{18.7}$$

where F stands for the strength of the force – a *constant strength*, since m and f are constant.

18.18 Now consider how it looks to observer A of Chapter 17. In his view, equation (17.14) of §17.26 gives the relation between W's velocity, v, and the time, T, taken to attain it. Multiply both sides by m and use (18.7) to substitute F for mf, getting

$$FT = \frac{mv}{\sqrt{(1 - v^2)}}. \tag{18.8}$$

Do you recognise the left-hand side?

It's the strength of force multiplied by the time during which it acts – which, according to (18.6), is equal to the change in momentum (force strength being constant, §18.17). But at zero time W was stationary relative to A, and so had zero momentum. Thus FT is W's actual momentum at time T. And using (18.8), we conclude that

If a body of rest mass m is moving at velocity v relative to a certain observer, then its momentum relative to him is

$$\frac{mv}{\sqrt{(1 - v^2)}}. \tag{18.9}$$

This expression involves only the rest mass of the body and its velocity. So our conclusion refers to *any* body with rest mass *m*, moving at velocity *v*, independently of how it reached that velocity.

18.19 By (18.1) the classical formula for momentum is *mv* – indistinguishable at low speeds from (18.9). But at relativistic speeds, clearly, something will have to be changed. We *could* say that the mass is always *m*, in which case we are forced to redefine momentum according to (18.9). *Or* we can retain the formula of (18.1) for momentum; and then (18.9) forces us to concede that W's mass, when moving at speed *v*, is $m/\sqrt{(1 - v^2)}$.

We shall find in the next chapter that the latter alternative leads to a very beautiful mathematical simplification, which in turn suggests a wonderful new physical hypothesis. So we choose this one, and say that

> If a body of rest mass *m* moves at speed *v* relative to a certain observer, then its mass according to him is
>
> $$\frac{m}{\sqrt{(1 - v^2)}} \,.$$ (18.10)

(*Speed*, not velocity. Mass can't depend on direction of motion. And (18.10) is the same whether *v* is positive or negative – §12.12). **Can you see now why I insisted in §18.2 that mass is not 'quantity of matter'?**

18.20 Experimental verifications of this prediction provide some of the best evidence in favour of Relativity. (Actually it is the momentum prediction of §18.18 that gets tested, rather than the mass prediction arising from our decision in §18.19. But the results are usually expressed in mass terms.)

When we speak of mass as a measure of resistance to change in motion, we can refer to change of direction as well as change in speed. Force is needed to deflect a moving body. And the above considerations can be extended to cover the relation of transverse force to change in direction of motion. Around 1900, experimenters were finding that in order to deflect streams of fast-moving electrons in cathode ray tubes they had to use bigger forces than they expected. So Einstein's prediction was verified in advance – the quantitative agreement with (18.10) was excellent.

The great 'atom-smasher' machines (particle accelerators) of today use enormously powerful magnetic forces to deflect streams of high-speed particles, so that they whirl round and round in circles, gradually speeding up. Obviously the strength of the magnet required to do the job depends on the mass of the particles. In the CERN machine near Geneva, particles called protons reach almost 30 times their rest mass at top speed. And the machine only works because the designers took this into account. Machines under construction will push this figure up by a factor of more than 10. Less massive electrons can be accelerated much more easily than protons – giving machines that only work because the design engineers allowed for a 40 000-fold mass increase.

In recent years this branch of Relativity has made its way even into the engineering workshop. Cutting, drilling and welding of metals can be done by well focussed beams of electrons moving at speeds which increase their masses by amounts of the order of 50 per cent. They work because (18.10) was used in designing the focussing arrangements. Even your TV set gives a clear unblurred picture because due allowance has been made for a per cent or so increase in the mass of the cathode ray electrons.

William Kingdon Clifford has reminded us that scientific thought 'is the guide of action; that the truth which it arrives at is not that which we can ideally contemplate without error, but that which we may act upon without fear'. And surely that applies to Relativity.

19

The mass–energy relation

19.1 Einstein's most famous prediction concerns a previously unsuspected relation between mass and energy, which led to far-reaching consequences that have abolished for ever any comfortable illusions about science – even its most abstract parts – being an ivory tower that can be left in the care of specialists that dwell therein.

In everyday life we say that a person has a lot of energy if he has a large capacity for doing work. And the same idea will serve for defining energy in its scientific sense, provided we say first what is meant by 'work' in this context.

A typical example of physical work, and an easy one to think about, is lifting weights. And the only reasonable formula for the amount of work done is

> work done
> = (weight lifted) × (height through which it is lifted).

I'm sure you'll agree **after a few minutes' thought.**

Doubling the weight lifted, or doubling the height, both double the amount of work. And these are the only factors involved. **Think on till you feel sure.**

19.2 Weight-lifting is just a particular case of using a force to move something in the direction of the force – which leads to a more general definition (not completely general, but good enough for us):

> If a force of constant strength moves a body in its own direction, the amount of *work* which it does on this body is given by
> work done = (force employed) × (distance moved).   (19.1)

And the *unit* of work will be the work which is done when a unit force moves a body through unit distance. ('Distance moved' means the distance which the body moves *while the force is acting on it.*)

19.3 And now we lay down the definition that

Energy is capacity for doing work. And the amount of energy possessed by anything is equal to the amount of work it is capable of doing.

And so energy is measured in the same units as work.

19.4 Thus energy takes many forms and resides in many things. A lump of coal has energy, since you can use it to drive a steam engine that does work. The Sun's rays have energy, since they lift water to great heights – as experience tells us when it comes down again. And anything that's moving has energy, since (for example) you can use its motion to lift its own weight by making it run up a slope. **Spend a few minutes thinking about other forms of this all-pervading energy.**

19.5 The world is full of examples of energy changing from one form to another. In your car the petrol burns, changing *chemical* energy into *heat* energy; the hot gases push back the piston, changing this to *mechanical* energy. **Think up a few more examples of energy transformations.**

Unfortunately energy transformations seldom happen efficiently – most of the heat in your car engine comes out of the exhaust instead of driving the car. But if you carefully consider all the energy that goes into each transformation and all the energy – *of all types* – that comes out, you reach a conclusion of extreme importance called

The Principle of the Conservation of Energy
Energy can be neither created nor destroyed.

– only changed from one form to another.

As so often happens, we can only test this directly in a few simple cases. But it is used as one of the basic assumptions of almost every physical theory, and it has never yet led to a false prediction. So the evidence for it is overwhelming.

19.6 The energy that a body possesses by virtue of its motion is called its *kinetic energy*. This is obviously relative to the observer.

So a body's kinetic energy is equal to the work we can get it to do in the process of stopping. But this would be exactly the same as the work that would have to be done to speed it up from zero to whatever speed it has got (speed, not velocity – cf. §18.19). **Can you prove this?**

If you could get more work out of the body in stopping it than you had to put in to produce its motion, then repeating the process would give as much work as you wish – as much energy as you wish. Apart from growing very rich by getting all this work done for nothing, you'd have breached the Conservation of Energy. **Work out the reverse case for yourself** (this time you have to be very rich to start with).

Now we apply these ideas to calculate the kinetic energy of the body W defined in §18.17.

19.7M We shall work this out by a *chain calculation,* in which we find that some quantity is equal to another, which is equal to another, which . . . till at last we reach our answer. Giving the full reasoning behind each step takes a lot of space, interrupts the continuity, and can be a nuisance if you can see the point for yourself anyway. So in future I shall lay out such calculations concisely, as in §19.8–glance at it and come back, please. On the extreme left are brief notes about the reason for asserting that the line which follows is equal to the line above. These notes will usually be equation numbers or article numbers, which you can look up if necessary; sometimes there will be a few words.

To see the sort of thing I mean, will you now read §19.8 in parallel with the following explanations. Don't forget the points of §§3.15 (last sentence), 9.19 (end of paragraph 2) and 8.19 (with examples in §§9.3, 9.17 and 11.18).

At the start we refer to §19.6. There you will find the two recipes for calculating kinetic energy, of which we use the second. Then the (19.1) tells us that we are using the definition of work to say that the quantity of the previous line is equal to the one that follows this (19.1) reference.

Now instead of generalities about force and distance, we want to introduce the actual force and distance that relate to the present problem. Well, it was at the end of §18.17 that we decided to use F to stand for the force in question. And (17.7) tells us that the distance we're concerned with is $X - 1/f$. (Of course if you are not clear about *why* this is the right distance, you must search earlier material.) So substituting these (§9.16) in the previous line gives $F(X - 1/f)$.

Then comes the reminder of (18.7) which enables us to transform this last into $mf(X - 1/f)$. Next a reference to the identity of §16.12 (and I shall usually leave you to work out what the three quantities are and which alternative to use). When you apply this identity, you actually get

$$mfX - mf\frac{1}{f}.$$

But the instructions to multiply and divide by f cancel each other and bring us back to m. (I shall usually leave such details to you.) And we do the transition to the last line by using (17.13) to substitute for fX in the line above. Now check that you can follow, step by step, the whole chain calculation of §19.8.

Naturally, when *I* refer to an article, *you* will have to decide which bit is relevant; and when *I* give you an equation, *you* will often have to do the 'with *a* changed to *t*' sort of thing. When you meet a future chain calculation, see if you can spot the reasons behind each '='. You'll often succeed. If not, my reference will put you right.

19.8 By §19.6,

> kinetic energy of W at speed *v*
> = work required to raise speed from 0 to *v*

(19.1)

> = (force used) × (distance W moves to attain speed *v*)

§18.17 & (17.7)

$$= F\left(X - \frac{1}{f}\right)$$

(18.7)

$$= mf\left(X - \frac{1}{f}\right)$$

§16.12

$$= mfX - m$$

(17.13)

$$= \frac{m}{\sqrt{(1 - v^2)}} - m \; .$$

So we conclude that

> If a body of rest mass *m* is moving at speed *v* relative to a certain observer, its kinetic energy according to that observer is

$$\frac{m}{\sqrt{(1 - v^2)}} - m \; . \tag{19.2}$$

(If you are at ease with the necessary mathematics, you might care to prove that the familiar $\frac{1}{2}mv^2$ is a slow-speed approximation to this formula.)

19.9 Comparing (19.2) with (18.10), **do you notice anything interesting?**

When we accelerate W from rest to speed *v*, its mass changes from *m* to $m/\sqrt{(1 - v^2)}$ – i.e., changes *by* $m/\sqrt{(1 - v^2)} - m$. Comparing this with (19.2), we see that *the kinetic energy and the mass are changed by exactly the same amount* (using natural units – otherwise there would be a conversion factor). **Can you think of a physical interpretation of this interesting relationship?**

19.10 The answer depends on a certain leap of intuition, not just pure logic. We give the body a push – *supply it* with kinetic energy. And we find that *as a consequence* its mass increases correspondingly. So it's plausible (though not logically necessary) to suggest that the kinetic energy has brought the additional mass with it. In other words,

> Kinetic energy has mass

– which gets added to the body's mass when we set it in motion. And of course it follows from the comparison in §19.9 that

> When expressed in natural units, the kinetic energy and its mass are equal – one unit of kinetic energy has a mass of one unit.

Does anything occur to you about extending the idea that kinetic energy has mass?

19.11 All forms of energy are interchangeable. It would be very surprising if kinetic energy had mass and chemical energy (say) did not. The queerest things would happen when you converted one to the other. And so we are led to widen the suggestion to:

> All energy has mass. And when expressed in natural units, the energy and its mass are equal.

19.12 The expression (19.2) looks clumsy. It seems to describe the difference (§16.11) of two quantities rather than a single quantity. What could these quantities be, considered separately?

First, what *sort* of a quantity is $m/\sqrt{(1 - v^2)}$? In the present context, kinetic energy is equal to this quantity diminished by m. In other words, *part* of this quantity is a bit of energy. So the *whole* of it must also be a quantity of energy, connected somehow with the body in question. Furthermore, it has a neat and simple form – compared with (19.2). And neat simple expressions are often found to stand for something important – not *very* often, it's true, but often enough to make the possibility worth considering.

Summarising, $m/\sqrt{(1 - v^2)}$ is a quantity of energy, possibly an important one. Part of it is W's kinetic energy, and the whole of it is associated with W. **Any guesses?**

Surely it's tempting to suppose that it represents the *total* energy of the body. We're suggesting that

> The *total energy* of a body whose rest mass is m, when moving with speed v, is
> $$\frac{m}{\sqrt{(1 - v^2)}}.$$ (19.3)

19.13 Apart from its kinetic energy, there are many forms of energy that belong to the body itself, independently of its motion – heat energy, chemical energy, nuclear energy, etc. We lump these together as its *intrinsic energy*. If that last guess is correct, **how much intrinsic energy does W possess?**

   The intrinsic energy is what remains when W is stationary; so we find it by putting $v = 0$ in (19.3). Thus

   A body of rest mass m has intrinsic energy m.

(in natural units).

19.14 We have met the quantity $m/\sqrt{(1 - v^2)}$ in §18.19 as W's *mass* at speed *v;* and again in §19.12 as W's *total energy* at speed *v*. And m has occurred as *rest mass* (§18.16) and as *intrinsic energy* (§19.13). So the equation

$$\frac{m}{\sqrt{(1 - v^2)}} = \frac{m}{\sqrt{(1 - v^2)}} - m \qquad + m$$

can be interpreted as

$$\frac{\text{Total energy}}{\text{at speed } v} = \frac{\text{kinetic energy}}{\text{at speed } v} \qquad + \text{ intrinsic energy}$$

or as

$$\frac{\text{Mass at}}{\text{speed } v} = \frac{\text{increase in mass due to}}{\text{motion at speed } v} + \text{ rest mass.}$$

Try to draw another plausible conclusion.

   We said in §19.11 that all energy has mass. Surely this latest comparison suggests that

   All mass is the mass of energy,

and that, just as (19.2) is the mass of kinetic energy, so also

   The rest mass is the mass of the body's intrinsic energy.

19.15 So (19.3) tells us how much the energy of a moving particle is increased, compared with its intrinsic energy as measured by its rest mass. An experiment reported in 1975 showed that if, by speeding it up, an electron's energy is increased 30 000 times, it is then moving slower than light by 5 parts in 10 000 million, which agrees with (19.3) – an immediate verification of all these guesses.

19.16 So is there any point in thinking of energy and mass as distinct things? Surely it would be better to regard mass as simply one of the ways in which energy makes itself felt. Remember that mass is merely resistance to change of motion. All we're saying now is that energy puts up a resistance to change in motion, and that we use the word 'mass' to describe this particular attribute of energy.

Don't let the shortness of this article deceive you. It's very important – a revolution in outlook.

19.17 So a lump of matter, inert though it may seem, has in fact got energy locked up in it. How much?

For this calculation, let's take the unit of mass to be 1 kilogram (§18.9) and the unit of time to be 1 year – so that the natural unit of distance is 1 light year (§1.8). As we calculated in §17.21, the unit of acceleration is then (roughly) 1*g*. Then by §18.13, the unit of force is the force that gives an acceleration of 1*g* to a mass of 1 kilogram – which is simply the *weight* of 1 kilogram. (When you let 1 kilogram fall freely, the force that produces the 1*g* acceleration is, of course, its own weight.)

19.18 Then by §§19.2–3 the unit of work or energy is the work done by a force equal to the weight of 1 kilogram moving the body through a distance of 1 light year, or roughly 9·5 million million kilometres (§1.8). But to avoid the variation of gravity with height, we'd better change that to the work done when a force equal to the weight of a million million kilograms moves a body 9·5 kilometres – that is to say, the work done in lifting 1000 million tonnes to a height of 9·5 kilometres. (A tonne is 1000 kilograms; at 2205 lb. it is very close to the old-fashioned ton.)

19.19 According to §§19.13–14 a lump of matter having a rest mass of 1 unit contains 1 unit of intrinsic energy. So we've calculated that

> One kilogram of matter contains enough intrinsic energy to raise 1000 million tonnes to a height of 9·5 kilometres – rather higher than Mount Everest.

A thousand million tonnes is the mass of a cube of water whose side is 1 kilometre ($\frac{5}{8}$ mile) long. Enormous! Before we ask the obvious question, let's do the calculation in a more orthodox fashion.

19.20 The unit-for-unit correspondence between mass and energy only holds good when we use natural units for distance (§7.11). But we can easily convert to some other system in which light travels *c* distance units in 1 time unit – so that *c* is the speed of light. Every time a distance occurs, we must multiply by *c* (substituting *c* of the new units for 1 natural unit). The calculation of §§19.17–19 is a fair example, and we notice that distance comes into it twice – in working out the acceleration (§§19.17, 17.21) and the work done (§19.18). So we have to multiply by *c* twice – by c^2. Thus

> In units such that the speed of light is *c*, the amount of energy corresponding to a given mass is equal to that mass multiplied by c^2.

In particular, if E stands for the intrinsic energy, this gives the famous equation

$$E = mc^2. \tag{19.4}$$

19.21 The internationally agreed system of scientific units takes the second, metre and kilogram as its units of time, length and mass, so that the speed of light is 300 million metres per second, and $c^2 = 90\,000$ million million. The corresponding energy unit is called the *joule*, and so 1 kilogram of mass corresponds to 90 000 million million joules of energy. However, joules are strangers to most of us. More familiar is the kilowatt-hour (the energy used in an hour by a normal one-bar electric fire). We convert on the basis that 10 million joules = 2·778 kilowatt-hours (because a watt is defined as 1 joule per second). And we find that 1 kilogram of matter has 25 000 million kilowatt-hours of energy tied up inside it.

19.22 So the amount of energy imprisoned in matter – *any* matter – is enormous. Can some or all of it be released?

Relativity can't answer that question. It can only assert that *if* intrinsic energy is released as free energy, there must be a corresponding decrease of rest mass – and the portion of matter to which this rest mass belongs must disappear. And conversely, if matter disappears, a corresponding amount of energy must be released. That clear, precise prediction demands experimental test. And Einstein said, at the end of his second Relativity paper:

> It is not impossible that with bodies whose energy content is variable to a high degree (e.g., radium salts) the theory may be successfully put to the test.

That must be the understatement to end all understatements! As you know, the prediction has been thoroughly tested, first in the laboratory experiments of the thirties, and then in the engineering successes of nuclear power stations (and, alas, bombs). Let's have some examples.

19.23 When an electron encounters a positron (a similar particle, but with positive electric charge instead of negative), both of them completely disappear, and two gamma-rays are given out, whose energy can be calculated from their frequencies (§19.30 below). The agreement with (19.4) is found to be excellent.

19.24 We don't know how to produce such total annihilation of matter on a large scale. But we can arrange situations in which the rest mass of a lump of matter decreases significantly and a corresponding amount of intrinsic energy is set free. (If some of the following words are unfamiliar, please bear with me.)

When a neutron collides with the nucleus (central portion) of a uranium-235 atom, a thorough bust up (called *fission*) leads to an end-product of two large fragments and a few neutrons – some of which, in suitable conditions, will strike further uranium nuclei, producing more fission . . . building up to a chain reaction. In the units used for such purposes the masses of neutron and uranium-235 atom are 1·01 and 235·04. A typical fission (for there are many variants) gives an output of two neutrons plus fragments of masses 94·91 and 138·91. (The masses are for complete atoms – nuclei and surrounding electrons.) Assuming that this case *is* typical, **how much energy is released by the fission of 1 kilogram of uranium-235?**

The total mass is 236·05 before fission and 235·84 after. So 0·21/236·05 of the kilogram mass disappears and the energy released must be that fraction of 25 000 million kilowatt-hours (§19.21) – about 22 million kilowatt-hours. It is upon such calculations (immensely more complicated in practice) that the success of nuclear power engineering depends. There's verification for you!

The Sun's energy comes from a process which indirectly fuses 4 protons (hydrogen nuclei) into a helium nucleus and two positrons. **How much energy is released per kilogram thus transformed?** (Relevant masses: proton 1·0073, helium nucleus 4·0015, electron or positron 0·000 55. Solution at end of chapter).

19.25   We have ample confirmation, then, that when energy is released from a body, rest mass disappears. (N.B. The *mass* does not disappear – it accompanies the released energy – but it is no longer *rest* mass.) If we completely drain away all the body's energy, then *all* its rest mass disappears, and so the body itself disappears. Its matter has no existence, apart from the energy locked up in it.

> Matter is just energy tied up in such a way that it has rest mass.

And it then follows, of course, that *rest* mass is a measure of the quantity of matter in a body. **So was I wrong to warn you in §18.2 not to think of mass as quantity of matter?**

The difference between mass and rest mass cannot be ignored. And if we'd started with the idea that *mass* measures amount of matter, we couldn't have done the reasoning which proves that *rest mass* fulfills this function.

It's wrong, by the way, to describe phenomena like those of §§19.23–4 as mass being transformed into energy. The total energy remains unchanged (§19.5). *Intrinsic* energy is released as *free* energy; and the *rest* mass of the former becomes the *non-rest* mass of the latter. So, using the indented passage, we can fairly say that *matter* is transformed into *free energy*.

19.26 Back to natural units from now on. If E is the energy and m the rest mass of something moving at speed v, it follows from (19.3) that $E\sqrt{(1 - v^2)} = m$ (§9.11). If the something is light (which has energy and therefore does have mass), then $v = 1$, and so $m = 0$. Thus *light has zero rest mass*.

That makes sense – light at rest does not exist. When light is stopped by hitting something, **what happens to its mass?**

It usually goes to increase the rest mass of whatever stopped the light and absorbed its energy. Commonly the light just slightly warms the body it hits, adding the mass of this increased heat energy. Sometimes a very energetic gamma-ray disappears and is replaced by an electron and a positron – the reverse of §19.23. The gamma-ray's energy and mass have become the intrinsic energy and rest mass of the new particles.

We've now covered the essentials of what Relativity has to say about energy. I'm going to make a few further points in more condensed style, using mathematics that you may or may not be ready to take. Feel free to skip (or skim) and come back when you're more experienced. But do read §19.34.

19.27 Let's use Doppler factor k (§§15.4–5,15.8) instead of speed to describe the motion of a body. Then from (19.3) and (16.15) we easily deduce that its

$$\text{Energy} = \tfrac{1}{2}mk + \tfrac{1}{2}m/k . \tag{19.5}$$

And from (18.9) and (16.16), its

$$\text{Momentum} = \tfrac{1}{2}mk - \tfrac{1}{2}m/k . \tag{19.6}$$

The form of these equations suggests that we might consider the body as being equivalent to a combination of something that has energy $\tfrac{1}{2}mk$ and momentum $\tfrac{1}{2}mk$ with something else that has energy $\tfrac{1}{2}m/k$ and momentum $-\tfrac{1}{2}m/k$. The former is moving in the positive direction and the latter in the negative (§18.5). **Can you think what this 'something' and this 'something else' could be?**

19.28 The mass of the 'something', by §19.11, is also $\tfrac{1}{2}mk$. So its momentum is equal to its mass. **Does that suggest anything?**

By (18.1) its velocity is 1. **So what is the 'something'?**

Since it moves at the speed of light, it must be a light signal (ignoring the possibility that it might be a weird particle called a neutrino) – a positive-going light signal. **And the 'something else'?**

A negative-going signal. So this body of rest mass m can be considered as equivalent to two light signals – two packets of light energy – going in opposite directions.

A neutral pion (§1.9) gives concrete embodiment to this idea. As it dies it becomes two gamma rays, which go off in opposite directions. And the energies and momenta fit equations (19.5) and (19.6). But perhaps we shouldn't take too literally the idea that every particle is some sort of frozen combination of two packets of light. Better just say cautiously that the mathematics of energy, mass and momentum is the same for the sum of the two signals as for the body.

19.29 Let A be the observer relative to whom the body has Doppler factor k. And let B be another observer travelling with the body. Reasoning on the lines of §§15.4–6 (where k was just called 'constant factor') leads to the conclusion that if A and then B observe the same continuous positive-going signal, then

its frequency according to A
$= k \times$ its frequency according to B. (19.7)

Now think of B observing the same body. For him $k = 1$ (§15.4), and so (19.5) reduces to

Energy $= \frac{1}{2}m + \frac{1}{2}m$.

So, confining attention for the present to the positive-going signal, we find that its

Energy according to A $= k \times$ energy according to B.

Compare this with (19.7) and draw a conclusion.

The energy of a light signal, as measured by various observers, is proportional to its frequency according to them. (Work out the negative-going case yourself.)

19.30 The branch of physics called quantum theory asserts, as its most basic assumption, that light always comes in packages – called *photons* – of definite sizes. From the previous article it follows that the energy of a photon is proportional to its frequency (the constant of proportionality being known as Planck's constant). This is one of the most fundamental laws of modern physics; but if you don't know it already, explaining its significance would take too much space.

19.31 If we take (18.8) and the corresponding equation for *FX* (calculated by a slight modification of §19.8), and convert these entirely to radar terms, we get $F\Phi = mk$ and $F\Theta = -m/k$. Can *you* suggest a physical interpretation of these *very* simple equations? I can't–except that the quantities involved are obviously radar equivalents, in some sense, of momentum and energy.

19.32 Changing our notation slightly, let bodies A and C be approaching observer B at the same speed from opposite directions. Let k stand for the Doppler factor between A and B, and also between B and C. Then by (16.20), the Doppler factor between A and C is k^2. If A and C both have rest mass m, and E stands for their (equal) energies as measured by B, (19.5) gives

$$\frac{2E}{m} = k + \frac{1}{k} . \tag{19.8}$$

Similarly if E_{AC} stands for the energy of body C as measured by an observer travelling with A, we have

$$\frac{2E_{AC}}{m} = k^2 + \frac{1}{k^2}$$

§26.28 below

$$= \left(k + \frac{1}{k} \right)^2 - 2$$

(19.8)

$$= \frac{4E^2}{m^2} - 2 .$$

And so

$$E_{AC} = \frac{2E^2}{m} - m . \tag{19.9}$$

If $m = 0{\cdot}938$ and $E = 28{\cdot}94$, **check that** $E_{AC} = 1785$.

So E_{AC} is not 4 times E (as you'd deduce from the classical formula – end of §19.8), but *more than* 60 *times*. I suppose you're past being shocked by now.

19.33 Those figures refer to a practical job. In modern research on the so-called fundamental particles, the main method is to use an accelerator (§18.20) to get particles moving very energetically, let them bash into other particles in a stationary 'target' and watch what emerges. The bigger the energy involved in the collision, the more exciting the results. But building more powerful accelerators is *very* costly. Hence arises the idea of diverting the particles into what are called 'storage rings', and then arranging for two streams of them, moving at top speed in opposite directions, to collide head on.

Our calculation refers to such an apparatus at the CERN establishment, Geneva. The energy unit used in this field of research is the *giga electron volt* (GeV) – a thousand million electron volts, where 1 electron volt is the energy you impart to a particle by accelerating it through a 1 volt drop in an electric field. The CERN machine gives 28 GeV to a proton. But to find E we have to add the intrinsic energy of 0·938 GeV that the proton started with. Hence the §19.32 figures – though in practice the energies obtained are not quite so high. This CERN project has paid handsome dividends in research results. Yet if one had trusted the classical formula instead of (19.9), it would not have seemed worth trying.

19.34 You may have had a feeling that Relativity is mainly concerned with things like the high-speed space travel of the distant future. And I may have reinforced this impression by using spacemen observers to make my descriptions more vivid. But these last two chapters must have convinced you that Relativity is actually concerned with very practical matters; it is an essential tool for effective engineering whenever high speeds or high energies are involved. Most of the applications so far are concerned with the design of machines for scientific research. But history shows the advanced scientific apparatus of one generation becomes the practical technology of the next. And the examples of §§18.20 and 19.24 show that this trend has already gone a long way.

Answer to §19.24 The positrons will meet, and mutually annihilate with, electrons. So the input is really 4 protons and 2 electrons, total mass 4·0303; and the final output is just the helium nucleus. Thus fraction 0·0288/4·0303 of the rest mass disappears and about 180 million kilowatt-hours of energy is released.

20

The effect of acceleration on time measurement

20.1 Twice before we've discussed the so-called Clock Paradox (Space-twin Paradox). Please revise §§4.14–20 and 14.22–4. After that, if you've not already tackled the problem at the end of §4.18, please do so now. Figure 14.22 should help. **Action!**

20.2 You are being asked to compare the stories told by different versions of Figure 14.22 in which the curves near P, O and Q remain always the same in shape and length, but the two intervening straight portions of D's world line may be of any length we wish. If the accelerations do have the effect (suggested in §§4.18 and 14.24) of increasing the time that passes for D, then (with the same accelerations used in every case) the amount of this increase is fixed. But on the inertial parts of D's journey the dilation of time (§§3.10, 13.7) is always operating to diminish his total time measurement compared with A's; and the longer the inertial portions of the journey, the bigger this effect will be. So, *even if there is* an 'acceleration effect', it could only compensate for the dilation of time on a journey of one particular length. On longer journeys less time would pass for D than for A; on shorter ones it would be the other way round.

20.3 With a theory based on the assumptions we've used so far (all inertial observers equivalent, etc.) we can't definitely decide whether accelerations do or do not have *some* effect on the traveller's time. To make progress we must introduce a *new assumption* on this question, work out its consequences, and test them as usual against experiment.

What assumption shall we use then? Surely, as a first shot, we'll try assuming that *the accelerations have no effect on the traveller's time measurements,* so that we need only take account of relative speed. We assume, in other words, that we can investigate how time passes for the traveller by simply applying the dilation-of-time formula, taking account of the instantaneous relative speed at every moment. We must use this formula to calculate A's opinion of D's clock times. It cannot be used to give D's opinion of A's clock times, since we have no idea yet of what an accelerated observer's observations and opinions would be like.

20.4 The obvious next step is to do the calculation for some specified journey and compare the prediction with experiment. But there's a snag. According to §13.7, dilation-of-time calculations involve the factor $\surd(1 - v^2)$, where v stands for the relative speed. That gives no trouble when v stays constant. But when v is changing from moment to moment, it is far from obvious how to do the calculation. To avoid (temporarily) this difficulty, **can you think of a case in which D is accelerated all the time and yet moves at constant speed?**

20.5 We get what we want if D travels at constant speed *in a circular path*.

But can we really say that D's motion is accelerated when his speed is constant? Yes. For us the word 'accelerated' means, at rock bottom, 'non-inertial' (§§17.2–3). That's the sense that matters – because we already know what happens in the case of inertial motion, and we are now starting an enquiry about what happens when the motion is non-inertial. And D's circular motion is certainly non-inertial (§5.2): his test particle leaves him – by 'flying off at a tangent'.

But even taking a more familiar point of view, at one instant D is going in a certain speed in one direction, and half a revolution later he is going at the same speed in the other direction. To get from the one to the other he must have accelerated. And indeed, when we get beyond simple straight line motion, acceleration has to be defined as changing *velocity* (§18.6), not just changing speed. (This is needed, e.g., to make sense of the relation between transverse force and acceleration.) So D's continuously changing direction of motion counts as acceleration – while leaving us with the mathematical convenience of constant speed.

20.6 Consider a laboratory experiment in miniaturised space travel in which D is a bunch of muons (§3.24) racing round a circular track, while A is a trackside observer. (**Does it matter that D merely flies past A at the start and finish, rather than beginning from and returning to rest relative to A?** No. By repeating the circular trip many times the effect of starting and stopping can be made negligible – cf. §20.2.)

Taking time-dilation in the form discussed in §§3.34–6 and adding the quantitative information of §13.7, we see that if a clock is moving at speed v relative to an inertial observer, he thinks it is running slow by a factor of $\surd(1 - v^2)$. And our new assumption (§20.3) says we can apply this formula to A's opinion of D's clock (but not the other way round).

20.7 In a very carefully designed experiment conducted at the CERN laboratory (details in *Nature*, **268** (1977) 301) the muons travelled in a circular path 14 metres in diameter at a speed rather more than 0·9994 of the speed of light, giving $1/\sqrt{(1 - v^2)} = 29\cdot33$ – so that their lifetime as observed by a laboratory-fixed observer should appear 29·33 times the lifetime of a muon at rest. *The experimental results agreed with the prediction to within* 0·2 *per cent.*

So now we have experimental evidence that the accelerations have no effect on the time that passes for the traveller; and that in calculating his time *one need only take account of his speed* relative to the stay-at-home and make use of the dilation of time.

Furthermore, if acceleration had even a slight effect, it would surely show up in this case – for the muons' acceleration works out at about a million million million times *g*.

20.8 In this experiment the traveller's clock seems to an observer fixed in the laboratory to be running slow all the time. So when it comes back to where it started, it must actually have recorded less time as having passed than is recorded by a stationary clock at that point. Less time *really does* pass for traveller than for stay-at-home.

Please note: we are *not* saying that the traveller's clock runs slow. The stay-at-home *thinks* it runs slow. Really, however, it records less time because it took a different – and shorter – route through space-time. If this is not clear, please read once again §§14.20–4.

20.9 Yet you'd feel happier if we could demonstrate this space-twin effect for what a human being would regard as a journey? What about a flight round the world? Rough calculation shows that the time difference ought to be something like 100 nanoseconds (a nanosecond being a thousand-millionth of a second). And that is well within the capacity of an atomic clock (whose ticks are the successive waves of light produced by rearrangements of the electrons in the outer parts of atoms). So it looks promising.

But there are complications. A point on the Earth's surface is already doing a circular trip relative to a hypothetical inertial observer. An east-going plane will be making a faster circular journey, and a west-going plane a slower one. So the dilation of time implies that a clock going eastward round the world will record less time than one that stays put, and a west-going clock will record more. *You* could now easily work out the times for planes flying steadily round the equator. But doing the experiment that way would be prohibitively expensive. So commercial services must be used, and the calculations have to allow for flights at various speeds in various directions. Again, time measurements are affected by gravity to roughly the same extent, but in a way that always increases the traveller's time by an amount proportional to his height (we'll come to this in Chapter 23). And there are problems of allowing for the residual unreliabilities of even these very reliable clocks.

Nevertheless the experiment was tried in October 1971 (details in *Science* **177** (1972) 116). The prediction emerging from the complex calculations was that the east-going clock should record 40 nanoseconds less than an Earth-fixed clock, and the west-going should give 275 nanoseconds more. The experimental results were 59 nanoseconds less and 273 nanoseconds more. Considering the difficulties, one couldn't ask for better agreement.

20.10 As we noticed in §20.4, the calculations are going to be harder when D's journey involves changing speed. To make progress we need to formulate our new assumption more precisely. And that's quite difficult. The rest of this chapter is mathematically rather tough. You may have to study it more than once – and see §16.23. In the end you may need to leave some of the new ideas half digested and come back when you're more experienced.

The new assumption, as we stated it in §20.3, is far too vague. We must find some precise way of specifying how time measured by an *accelerated* clock compares with time measured by an *inertial* clock. **Any suggestions?**

20.11 One idea might be to take an inertial observer F who at event Q is momentarily moving with accelerated traveller D (use the slot to check that this is what Figure 20.11 shows) and say that D's clock will keep

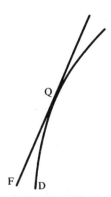

Figure 20.11

time with F's while they are moving together at Q. (And this is to apply, of course, for *any* point Q on D's world line.) That's a brave try. **But why won't it do?**

To decide whether the clocks keep time with each other, we should have to compare them *twice*. And if they're only together for an instant, we can't do that. On the other hand, if we make the comparison over some non-zero time-interval, then the clocks are not moving together (slot again!). Either way we seem to be stuck.

Actually we met a rather similar dilemma in §§17.12–16, and we got out of it by using the as-accurately-as-we-wish technique. Do study that again, and **ponder on how we might apply similar thinking in this case.**

20.12 Take *two* ticks, P and Q, of D's clock (Figure 20.12) and consider an inertial observer E who is present at both (slot!). Let t stand for the time between P and Q by D's clock, and t_i ('inertial time') for the time between them by E's clock. Then D plays mini-space-traveller to E's stay-at-home, and so the equation

$$t = t_i \tag{20.1}$$

is false. But – surely this is what we want our new assumption to say – if we make t small enough (i.e., take Q near enough to P), then the error in this equation will be small – because only low relative speeds will be involved, so that there will be little time-dilation effect, and (we are assuming) *no* acceleration effect. The smaller we take t to be, the nearer will (20.1) come to being true; and by taking t small enough† *we can make the equation $t = t_i$ as nearly true as we wish.*

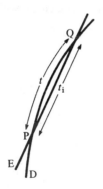

Figure 20.12

---

† Strictly we should say 'by taking the absolute value (§13.20) of t small enough' – we have to allow for the possibility of t being negative (Q happening before P). You probably took that for granted, and I shan't mention it again.

20.13 We must state clearly what we mean by those italicised words. If we assert that $t = t_i$ – that t is actually equal to t_i – we introduce an error, whose size is, of course, $t_i - t$ (the difference between t_i and t). As we take P and Q closer together, both t and t_i get nearer to zero, and so the difference $t_i - t$ gets smaller – and we're *automatically* making t and t_i as nearly equal as we wish. For example, if I assert that $t = 2t$ (i.e., take t_i to be $2t$), then clearly I'm wrong. And surely we'd say that no matter how small I make t, I'm still just as wrong as before. Yet the error – the actual difference between the sides – is t; and we can make that as small as we wish by taking t small enough (obviously!). So when we say 'make the equation $t = t_i$ as nearly true as we wish, we don't just mean 'make the difference between t and t_i as small as we wish'. We need a stronger statement.

What really matters is the size of the error *considered in relation to the size of the quantities involved*. And surely we'd measure that by means of the difference between t_i and t *divided by one or other of these quantities* – say by t_i. We can call this the *proportionate difference*† between t_i and t, and take it to be the measure of how nearly (20.1) comes to being true. So

$$\text{proportionate difference} = \frac{t_i - t}{t_i} . \tag{20.2}$$

And the italicised passage of §20.12 acquires real significance when we interpret it as 'we can make the proportionate difference between t_i and t as small as we wish'. Reread §20.12 thus amended.

20.14 A numerical example may help. **Study Table** 20.14 (ignoring last column) **to see how it illustrates** §§20.12–13.

The difference $t_i - t$ gets less as we work our way down the table. But that only says that as t and t_i both get nearer to zero, they get

Table 20.14

t_i	t	Difference $t_i - t$	Proportionate difference $(t_i - t)/t_i$	Ratio t/t_i
9·90	4·61	5·29	0·53	0·47
4·80	3·22	1·58	0·33	0·67
1·500	1·386	0·114	0·076	0·924
0·190 91	0·190 62	0·000 29	0·001 5	0·998 5
0·019 901 0	0·019 900 6	0·000 000 4	0·000 02	0·999 98

† N.B. 'Proportion*ate*', not 'proportion*al*'.

nearer to each other. It's the proportionate difference of the fourth column that gives a useful measure of how near (20.1) is to being true – since it takes the error you commit in saying that the equation *is* true, and presents it in proportion to (as a fraction of) the quantities involved. (If you multiplied that fourth column by 100, you would have the *percentage* difference. Maybe you'd find it easier to think in those terms.)

But notice that *we can't allow t and t_i to be actually zero*. That would take us back to the useless §20.11 situation, and the proportionate difference would be 0/0 – which is meaningless. (If you're struggling, this would be a good time to work through §§20.10–14 again.)

20.15 So our new assumption about the acceleration having no effect can be put in the form that

> By taking *t* to be small enough (but not zero) the proportionate difference between t_i and *t* can be made as small as we wish.

And this is a more precise way of stating (§20.12) that by taking *t* small enough we can make the equation $t = t_i$ as nearly true as we wish.

There's another way of putting it. From (20.2) (since subtracting and then dividing gives the same result as dividing and then subtracting) we have

$$\text{proportionate difference} = \frac{t_i}{t_i} - \frac{t}{t_i} = 1 - \frac{t}{t_i},$$

whence (§9.12)

$$\frac{t}{t_i} = 1 - (\text{proportionate difference}).$$

So as the proportionate difference gets nearer to zero, the ratio (§9.21) t/t_i gets nearer to 1. Thus the alternative form is

> By taking *t* to be small enough (but not zero) the ratio t/t_i can be made as close to 1 as we wish.

This is illustrated by the last column of Table 20.14.

On the other hand, making *t* smaller and smaller does not diminish the acceleration. So if this genuinely had an effect, the proportionate difference would stay clearly different from zero (and the ratio different from 1).

So what started off as a vague idea about the acceleration having no effect (§20.3) has now been transformed into a precise statement. And we'll soon see how it can be used to make actual calculations.

A small 'm' appended to the number of an article – as in §20.16 below – means that it deals with a mathematical point, but using such an unorthodox approach that even expert mathematicians will have to pay some attention.

20.16m It's handy to have a special sign for this new idea. I'm going to write

$$t \sim t_i \tag{20.3}$$

as shorthand for 'By taking t to be small enough. . .as we wish' in either of the forms indented in §20.15.

20.17m So every time you meet (20.3) you would, if you're following my read-it-in-full advice, say or think one or other of those indented passages. But you can't be expected to go through that rigmarole, except for the first few times while the idea is sinking in. I suggest you verbalize '$\sim$' as 'nears' (in the sense of 'gets nearer to') – so that you'll read (20.3) as 't nears t_i'. This is meant to remind you of the idea that as t gets smaller, the equation $t = t_i$ gets nearer to being true (in the more precise sense of §20.15). (Mathematical experts! Don't confuse this with the normal use of 'approaches' referring to a limit.) And just as we call (20.1) an equation, so we can refer to (20.3) as a *near-equation*.

This is *not* standard usage – just my own invention for present purposes. In other books '$\sim$' may be used (for example) where I use '$\approx$' (§13.11). Note carefully the difference: '$\approx$' says that two quantities are approximately equal – for the purpose you have in hand at the moment you can use one in place of the other; but '$\sim$' says (in the carefully qualified sense of §20.15) that we can make them near enough equal to suit our needs *for any purpose whatsoever*.

20.18m It is most important to understand that near-equation $t \sim t_i$ is *not* a statement about the relation between t and t_i 'when t is small'. It is a statement about *what happens* to this relation *as t gets smaller*. Please keep reminding yourself of this.

20.19 Now we can make a neater statement of our

Additional Assumption
If t and t_i are defined as in §20.12, then $t \sim t_i$.

And this, of course, is to apply *anywhere* on the observer's world line – that's how we insist that his time measurements are *always* unaffected by his accelerations.

20.20m We shall want to use the 'nears' idea for other quantities besides t and t_i. So we make the more general *definition:*

If Y and Z stand for two quantities, the near-equation $Y \sim Z$ means that by taking some specified quantity to be small enough (but not zero) the proportionate difference between

Y and Z can be made as small as we wish, or (as alternative
form) the ratio Y/Z can be made as near to 1 as we wish.

The proportionate difference means the difference between Y and Z
divided by either of them (cf. §20.13). It is sometimes useful to think
of this quantity as the *proportionate error* that would be introduced by
saying that actually Y = Z. That proportionate error, we are saying,
can be made as small as we wish by taking the specified quantity small
enough. And indeed when we are only trying to get a general impression of what is being said, rather than present a strict argument, it will
often be good enough to think of Y ~ Z as meaning that by taking the
specified quantity small enough we can make the equation Y = Z as
nearly true as we wish (cf. §20.12).

The contents of §§20.17–18 clearly apply (with obvious minor
changes) to near-equation Y ~ Z.

.21m Near-equations take us on another giant stride into that new realm that
we timorously entered in §§17.12–16. They will grow increasingly important as we go on. So please make sure you master the line of
thought that we've been developing since §20.10. If it's not yet fully
clear, don't let that prevent you from moving on; but keep coming
back till you're completely satisfied. And in future, every time you
meet a near-equation, use §20.20 (at least) to make sure that you
understand its meaning – and keep on doing so until this way of thinking becomes second nature.

20.22 Let's return to the case of inertial stay-at-home A and traveller D, who
part at event R and reunite at S (Figure 20.22), and see how our
Additional Assumption enables us to compare their times. We assume
that A has made a complete set of measurements giving D's position
relative to him at every instant.

Divide D's world line into a number of portions (marked by short
cross lines), one such portion running from P to Q. With t and t_i
defined as in §20.12, and with T standing for the time from P to Q
according to A (which is actually the time by A's clock between events
L and M which A regards as simultaneous with P and Q), **can you
work out a near-equation connecting t and T?**

20.23 By the dilation of time (§13.7 – check which times are proper and improper, §3.8), we have

$$t_i = T\sqrt{(1 - v^2)} ,$$
(20.4)

where v stands for the relative speed of A and E. And then the Additional Assumption (§20.19) gives $t \sim t_i$, so that

$$t \sim T\sqrt{(1 - v^2)} \, . \tag{20.5}$$

Make sure you know what that means – by changing Y to t and Z to $T\sqrt{(1 - v^2)}$ in §20.20 (with t as the 'specified quantity'). An informal interpretation would be that the equation

$$t = T\sqrt{(1 - v^2)} \tag{20.6}$$

is only *approximately* true; asserting it to be actually true would introduce an error, but the size of the proportionate error is under our control (in the sense specified by the definition of the $\sim$ sign).

20.24   We can carry out this procedure on each one of the portions into which D's world line is divided. For each of them we have a near-equation like (20.5). Put otherwise, for each portion an expression like $T\sqrt{(1 - v^2)}$ is an approximation to D's time, t. And so the sum of these expressions for *all* the portions will be an approximation to D's total time between R and S.

And now comes the vital question: Can we make that approximation as good as we wish – keep the total error as small as we wish? Consider the possible snag that though we're making each portion shorter (reducing t) in order to make *its* error smaller, we're also increasing the number of portions – increasing the number of errors that have to be added to get the total error. Paying careful attention to the exact definition of '$\sim$' in §20.20, **decide whether we can or can not make the approximation as good as we wish.**

20.25   We can. If the *proportionate* error in each portion is less than some stated figure, then the *proportionate* error for all the portions added together must still be less than that figure, no matter how many portions there may be. (Analogy: if you have any number of debts, on each of which you pay interest at less than 8 per cent, then the interest on your total borrowing must also be less than 8 per cent.) Let's set that out more formally.

We begin by deciding what size of error we shall permit in the total time from R to S. Dividing this by the total time (or by our approximation to it) gives the maximum permissible *proportionate* error (§20.20) – call it p for short. The Additional Assumption (§20.19) assures us that for each portion we can, by taking t small enough, make the proportionate error less than p. And if we do so for *each* portion, then the *proportionate* error in the sum of all the portions will also be less than p. We have thus given a recipe for making the total error less than whatever size we chose at the start of this paragraph – which could have been as small as we please. And so we have justified the conclusion that

The procedure described in the first paragraph of §20.24 enables us to calculate the time from event R to event S by D's
clock to as high an accuracy as we may wish.

Although Figure 20.22 is drawn in two dimensions, confining D's
motion to the 1D universe, you'll find that the reasoning holds good
even when D's journey uses all three space dimensions. That point will
be important later. (If you're worried about what D's world line will be
like in these conditions, see §20.30.)

Articles 20.24–5 use one of the key processes of the as-accurately-as-
we-wish technique. They merit careful study, particularly the paragraph in which we put the argument formally. Notice there how vital it
is that we're using proportionate error, not just error.

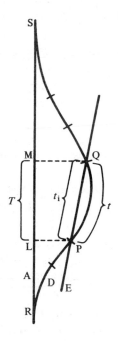

Figure 20.22

20.26 Now we have a recipe for calculating the traveller's time on a
straightforward out-and-back journey. We'll meet a fully worked out
example in the next chapter, and so I hope to be excused for not
doing the calculations for the only case that has, to date, been
tested experimentally.

This concerns a very short journey indeed – submicroscopic, in fact. An atom in a crystal is always vibrating back and forth about a fixed point – performing repeated tiny space trips. So its clock should (we predict) record less time than a stationary one. The higher the temperature, 'the more furiously does the atom vibrate – moving faster, going farther – and so the greater the time difference should be. And we could calculate what relation the theory predicts between temperature and time measured by the atom's clock.

By relativistic standards even a nimble atom is moving slowly. The experimental test would have to cope with time differences around one part in ten million million. But this would be well within the capability of a *nuclear* clock – rather like the atomic clock of §20.9, except that the ticks are successive waves of a gamma-ray given out when certain rearrangements take place in the nucleus (central portion) of an atom. A phenomenon called the Mössbauer effect – which I regret having no space to explain – allows two nuclear clocks to be compared to an accuracy of one part in a thousand million million or even better. And when it was used to test the predicted relation between time measured and temperature, the agreement was excellent (*Physical Review Letters,* **4** (1960) 274). This time the accelerations were around 100 million million times g. So the Additional Assumption (§§20.15, 20.19) is again verified. We shall make it the basis of future developments.

20.27 The conclusion, then, is that the accelerations do not affect the traveller's time measurements. The difference between the times experienced by traveller and stay-at-home arises solely from their relative speed. Yet if there were no accelerations, there could be no relative speed and therefore no difference in times. So the accelerations do play a vital role. **Do you find that puzzling?**

Well reread §§14.20–4. When we are considering Figure 14.22 as a piece of geography – as in §14.23 – the difference in distance arises because there are parts of D's route which go in different directions from A's. To produce these differences of direction there have to be curves. Yet the greater length of D's route does not lie in the curves as such.

Similarly, when we think of Figure 14.22 as a spacetime diagram, the difference between A's and D's times arises because there are stretches in which D has non-zero speed relative to A. To produce the speed differences there have to be accelerations (curves in D's world line). Yet the shortness of D's time measurement compared with A's does not lie in the periods of acceleration (curves) as such. I hope the puzzle has disappeared.

20.28 Besides inertial observer A, there could be any number of travellers (like D, Figure 20.22) who part at R and come together again at S. **Can you suggest a very broad generalisation concerning the time between R and S as measured by A and these various travellers?**

The quantity $T\sqrt{(1 - v^2)}$ which occurs in (20.4)–(20.6) is *equal* to T if $v = 0$. In all other cases it is *less* than T. So when we carry out the procedure described in the first paragraph of §20.24, the resulting approximation to the traveller's total time from R to S will be *less* than A's. And this will remain true no matter how small the portions into which we divide the traveller's world line in making the approximation as close as we wish (§§20.24–5). Calculated to whatever accuracy we demand, the traveller's time is less than A's – the inertial observer's. So we conclude that

> Among all the observers who are present at two specified events, the inertial observer is the one whose clock records the longest time between them.

Or using the language of routes to emphasise the analogy with a familiar piece of geometry,

> Of all the spacetime routes between two events, the inertial route takes the longest time.

And this applies for journeys in all three dimensions of space (see second-to-last paragraph of §20.25).

20.29 This last displays the space-twin effect in all its generality. No matter what journey the traveller undertakes, his clock on return will have recorded less time than the stay-at-home's. **But will he have aged less?**

First reread §4.16. If an 'ageing clock' keeps good time with a wrist watch or atomic clock (as it does in ordinary conditions), then we should expect the travelling twin to age less.

In these ordinary conditions we are non-inertial to a small extent (§5.4). The traveller might at some parts of his journey be inertial (weightless) and at other parts have greater accelerations than our 1g. **Would this make his ageing clock behave differently from his ordinary clocks?**

I'll only make some brief remarks to stimulate your further thoughts:

(1) The ageing clock ultimately depends on the same physical laws of nature as any other clock. So there seems to be no reason why it should behave differently, except that

(2) It is delicately built; if subjected to large accelerations (and probably if subjected for longish times to accelerations significantly differ-

ent from 1g), it stops altogether. But within the limits at which life can go on,

(3) Experience so far has not revealed any gross effect. Cosmonauts subjected to several g for a few minutes or living weightless for some months have not shown obvious signs of accelerated ageing. In any case,

(4) If subjecting a human traveller to greater or smaller accelerations does in some way affect his rate of ageing, then (cf. §20.2) it still can't make both twins' ageing the same for *all* journeys.

(5) But if you're thinking in terms of very long distance space journeys that might sometime become possible, then this problem doesn't arise. **Can you think why?** Try to answer that question before you get to §21.33.

20.30 We've been discussing the affairs of observers who move in all three dimensions of space. But how, you may be wondering, can we draw a world line in four dimensions – three of space and one of time? Well of course we can't. We have to think of 'world line' as a shorthand phrase for 'a complete description of the history of an observer, giving the time and position of every event in his life'. The description may be in words or in mathematical symbols. And when we're concerned with only one or two dimensions of space, we can actually *draw* the world line in two or three dimensions.

In fact, in all the applications that will concern us, we can stick to two spatial dimensions (a plane) plus time. So we could draw space-time diagrams in three dimensions – using, for example, vertically upwards for time and two perpendicular horizontal directions for the space dimensions. This is not very helpful in a book – which could only offer two-dimensional perspective drawings of three-dimensional diagrams. If you have a strong visual imagination, you may find it worthwhile trying to form mental pictures of world lines in this two-plus-one-dimensional diagram. But if you're not particularly good at visualising, you'll probably do better by thinking in 'complete description. . .' terms.

Our reasoning never depends on being able to draw anything properly. The diagram serves simply as an easy-to-read summary of points we need to keep in mind – the meanings of symbols, for example. And so a rather schematic diagram on a sheet of paper is usually good enough. And the logic of our reasoning has to be independent of how adequately or inadequately we can draw.

21

Time as experienced by a constant acceleration traveller

21.1 Special interest attaches to the time measurements of a traveller moving with *constant* acceleration. So we return to our old friend W of Chapter 17, who has constant acceleration f. We'll compare his times with those of inertial observer A as defined in that chapter, and retain the symbols used there (summarised in §17.19), *except that t* will now stand for the time from P to Q by W's clock, as marked in Figure 21.1.

The straight line through O and Q is the distance axis of inertial observer B, whose world line (labelled 'B') is fixed by the equal-angles-with-a-signal-line rule of §8.26. (His axes do not carry the usual labels, since we are now using t for W's time, not for B's.) Then W's

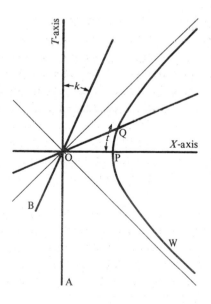

Figure 21.1

instantaneous speed at event Q is equal to B's constant speed (end of §17.5).

21.2 Let k stand for the Doppler factor between A and B (§§15.4–5, 15.8). If the value† of k is given, that fixes B's motion (relative to A) and so fixes the slope of his world line and therefore also the slope of his distance axis, which – finally – fixes the position of Q on W's world line. So event Q can be specified by stating the value of k.

We shall call k the 'parameter' of Q. And in general we define the *parameter* of an event on W's world line to be the Doppler factor between A and the inertial observer whose constant speed is the same as W's instantaneous speed at that event. Our present object is to find a rule for calculating t in terms of k.

21.3 We now take another point S on W's world line (Figure 21.3) and add inertial observer D, whose constant speed is the same as W's instantaneous speed at event S. **Make sure you know how to place D's world line** (cf. B's in §21.1). To avoid clutter the distance axes are omitted.

Let‡ τ stand for the time between Q and S by W's clock, and let κ

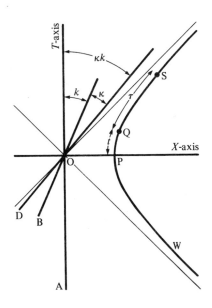

Figure 21.3

† Mathematicians usually speak of the 'value', rather than the 'size' of a number or quantity, but it means the same thing.

‡ τ (tau) and κ (kappa) are the Greek t and k.

stand for the Doppler factor between B and D. Then equation (16.20) (with what changed to what?) gives

Doppler factor between A and D = κk .   (21.1)

So κk is the parameter (§21.2) of event S.

21.4 Figure 21.4 adds a further event R, the time from P to R being τ – the same as from Q to S. (*Lengths* PR and QS are unequal because scale varies with slope). We also put in inertial observer C, whose constant speed is the same as W's instantaneous speed at event R. Then the Doppler factor between A and C is κ – the same as between B and D. **Can you prove that?**

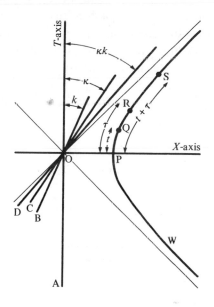

Figure 21.4

Very briefly, W's world line, being an isoval, is related in the same way to all the observers. And A, C, P and R are related to each other in the same way as B, D, Q and S. **Finish the proof for yourself** (perhaps draw another diagram giving B rectangular axes).

So the parameter (§21.2) of R is κ. **Now can you see an interesting relation between the various times and parameters?**

Measuring all times from P, the times t and τ of Q and R are *added* to give the time $t + \tau$ of S. And the parameters k and κ of Q and R are *multiplied* to give the parameter κk of S.

21.5 Since Q and R could be anywhere on W's world line, this can be generalised to say that

If t is the time of the event which has parameter k and τ is
the time of the event which has parameter κ, then $t + \tau$ is the
time of the event which has parameter κk

—where 'time' always means time by W's clock from P to the event in
question. Or putting it more briefly,

Multiplying parameters corresponds to *adding* times.

21.6 **I wonder can you spot the importance of this?**

Suppose we had a table of the values of t that correspond to various values of k between (say) 1 and 2. Then we could find what value
of t goes with $k = 3$. For $3 = 2 \times 1.5$; and since 'multiplying
parameters corresponds to adding times', we deduce that the time for
parameter 3 is the sum of the times for parameters 2 and 1.5. By
extending this method, we could build up a table for as large values
of k as we want, provided we've got a little bit to start from. So our
next job is to work out that little bit. And for this purpose we'll be
using near-equations, so that perhaps you should carefully rework
§§20.12–21.

21.7 Starting again with Figure 21.7, let E be the inertial observer who is
present at P and Q (cf. §20.12), and let t_i be the time from P to Q by

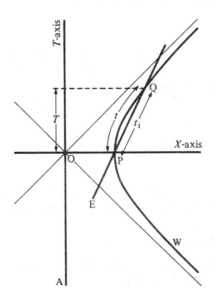

Figure 21.7

E's clock. If v stands for the speed of E relative to A, the dilation of time (§13.7) gives

$$t_i = T\sqrt{(1 - v^2)} \, . \tag{21.2}$$

By taking t small enough (giving W very little time to accelerate) we can arrange for v to be as small as we wish, and so for $\sqrt{(1 - v^2)}$ to be as near to 1 as we wish. So

$$t_i \sim T \, . \tag{21.3}$$

21.8m **Do you feel certain that we've really established the truth of (21.3)?**
To prove it, we go back to the definition (§20.20). The ratio of the sides is t_i/T, and by (21.2) this is equal to $\sqrt{(1 - v^2)}$ – which we *can* make as near to 1 as we wish by taking t small enough. So (21.3) conforms to the definition.

21.9 By the Additional Assumption (§20.19), $t \sim t_i$. And so, using (21.3),

$$t \sim T \, . \tag{21.4}$$

In other words, the smaller the time-interval from P to Q, the nearer do A and W come to agreeing about it – as you'd expect. From (21.4) we deduce that

$$ft \sim fT \, . \tag{21.5}$$

1.10m Two steps in that last article need justification. In deriving (21.4) we used a proposition that could be stated more generally as

If $Y \sim Z$ and $Z \sim U$, then $Y \sim U$.

Make sure you understand what this means (using §20.20); and think about its resemblance to a more familiar proposition that you'd get by changing $\sim$ to $=$. **Now can you prove it?**
The two near-equations that we start with say that by taking the specified quantity small enough, we can make Y/Z and Z/U as near to 1 as we wish. So we can also make their product – which works out at Y/U – as near to 1 as we wish, thus justifying the conclusion that $Y \sim U$.

Most of the manipulations we use for equations apply also to near-equations. This is a case in point. But there are exceptions, and so we must keep checking.

1.11m In deriving (21.5) from (21.4) we assumed that

Multiplying both sides of a near-equation by the same amount yields another near-equation.

Is that true?
Multiplying both sides by the same thing does not alter their ratio.

Need I say more? And the same obviously applies to dividing both sides.

21.12 As usual, X stands for the distance of Q from A according to A. By taking t small enough (Q close enough to P) we can make X as near as we wish to the distance OP, which is $1/f$ (§17.19). So

$$X \sim \frac{1}{f} . \tag{21.6}$$

(See §21.13 for justification.) Hence (using §21.11)

$$fX \sim 1 . \tag{21.7}$$

Adding corresponding sides of (17.13) and (17.14) gives†

$$fX + fT = \frac{1 + v}{\sqrt{(1 - v^2)}} .$$

Then (16.13) gives

$$fX + fT = k , \tag{21.8}$$

and so (§3.19)

$$fT = k - fX , \tag{21.9}$$

and by (21.7)

$$fT \sim k - 1 . \tag{21.10}$$

Finally, by (21.5)

$$ft \sim k - 1 . \tag{21.11}$$

Make sure you're clear what this means (§20.20).

21.13m As we make t smaller, the two sides X and $1/f$ of (21.6) do not get closer and closer to zero (as happened in our previous near-equations). So if the actual difference between them can be made as small as we wish, the same will be true of the *proportionate* difference. That is why I was able to assert (21.6) on the basis of the two preceding sentences, without checking the usual details.

21.14m **Using the definition (§20.20) can you justify the step in which we derived (21.10) from (21.7) and (21.9)?**

If we had a true equation $fX = 1$, instead of (21.7), then we could change fX to 1 in (21.9), getting $fT = k - 1$ instead of (21.10). It seems fairly obvious that this will also work with near-equations. But we'd better make sure. Using (21.9) to substitute (§9.16) for fT, the ratio of the two sides of (21.10) becomes $(k - fX)/(k - 1)$. The nearer

---

† Adding two fractions with the same denominator (§16.22) is just a matter of adding their numerators and keeping the same denominator. I'm sure you can see why.

fX gets to 1, the nearer does $k - fX$ get to $k - 1$, and so the nearer does the ratio get to 1. Thus (21.10) does comply with the definition.

21.15 The relation of κ to τ (§21.3) is exactly the same as the relation of k to t (check that you can get from one to the other by merely changing the names of a few quantities and events). So, analogously to (21.11), we have

$$f\tau \sim \kappa - 1 .\qquad\qquad (21.12)$$

Combining this with what we proved in §21.5 will give us the machinery for starting that table of t against k that we are seeking (§21.6).

21.16 First let's take the set-up illustrated in Figure 21.3 and change the names of some of the quantities, as shown in Figure 21.16. The time and parameter of Q will be called t_0 and k_0 instead of t and k. Similarly t_1 and k_1 will stand for the time and parameter of S (with τ and κ unchanged in meaning). We make these changes in order to stress what happens when we take the 'step' (as we shall often call it) from event Q to event S. Subscript $_0$ refers to any quantity at the beginning of the step, subscript $_1$ to the same quantity at the end. And we seek a near-equation relating $t_1 - t_0$ to $k_1 - k_0$ – relating the change in time (by W's clock) to the change in parameter.

Each parameter is the Doppler factor between the appropriate

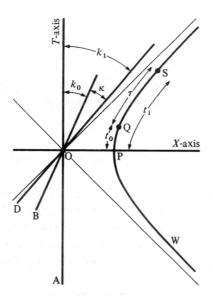

Figure 21.16

pair of observers, as marked in Figure 21.16. So (21.1), appropriately amended, gives $k_1 = \kappa k_0$, and therefore $\kappa = k_1/k_0$. And obviously $\tau = t_1 - t_0$. Substituting these in (21.12) gives

$$f(t_1 - t_0) \sim \frac{k_1}{k_0} - 1 \ .$$

This is better written in the form

$$ft_1 - ft_0 \sim \frac{k_1 - k_0}{k_0} \tag{21.13}$$

(using §16.12; and $k_1/k_0 - 1 = k_1/k_0 - k_0/k_0 = (k_1 - k_0)/k_0$ – see footnote to §21.12). Note that τ and κ have disappeared (imagine them deleted from Figure 21.16). And so we have reached our immediate objective: a near-equation showing how the change in W's time is related to the change in the parameter over the step from Q to S.

The 'specified quantity' (which has to be made small enough, §20.20) started off as t (§21.7). In (21.12) it became τ, which now appears as $t_1 - t_0$. But (21.13) shows that as we make $t_1 - t_0$ closer to zero, we do the same for $k_1 - k_0$. So we can now take $k_1 - k_0$ as the specified quantity. And then (21.13) says (check this with §20.20) that

> By taking $k_1 - k_0$ small enough, the proportionate difference between $ft_1 - ft_0$ and $(k_1 - k_0)/k_0$ can be made as small as we wish

– and taking $k_1 - k_0$ small enough means taking a short enough 'step' from Q to S. (In Figure 21.16 this step is drawn quite big; but see §20.18).

21.17 Near-equation (21.13) is the climax of this chapter. Keep coming back to it as we develop its consequences. And now would be a good time to consolidate. You might find it helpful to follow a strategy like the one outlined in §§9.20–1. The main blocks this time are §§21.1–2 (the set-up), 21.3–6 (proving the proposition of 21.5), 21.7–15 (the toughest, proving that $f\tau \sim \kappa - 1$) and 21.16 (gathering it all together).

21.18 Let us use (21.13) to calculate the first few lines of a table of ft against k. We start from P, where $ft_0 = 0$ and $k_0 = 1$ (since W is stationary relative to A, §15.10), and take the first step from there to (say) $k_1 = 1 \cdot 01$. Then (21.13) gives

$$ft_1 - 0 \approx \frac{1 \cdot 01 - 1}{1} = 0 \cdot 01 \ .$$

And so at the end of the step, $ft_1 \approx 0 \cdot 01$.

Note the switch from $\sim$ to $\approx$. Once we've fixed the size of step, we can no longer speak of taking $k_1 - k_0$ small enough. Our calculation now involves a fixed error.

The starting values k_0 and ft_0 of the second step are the values we've just calculated for the end of the first step. So the second step starts from $k_0 = 1.01$, $ft_0 \approx 0.01$, and ends with $k_1 = 1.02$. Then (21.13) gives

$$ft_1 - 0.01 \approx \frac{1.02 - 1.01}{1.01} = 0.0099 ,$$

whence $ft_1 \approx 0.0199$. And the table so far reads

k	ft
1.00	0.0000
1.01	0.0100
1.02	0.0199

21.19 Threatened by a tedium of calculation, we'd better investigate lab-our-saving devices. If we use larger steps, there'll be less work; but the results will be less accurate. So let's examine how step size affects accuracy. Consider calculating the change in ft between $k = 1.2$ and $k = 1.25$ by using (a) five steps of 0.01, and (b) a single step of 0.05. Process (a) will give the increase over the whole interval as

$$\frac{0.01}{1.2} + \frac{0.01}{1.21} + \frac{0.01}{1.22} + \frac{0.01}{1.23} + \frac{0.01}{1.24} = 0.04099 \qquad (21.14)$$

−each term arising from the application of (21.13) to a single step (**please check details**). The (b) calculation gives

$$\frac{1.25 - 1.2}{1.2} . \qquad (21.15)$$

This is equivalent to repeating the *first* term of (21.14) five times, and is therefore too large − it comes to 0.04167.

Calculation (21.15) is biassed by using a denominator that refers only to the start of the step, whereas the denominators of (21.14) are more evenly spread. **Any ideas for improving the one-step calculation?**

Reduce the bias by using as denominator, not k_0, but the value of k in the middle of the step, which is $\frac{1}{2}(k_1 + k_0)$ (cf. §16.1). Then the one-step calculation would be

$$\frac{1.25 - 1.2}{\frac{1}{2}(1.25 + 1.2)} = 0.04082$$

−much nearer than (21.15) to the (21.14) result which we know to be superior.

21.20 So we can use bigger, and therefore fewer, steps if in place of (21.13) we use the near-equation

$$ft_1 - ft_0 \sim \frac{k_1 - k_0}{\frac{1}{2}(k_1 + k_0)} \qquad (21.16)$$

to make the calculation. This last will lead us to the same results as (21.13), but requires less work to reach the same accuracy.

21.21 Let's set out the calculation in tabular form, assuming provisionally that steps of 0·1 in k will be good enough. The first few lines are given in Table 21.21.

Table 21.21

k	$\frac{1}{2}(k_1 + k_0)$	$ft_1 - ft_0$	ft	ft to 2 decimal places
1·0			0·0000	0·00
	1·05	0·0952		
1·1			0·0952	0·10
	1·15	0·0870		
1·2			0·1822	0·18
	1·25	0·0800		
1·3			0·2622	0·26
	1·35	0·0741		
1·4			0·3363	0·34
	1·45	0·0690		
1·5			0·4053	0·41

Note that the second and third columns, with their entries set in the half-line positions, refer to quantities involving subscripts: k_0, k_1, ft_0 and ft_1. Here the subscript $_0$ refers to the line above, and subscript $_1$ to the line below. For example, $\frac{1}{2}(k_1 + k_0) = 1\cdot15$ in line $2\frac{1}{2}$ is the average of $k_0 = 1\cdot1$ in line 2 and $k_1 = 1\cdot2$ in line 3.

Since our step $k_1 - k_0$ is always 0·1, (21.16) tells us that we get the third column entry for $ft_1 - ft_0$ by simply dividing 0·1 by the value of $\frac{1}{2}(k_1 + k_0)$ in the same line of the second column.

This value of $ft_1 - ft_0$ is the difference between the ft entries (fourth column) in the lines immediately above and below. So each ft is obtained from the one above by adding the entry for $ft_1 - ft_0$ (third column) which lies halfway between. And in the last column this value of ft is rounded off to two decimal places (to the *nearest* two-decimal-place number).

Now please make sure that you understand how this table is constructed by using (21.16) – in other words, **go through the calculations in detail.**

21.22 But do steps of 0·1 give sufficient accuracy? We can check that by repeating the calculation in smaller steps and seeing what difference that makes. Let's compare the above result for $k = 1\cdot1$ with what we get by using two steps of 0·05. For the first, $k_0 = 1$, $k_1 = 1·05$ and $ft_0 = 0$; and **you can check that** (21.16) gives $ft_1 \approx 0·0488$. And the second step has $k_0 = 1·05$, $k_1 = 1·1$ and $ft_0 \approx 0·0488$, so that $ft_1 - 0·0488 = 0·0465$, giving 0·0953 as the value of ft corresponding to $k = 1·1$.

Now the calculation in two small steps is more accurate – that's what the ~ sign says. And yet there's only a trivial difference. So it seems extremely likely (though a highbrow mathematician would demand stricter proof) that even if we worked with very much smaller steps, the result would not be greatly altered. At the worst perhaps that fourth decimal place would be changed a little more. And as we're ending with a *two*-decimal-place table, surely we have all the accuracy we want. (How about *you* doing a similar check on another step?)

We must, however, continue to work to four decimal places in columns three and four – otherwise we may introduce what are called 'rounding off errors'. For example, if we had reduced the first two entries in column three to 0·10 and 0·09 and then added, we should have got $ft = 0·19$ in the third row, and that 0·01 error would be carried on right through the table.

21.23 Now will you please **extend Table 21.21 as far as** $k = 2·5$. Keep your workings – we'll use them again. Your final results should be as in Table 21.23

Table 21.23

k	ft	k	ft	k	ft
1·0	0·00	1·6	0·47	2·1	0·74
1·1	0·10	1·7	0·53	2·2	0·79
1·2	0·18	1·8	0·59	2·3	0·83
1·3	0·26	1·9	0·64	2·4	0·88
1·4	0·34	2·0	0·69	2·5	0·92
1·5	0·41				

21.24 Let's do some space travel calculations, choosing the (for us) particularly interesting almost-1g traveller that we get by putting $f = 1$ and taking the time unit as 1 year (§§17.20–1). Let A_1 (Figure 21.24) be the inertial observer who is stationary relative to A and present at event P (cf. §17.29). **Slot please!** We compare A_1's and W's opinions about the journey from event P (where W takes off) to event Q whose parameter is k.

Combining (17.14) and (16.16) and putting $f = 1$, we find that

$$\text{Time taken according to } A_1 \text{ (or A)} = T = \frac{1}{2}\left(k - \frac{1}{k}\right).$$

Similarly (17.8), (17.13) and (16.15) give

$$\text{Distance travelled according to } A_1 \text{ (or A)} = X - 1 = \frac{1}{2}\left(k + \frac{1}{k}\right) - 1.$$

Next (16.18) or (17.2) gives W's speed relative to A_1, at the end of the journey. And finally the time taken according to traveller W is given by Table 21.23, with $f = 1$.

21.25 Table 21.25 gives the figures for three sample journeys, each defined by the value of k at the end. The first column's trip of almost 800 000 million kilometres in about 5 months just begins to show a difference between A_1's and W's times at this accuracy. For the quarter-light-year journey of column 2, the difference has grown to about 22 days in 9 months – about 8 per cent. **Interpret the last column for yourself and calculate another journey or two.**

Table 21.25

k	1·5	2·0	2·5
Time for journey by W's clock (years)	0·41	0·69	0·92
Time for journey according to A_1 (years)	0·42	0·75	1·05
Distance travelled according to A_1 (light years)	0·083	0·25	0·45
Terminal speed relative to A_1 (fraction of speed of light)	0·38	0·60	0·72

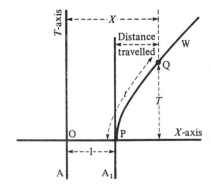

Figure 21.24

21.26 To arrange for W to do a round trip, we have merely to reverse the direction of his rocket thrust on two occasions, Q and S (Figure 21.26 – slot!). He accelerates towards the right between P and Q and between S and U, and towards the left between Q and S. Clearly the farthest distance reached (at R) is twice the 'Distance travelled' of Table 21.25; and the times are quadruple the ones in that table – giving results set out in Table 21.26.

After making that last journey a dozen times the traveller would be thinking about changing his or her spouse for a younger one. For trips that are biologically simple (§17.20) and cosmically modest (if only the engineers would make the rockets!), we are already beginning to find quite sizeable time differences. And *you ain't seen nothing yet!*

Table 21.26

Distance of R according to A_1 (light years)	0·17	0·50	0·90
Round-trip time by A_1's clock (years)	1·68	3·00	4·20
Round-trip time by W's clock (years)	1·64	2·76	3·68

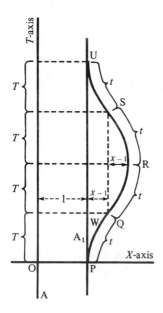

Figure 21.26

21.27 We can extend Table 21.23 by using the proposition of §21.5. For example, $k = 1.5$ corresponds to $ft = 0.41$ and $k = 2$ to $ft = 0.69$. So $k = 1.5 \times 2 = 3.0$ will correspond to $ft = 0.41 + 0.69 = 1.10$. But using the 2-decimal-place Table 21.23 would risk the dangers mentioned at the end of §21.22. **So repeat that calculation using the 4 decimal places of Table 21.21 and your extension of it.**

You get $ft = 0.4053 + 0.6929 = 1.0982$ – which *does* round off to 1.10. **Now please do similar calculations for $k = 4, 5, 6, 7 (= 5 \times 1.4)$, 8, 9 and** 10. After rounding off, you should get Table 21.27.

Table 21.27

k	ft		k	ft
3	1·10		7	1·95
4	1·39		8	2·08
5	1·61		9	2·20
6	1·79		10	2·30

21.28 We can calculate ft for higher values of k by using the multiply-parameters/add-times rule. Thus if $k = 230 = 2.3 \times 10 \times 10$, $ft = 0.83 + 2.30 + 2.30 = 5.43$. (We'll have to accept round-off errors from now on.)

21.29 If k is not much more than 1, (21.11) says that it will be good enough to take $ft = k - 1$. And for our two-decimal-place calculations this approximation will serve up to $k = 1.1$ – as the $k = 1.1$ entry of Table 21.23 shows.

That puts us in a position to calculate ft when k lies between the tabulated values. For example, we find by successive divisions that $3.73 = 3 \times 1.24 = 3 \times 1.2 \times 1.04$, so that the corresponding ft is $1.10 + 0.18 + 0.04 = 1.32$.

And now, if you're given *any* value of k, a suitable combination of these various devices will enable you to calculate the corresponding ft to as good an accuracy as we shall want in this book.

21.30 So let's continue our calculations about almost-1g ($f = 1$) space journeys – the straightforward outward journey of §§21.24–5. This time let's assume that we're told the distance (according to A_1) that W travels between events P and Q. By (17.8), adding 1 to this distance gives X. And from (17.9) we have $T^2 = X^2 - 1$, whence (§13.4)

$$T = \sqrt{(X^2 - 1)} . \tag{21.17}$$

Then (21.8), with $f = 1$, gives

$$k = X + T.$$ (21.18)

From k we can find t by using the tables and tricks of the last few pages. And W's terminal speed is given by (17.2).

Check the details on the following example. Distance travelled = 1 light year. So $X = 2$, and $X^2 - 1 = 3$. Thus $T = \sqrt{3} = 1{\cdot}73$, and $k = 3{\cdot}73$. Hence $t = 1{\cdot}32$ (§21.29, with $f = 1$). And terminal speed = $1{\cdot}73/2 = 0{\cdot}87$.

The upper part of Table 21.30 gives the results of such calculations for journeys of different lengths, with the first column referring to the case of the previous paragraph. **Check up one or two more for yourself. Then see what conclusions you can draw,** and read §21.31 before considering the bottom of the table.

21.31 You'll have noticed that the time discrepancy grows much faster than journey length.

To get the astronomical perspective right, note that shortly before ending the journey of column 3, W would (if aimed correctly) dash past Sirius, the brightest star in the sky. The time difference here is striking enough – 3 years against 10. But in the last column's journey, only 11 times as far, it's a question of 5 years 3½ months against a century. (One problem that might be troubling you can be solved by a glance at §3.38.)

21.32 The round-trip statistics in the last three lines of Table 21.30 are calculated as in §21.26. Remember that this is not just a matter of one observer's opinion of another's clock. Stay-at-home and traveller can actually compare clocks as they part, and again as they reunite. And that makes the table distinctly startling.

A journey to the nearest known star (other than the Sun) would be a little longer than the one described by column 2 – giving the traveller a four-year advantage over her Earth-bound friends. The next column, in which the Earth-folk grow 40 years older while the traveller ages only 12 years, is just what was intended by the love-lorn swain of §4.20. More shocking still, the last column's traveller, after passing 21 years in space, comes back in time to marry his original girl friend's great-great-great-great-great-great-great-great-great-great-granddaughter.

21.33 That last paragraph assumes that these calculations will apply to biological ageing as well as to physical clocks. **Is this justified?**

Better reread §20.29. Our calculations apply to times measured by any clock which is known by experiment to behave just as Relativity Theory predicts – the atomic and nuclear clocks of §§20.9 and 20.26, for example. Such clocks are not affected by acceleration – by the physical state of being non-inertial. But we can still wonder about the effects on an 'ageing clock' – even though we know, from §20.29(4), that this can't produce equal ageing in *all* cases. **Have you answered the challenge at the end of §20.29?**

The 1g acceleration cuts the Gordian knot. Traveller and stay-at-home are equally non-inertial. Any acceleration effect will apply equally to both. Now we know that in terrestrial conditions an ageing clock keeps time with an ordinary standard clock (at any rate to a fair approximation). So the same will apply to the traveller's clock. Thus the figures of Table 21.30 *can* be interpreted in terms of ages.

True, there are four occasions – P, Q, S and U of Figure 21.26 – when W's physical conditions differ significantly from A_1's. But these cannot make any great difference – see §20.29(3). There really is no escaping the conclusion that the differential ageing will be real, if and when we acquire the rocket power.

21.34 In the last column of Table 21.30, T is effectively equal to X. And (21.17) shows that this will always be so if X is large. Also, taking distance travelled to be equal to X (instead of $X - 1$) would introduce negligible error. So we may as well take T = distance travelled, and k = 2 × (distance travelled) by (21.18). With these simplifications we produce Table 21.34.

21.35 Column two makes a very generous estimate for a trip across our Galaxy and back – taking a longish working life of 46 years for the traveller, but bringing him back to an Earth where 200 000 years have passed since he left. With the fourth column, the round trip takes us out to the famous spiral galaxy in Andromeda (M31), the nearest spiral to our own. It might be rather discouraging to come back to an Earth where your friends had all died 4 million years earlier. But the one-way migration of 29 years would be worthwhile if you could be sure of a good house at reasonable rent on arrival.

The last journey tabulated goes almost as far as our present radio telescopes can reach. And the return journey would *certainly* not be worthwhile – for there would be no Earth to come back to. But the one-way trip could be done in time to go into retirement on a planet in a distant galaxy. A 1g traveller could certainly get around!

Note that the one-way journey to the farthest known galaxies takes the same time, from the traveller's point of view, as the rather local journey across our own Galaxy and back. By contrast, the former takes 50 000 times longer than the latter according to an Earth-dweller.

The rocketry for these journeys – even for those of Table 21.25 – seems utterly impossible from the standpoint of present knowledge.

Table 21.30

One-way journey, always accelerating outwards

Distance according to A_1 (light years)	1	2	9	99
Time according to A_1 (years)	1·73	2·83	9·95	99·995
k	3·73	5·83	19·95	200
Time by W's clock (years)	1·32	1·77	2·99	5·29
Terminal speed relative to A_1	0·87	0·94	0·995	0·999 95

Round trip

Furthest distance according to A_1	2	4	18	198
Time by A_1's clock (years)	6·9	11·3	39·8	400
Time by W's clock (years)	5·3	7·1	12·0	21·2

Table 21.34

One-way journey, always accelerating outwards

Distance according to A_1 (light years) Time according to A_1 (years)	500	50 000	500 000	1 000 000	5 000 000 000
k	1000	100 000	1 000 000	2 000 000	10 000 000 000
Time by W's clock (years)	6·90	11·5	13·8	14·5	23·0

Round trip

Farthest distance according to A_1 (light years)	1000	100 000	1 000 000	2 000 000	10 000 000 000
Time by A_1's clock (years)	2000	200 000	2 000 000	4 000 000	20 000 000 000
Time by W's clock (years)	27·6	46	55	58	92

But when you consider the things we can do today that would have seemed impossible in 1890, can you categorically assert that these journeys will never be made?

21.36 Ignoring column four of Table 21.34, **can you spot a very simple rule for calculating the round-trip time by W's clock?**

> If the furthest distance reached is given (in light years) by a number which is written as 1 followed by three or more zeros, then the time (in years) by the 1g traveller's clock is 9·2 multiplied by the number of zeros.

Multiplying the journey length by 10 merely *adds* 9·2 years to the time. And for comparison,

> The time (in years) by the stay-at-home's clock is twice the furthest distance (in light years).

You should have no trouble in **proving these rules** (see §21.28 for the former; for the latter think about W's speed).

21.37 The calculations for very long journeys should be taken with several large pinches of salt. For here Relativity gets mixed up with Cosmology. The Expanding Universe (or any alternative explanation of the cosmological red shift) could upset our time and distance measurements. But so far as that consideration is concerned, our calculations should be pretty good up to the penultimate column of Table 21.34, and maybe well beyond.

A deeper doubt arises because the experiments verifying Relativity Theory have lasted, at the most, a few days (§20.9). There are strong arguments suggesting that we can still apply the theory to a few years, a few decades, a few centuries . . . I don't know how long. But there could well be laws of nature, unknown to us because they have negligible effects over small time spans (just as relativistic effects are unobservable at low speeds), which would entirely upset our calculations for some of the longer journeys of Table 21.34. The interesting thing, however, is that it's the stay-at-home's measurements we are now doubting. Our calculations for the traveller's few decades stand a good chance of being right.

21.38m This work on the traveller's time measurements is not mere science fiction. The calculations about very long journeys *are* fiction, of course – unless they are accurate prophecy. But now that we know how time goes for the traveller, we can start asking questions about how spacetime appears to an accelerated *observer*. And that is the basis of the *General* Theory of Relativity.

The king-pin is near-equation (21.13). Let's study it further. The quantity $k_1 - k_0$ is simply the amount by which k increases between Q and S (Figure 21.16). We call it the *increment in k*. Similarly, $ft_1 - ft_0$ is the increment in *ft*. So (21.13) is a recipe for calculating (approximately) the increment in *ft*, when the increment in k is known. You may find it easier to remember it in the form

$$\text{increment in } ft \sim \frac{\text{increment in } k}{k \text{ at beginning of step}}. \tag{21.19}$$

Rephrase the indented passage of §21.16 in terms of increments.

1.39m It's useful to have a symbol for 'the increment in'. We use δ – the Greek d, called 'delta'.

Thus δk means 'the increment in k'; and δx means 'the increment in x' (whatever x may stand for).

1.40M I'm sure you won't let yourself get misled into thinking of δk as standing for some quantity δ multiplied by some quantity k. The symbol δ does not represent a quantity, and has no meaning at all except when attached to a quantity-symbol to signify 'the increment in . . .'. And when it is so attached, you must think of δk or δx as a *single symbol* for the increment in the quantity concerned.

Good reading habits (as always) are the way to avoid confusion. Even if you have found (as you probably have by this time) that you can safely read t as 'tee' instead of 'the time by W's clock' (yet keeping that concept at the back of your mind), I would still counsel you to read δt as 'the increment in t' – *not* as 'delta tee'. (And of course you can still say 'the increment in the time by W's clock', if that suits you better.)

Brackets are used along with δ in such cases as $\delta(y + z)$ and $\delta(az)$ to mean the increment in $y + z$ and in az.

1.41m With this notation (21.13) takes the neater form

$$\delta(ft) \sim \frac{\delta k}{k}. \tag{21.20}$$

You may find it helpful to use (21.19) as a transition from (21.13) to (21.20).

Two *important conventions* are involved here:

(i) In any such near-equation, $\delta(ft)$ means the increment in *ft* which *corresponds to* the increment δk in k – that is, the amount by which *ft* increases when k is increased by amount δk; and

(ii) The symbol k itself, when not qualified with a δ (the denominator on the right, for instance) refers to k *at the start of the step* – what we formerly called k_0.

These two conventions *always* apply when we have equations or near-equations connecting increments written in the δ form. **Using them, check that (21.20) is a correct translation of (21.13).**

Now translate into the δ notation the indented passage of §21.16. I'll not insult you with a 'crib' answer.

21.42m The definition of the meaning of '∼', as we gave it in §20.20 had an obviously unsatisfactory aspect – the need to specify every time the quantity that is to be made small enough. But when we're dealing with near-equations involving increments, the difficulty vanishes. It's the increments themselves that must be made small enough. When there are only two – as in (21.20) – then we have to make *one* of them small enough; the other will automatically become small as we do so. So we can add a supplement to the definition of §20.20:

> When the near-equation connects two variable quantities and their increments, the 'specified quantity' that has to be made small enough is one of the increments.

21.43m If your mathematical education went far enough, you will have recognised this stuff about near-equations connecting increments as roughly equivalent to the branch of mathematics known as 'the calculus'. If you do *not* know the calculus, I'm pretty sure you'll find the methods of this book easier than the more orthodox treatment (at least for our purposes).

On the other hand, if you *do* know that subject, you *really must resist the temptation* to translate what I write into orthodox calculus terms. The mixture can lead to confusion and ultimate incomprehensibility. So throw away your calculus and start again.

Incidentally, if your former calculus education went far enough, you've probably recognised that *ft* is the natural logarithm of *k* – that follows from (21.20). We've been calculating for ourselves the tables that you formerly took trustingly from a book.

22

Time and distance measurements of a constant acceleration observer

22.1 Now that we know how W's clock behaves, we can think of him as carrying out all the observations that an inertial observer might make (§5.20). So we can now give him the full status of an *observer with constant acceleration*. In particular, we can investigate his opinions about the time and distance of an over-there event. In preparation, please do any necessary revision of Chapter 7.

22.2 So we consider again traveller W with constant acceleration f (Figure 22.2), related in the same way as before to inertial observer A, P being the event of W's closest approach to A. And we seek W's opinions of some over-there event, Q.

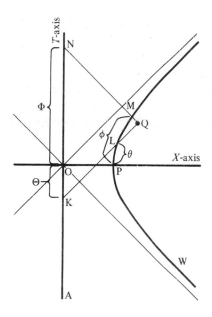

Figure 22.2

Let L and M be the events of W sending a light signal to event Q and receiving one back from Q (**slot please!**). Then the *only* observations that W can make concerning Q are the times of L and M. **Make sure you agree** (cf. §7.3).

22.3 **So what can we mean by 'the time of event Q' and 'the distance of Q from W' – both according to W?**

The situation is just like that of Chapter 7. The reasoning of §§7.9 and 7.13 would lead to the definitions:

> The time of event Q according to W is halfway between the times of events L and M by W's clock; and
>
> The distance of Q from W according to W (in natural units) is half of the time by W's clock between L and M.

22.4 Naturally you have doubts. Some will be dispelled by rereading §§2.15–18 and 7.4–8. Or you might object that W *knows* (by the particle test) that he is non-inertial, and must therefore make corrections to bring his ideas into line with those of an inertial observer. **But** *which* **inertial observer?**

Since W's world line is an isoval, he is related in the same way to *all* inertial observers with world lines through O (and any other agrees with one of these). There is no way of choosing a particular one as standard.

22.5 Here's a more plausible objection. The reasoning of Chapter 7 depended fundamentally on the assumption that the speed of light is invariant (§1.22). Acceleration being an absolute (§17.3), can we be sure that this invariance will still hold good for an accelerated observer? Might he find, for example, that the speed of light is smaller in the direction of his acceleration than in the opposite direction? We'd better hedge our bets. We can do so by the simple process of agreeing that 'time' and 'distance' according to W are merely *names that we've decided to use* for the quantities defined in §22.3 – names reminding us that these quantities are at least akin to time and distance as normally conceived. That leaves open the question of whether these are or are not exactly the quantities that we should obtain by some particular practical measurement.

In the long run it doesn't matter. Any experimental test of the theory must refer ultimately to quantities that can be measured on the spot. The ones that we have just agreed to call time and distance will only be intermediaries in the process of calculating a prediction – they will disappear from the prediction itself. So it doesn't matter whether they are 'true' time and distance or not.

22.6 **Can you prove the following proposition?**

> Events that W regards as happening at the same time are
> represented by points of a straight line through O.

Consider the case where Q is on A's distance axis. Then L is as far below that axis as M is above it. So the tick of W's clock lying halfway between L and M is P. Thus (§22.3) Q happens, according to W, at the same time as P – and that applies with Q anywhere on the X-axis (its positive half). Furthermore, this will still hold good if we redraw the diagram taking A's distance axis as *any* line through O less steep than 45°. (Compare §8.12.) If you are strong in mathematics, you might also prove that

> Events that W regards as happening at the same distance are
> represented by the points of an isoval centred on O (§13.23
> will help).

22.7 By analogy with §15.2, we use θ to stand for the time at which W sends or receives a positive-going signal which is present at Q; and ϕ for the time at which he sends or receives a negative-going signal that is present at Q. Strictly, we ought to alter the distance definition of §22.3 along the lines of §16.2. But we'll not bother, since we shall always draw our diagrams with Q on the positive side of W, so that θ and ϕ are the times of L and M, respectively (Figure 22.2).

We'll use capital letters, as usual, for A's quantities, and the corresponding small letters for W's. So t and x will stand for the time and distance of Q according to W. **Write down the equations that express t and x in terms of θ and ϕ.**

Starting from §22.3, reasoning like that of §§16.1–2 gives

$$t = \tfrac{1}{2}(\phi + \theta) , \tag{22.1}$$

and

$$x = \tfrac{1}{2}(\phi - \theta) . \tag{22.2}$$

22.8 Let K and N (Figure 22.2) be the events at which A sends a signal to Q and receives one from Q. Then his radar co-ordinates (§§15.1–2) are the times, Θ and Φ, of K and N by A's clock.

If W is to be able to make a full set of observations on Q, it follows from §17.28 that Q must be in the right quadrant (Figure 10.1). So Φ will always be positive and Θ always negative (§6.23).

We want to find a relationship between W's and A's measurements. But we only know how to handle W's clock times in the form of increments – using (21.20). So more research in the mathematics of increments is indicated.

22.9m We prove that

> If two variable numbers or quantities are always equal, then the corresponding increments are equal.

Or in symbols,

$$\text{if } Y = Z , \tag{22.3}$$
$$\text{then } \delta Y = \delta Z . \tag{22.4}$$

In fact, by *corresponding* increments we mean that when Y changes to $Y + \delta Y$, then Z changes to $Z + \delta Z$, and these altered values still make (22.3) true. That is

$$Y + \delta Y = Z + \delta Z .$$

From each side of this subtract the corresponding side of (22.3), and we get (22.4) – which proves the proposition.

Practically everything we need for manipulating increments is like that – very short and simple, but needing careful thought to make sure your intuition is not deceiving you.

22.10m We shall also use the proposition that

$$\delta(u + v) = \delta u + \delta v . \tag{22.5}$$

(You won't want me to keep repeating 'If u and v are variable quantities, then. . .'; take that as read in future.) This states that the increment in the sum of two quantities is equal to the sum of their increments. Intuitively obvious? But one can be misled; so let's have a proof.

If we increase u to $u + \delta u$, and v to $v + \delta v$, then we increase $u + v$ to $u + \delta u + v + \delta v = u + v + \delta u + \delta v$.

But $\delta(u + v)$ is the amount by which this last exceeds $u + v$ – i.e., it is $\delta u + \delta v$. Similarly

$$\delta(u - v) = \delta u - \delta v . \tag{22.6}$$

See if you can prove that if c stands for a *constant* number or quantity, then

$$\delta(u + c) = \delta u . \tag{22.7}$$

The increment of a constant is clearly 0. And so (22.5) gives $\delta(u + c) = \delta u + \delta c = \delta u + 0 = \delta u$.

22.11m If c is a constant number or quantity, but z is variable, then

$$\delta(cz) = c\delta z . \tag{22.8}$$

For if z increases to $z + \delta z$, then cz increases to $c(z + \delta z) = cz + c\delta z$ (§16.12). So cz has increased by $c\delta z$, which is thus the required increment. *But note well:* (22.8) is false if c is variable.

.12m We shall want to write 'the increment in $\frac{1}{2}(\phi + \theta)$'. To indicate that it is the *whole* of this expression whose increment we mean, we shall have to enclose it in brackets (cf. §§16.6, 21.40). A second pair of (. . .) would be confusing. So we use 'curly brackets'; and $\delta\{\frac{1}{2}(\phi + \theta)\}$ means the increment in $\frac{1}{2}(\phi + \theta)$. Curly brackets are always used in this way to cope with a brackets-within-brackets situation.

22.13 **Can you work out how to express the increments in t and x in terms of the increments in θ and ϕ?**

Applying §22.9 to equation (22.1), we get

$$\delta t = \delta\{\frac{1}{2}(\phi + \theta)\}^\bullet$$
$$= \frac{1}{2}\delta(\phi + \theta) ,$$

by (22.8). Then (22.5) gives

$$\delta t = \frac{1}{2}(\delta\phi + \delta\theta) . \tag{22.9}$$

Similarly (**you prove**)

$$\delta x = \frac{1}{2}(\delta\phi - \delta\theta) . \tag{22.10}$$

Starting from (16.1) and (16.2), we get the corresponding equations for A, namely

$$\delta T = \frac{1}{2}(\delta\Phi + \delta\Theta) , \tag{22.11}$$

and

$$\delta X = \frac{1}{2}(\delta\Phi - \delta\Theta) . \tag{22.12}$$

2.14 Figure 22.14 (less complicated than it looks) should clarify the significance of these equations. Concentrate first on event Q. The figure marks each observer's radar co-ordinates – Θ, Φ, θ and ϕ – along his world line. (Please check details.) Also shown are A's time and distance co-ordinates, T and X; but we can't do the same for t and x, unfortunately.

Next we bring in another event Q'. (Read it as 'Q dashed'; this is a common way of naming a close relative, as it were, of something you already have a symbol for.) And K', L', M' and N' are related to Q' exactly as K, L, M and N are related to Q (§§22.2, 22.8). We are concerned with the increments in the various co-ordinates – the amounts by which they change – as we pass from Q to Q'

Now A's phi co-ordinate for Q' is the time from O to N'. So the increment in Φ is the amount by which this time exceeds Φ. Thus $\delta\Phi$ is the time from N to N'. **Please check that the other increments are correctly marked in the figure.**

Equations (22.9)–(22.12) tell us, in terms of increments, the relations between the radar co-ordinates and the time-over-there and distance co-ordinates for each observer. Suppose, for example, that some body were moving in such a way as to be present at Q and Q' (world line through points Q and Q'). Then (22.12) means that if A's radar measurements on this body change by $\delta\Theta$ and $\delta\Phi$, his version of its distance will change correspondingly by $\frac{1}{2}(\delta\Phi - \delta\Theta)$ – with similar interpretations for (22.9)–(22.11), **which you can work out for yourself.**

22.15 If we could find a connection between A's and W's radar co-ordinates (in increment terms), then §22.13 would enable us to connect up their time and distance co-ordinates. Let T_M and X_M stand for the time and distance of M according to A. Obviously Φ is the phi co-ordinate of M as well as of Q (§15.2). And so (16.3), with T and X changed to T_M and X_M, gives

$$\Phi = T_M + X_M . \tag{22.13}$$

Let k be the parameter (§21.2) of M. Then (21.8), applied to M instead of Q, gives†

$$k = fX_M + fT_M$$

§16.12

$$= f(X_M + T_M)$$

(22.13)

$$= f\Phi .$$

Or summarising,

$$k = f\Phi . \tag{22.14}$$

Using §§22.9 and 22.11, this gives

$$\delta k = f\delta\Phi . \tag{22.15}$$

And dividing each side by the corresponding side of (22.14) yields

$$\frac{\delta k}{k} = \frac{\delta\Phi}{\Phi} . \tag{22.16}$$

Can you connect δk with $\delta\phi$?

Remember that ϕ is the time of M by W's clock – just the t of Chapter 21 under another name. **Try again.**

In (21.20) change t to ϕ, and use §22.11, getting $f\delta\phi \sim \delta k/k$. When we combine this with (22.16), we have

$$f\delta\phi \sim \frac{\delta\Phi}{\Phi} \tag{22.17}$$

† For the way this chain calculation is laid out, see §§19.7–8. I shan't remind you any more. But watch out for §22.18.

−the required connection between A's and W's phi co-ordinates. Refer to Figure 22.14 and use §§20.20 and 21.42 to make sure exactly what (22.17) says.

22.16 Whatever event Q we are interested in, we can always pick the observer (with world line through O) who thinks Q simultaneous with O, and take him to be our observer A. And in future we shall always use this simplified situation. Then Q lies on the X-axis, as in Figure 22.16; and $T = 0$, so that (16.3) reduces to $\Phi = X$. And substituting in (22.17) gives

$$f\delta\phi \sim \frac{\delta\Phi}{X}.$$ (22.18)

22.17 To avoid some rather complicated mathematics, we'll use a simple trick to discover a similar relation between the theta increments. Place Figure 22.16 flat on a table. Hold a rectangular mirror vertically, with

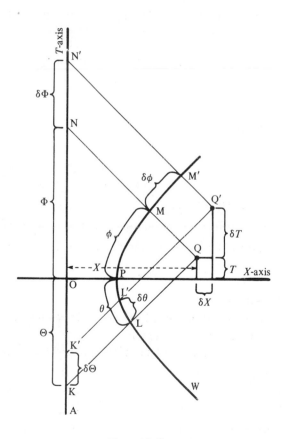

Figure 22.14

its bottom edge resting on the broken line across the top of the figure. Look at what you see in the mirror. (Or imagine all this.)

We have changed positive times into negative ones and vice versa. And in the reflected diagram Θ and θ play the roles that formerly belonged to Φ and ϕ. So the equivalent of (22.18) will apply to the theta increments, giving

$$f\delta\theta \sim \frac{\delta\Theta}{X}. \qquad (22.19)$$

Notice that the theta increments, as seen in the mirror, are both negative (indicating *decreases* in Θ and θ). If you don't feel comfortable with such negative increments – though they're perfectly permissible – redraw the diagram, before using the mirror, with Q′ so placed that K′ is below K. Then $\delta\Theta$ and $\delta\theta$ will both be positive.

When we go back to the original diagram, we again interchange positive and negative times (leaving X unchanged). So $\delta\Theta$ and $\delta\theta$ will still be both positive or both negative, and (22.19) will still hold good. Thus this is the required relation between the theta increments.

22.18 We can now put various pieces together to show how W's time-over-there and distance co-ordinates are related to A's. From (22.9),

$$f\delta t = \tfrac{1}{2}f(\delta\phi + \delta\theta)$$

§16.12

$$= \tfrac{1}{2}(f\delta\phi + f\delta\theta)$$

(22.18), (22.19)

$$\sim \tfrac{1}{2}\left(\frac{\delta\Phi}{X} + \frac{\delta\Theta}{X}\right)$$

§21.12, footnote

$$= \tfrac{1}{2}\left(\frac{\delta\Phi + \delta\Theta}{X}\right)$$

$$= \frac{\tfrac{1}{2}(\delta\Phi + \delta\Theta)}{X}$$

(22.11)

$$= \frac{\delta T}{X}.$$

Or in summary,

$$f\delta t \sim \frac{\delta T}{X}. \qquad (22.20)$$

Prove for yourself that

$$f\delta x \sim \frac{\delta X}{X}. \qquad (22.21)$$

Study the meaning of these near-equations with the help of Figure 22.16. Unfortunately δt and δx cannot be marked on the diagram – though you could think of them as represented by the same lengths as for δT and δX, but on different scales.

What we've got is two near-equations that tell us the relations between A's and W's versions of the time and distance from Q to Q'. You'll realise that these are the keys to a new world, since they will enable us to translate our knowledge about inertial observers into equivalent knowledge about accelerated observers. We can also put them (§9.11) in the form

$$\delta T \sim fX\delta t \qquad (22.22)$$

and

$$\delta X \sim fX\delta x . \qquad (22.23)$$

But *remember:* (22.20)–(22.23) hold good *only if* $T = 0$ (§22.16).

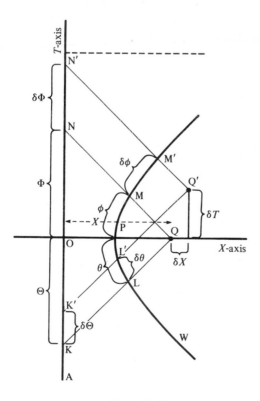

Figure 22.16

22.19 Near-equation (22.21) shows that A and W disagree about the increment in distance, and therefore about distance itself. And unfortunately the near-equations of §22.18 all involve A's version, X, of the
distance – which we shall therefore carry with us unwanted when we
make the transition from inertial to accelerated observer. So we are
prompted to enquire whether the disagreement between A and W is
big enough to matter in realistic conditions. Changing $\delta X/X$ to $f\delta X/fX$
(§16.22) and using (22.8), we can rewrite (22.21) as

$$\delta(fx) \sim \frac{\delta(fX)}{fX}. \tag{22.24}$$

Compare this with (21.20). **Can you draw a conclusion?**

The two have exactly the same form, with k and ft of the one
changed to fX and fx in the other. And just as $ft = 0$ when $k = 1$
(§21.18), so also $fx = 0$ (being measured from P) when $fX = 1$ (§17.19,
first sentence). So a table of fx against fX would be exactly the same as
Table 21.23 of ft against k and its continuations. And more generally,
everything we proved in Chapter 21 about the relation between k and
ft will also be true of the relation between fX and fx. In particular, it
follows from (21.11) that

$$fx \sim fX - 1. \tag{22.25}$$

Now let D stand for the distance of Q from W – distance PQ – according to A. Since distance OP $= 1/f$ (§17.19), we have $X = 1/f + D$,
and so

$$fX = 1 + fD. \tag{22.26}$$

Substituting in (22.25) gives $fx \sim fD$, or $x \sim D$ (§21.11).

This confirms that if we stick to reasonably short distances, it will be
good enough to take $x = D$ – to assume that A and W agree about
distance.

22.20 But how big an error should we thus introduce? This is the same as
asking about the error involved in replacing (21.11) by $ft = k - 1$.
When $k = 1\cdot1$, this last would give $ft = 0\cdot1$, against a correct value
of $0\cdot095$ (§21.21) – an error of 5 per cent. Similarly if fD is $0\cdot1$,
giving $fX = 1\cdot1$, the error involved in assuming $x = D$ will be about
5 per cent.

Taking the year and the light year as time and distance units, a 1g
acceleration is roughly equivalent to $f = 1$ (§17.21) – giving $D = 0\cdot1$, or
about a million million kilometres (§1.8). So an inertial observer and a
1g-acceleration observer would differ by 5 per cent or less in measurements up to a million million kilometres.

If we had calculated Table 21.21 for smaller steps and to higher accuracy, then (by taking $k = 1 \cdot 001$, which corresponds to $fD = 0 \cdot 001$) we could prove that for distances up to about 10 000 million kilometres these two would disagree by less than $0 \cdot 05$ per cent. And an observer with acceleration $100g$ would differ by less than $0 \cdot 05$ per cent from an inertial observer for distances up to 100 million kilometres. All the cases we shall deal with will lie well inside these limits, and so we may as well change permanently from A's to W's distance measurements, writing x in place of D. Then (22.26) gives $fX = 1 + fx$. Substituting this in (22.22) and (22.23) leads to

$$\delta T \sim (1 + fx)\delta t , \tag{22.27}$$

and

$$\delta X \sim (1 + fx)\delta x , \tag{22.28}$$

which – like (22.22) and (22.23) – are only valid if $T = 0$.

These two near-equations – relating the measurements of inertial and accelerated observers – are the basis from which we shall develop the General Theory of Relativity. Think carefully about their meaning (some lines of thought are suggested towards the ends of §§22.14, 22.15 and 22.18; and use Figure 22.16).

We have to emphasise that (22.27) and (22.28) have been obtained by assuming $x = D$, which is only approximately true. The validity of the approximation depends on fx (written in place of fD) being small compared with 1. And if we require high accuracy, fx must be *very* small compared with 1 – a very small fraction (e.g., the case $fD = 0 \cdot 001$ above). (Notice that the once-for-all approximation that we've introduced here is completely independent of the as-accurately-as-we-wish approximations indicated by the $\sim$ sign.)

22.21 If you are expert enough mathematically to know about the exponential function, $\exp x = e^x$, you will see that we could use it to avoid the approximations of §§22.20. The *exact* equivalents of (22.22) and (22.23), using x instead of X, are

$$\delta T \sim \exp (fx) \cdot \delta t, \tag{22.29}$$

and

$$\delta X \sim \exp (fx) \cdot \delta x, \tag{22.30}$$

where the dots indicate that $\exp (fx)$ is multiplied by the respective increments. And (22.27) and (22.28) then appear as first-order approximations to these.

If we developed the theory using these last instead of the §22.20 versions, naturally we should arrive at different theoretical predictions. The differences would be too small to detect by any test that could be

done at present, but in certain extreme conditions they could be significant. I'll come back briefly to this point before we end.

22.22 Let A_2 be an inertial observer who is stationary relative to A (Figure
22.22) and let x be the distance according to W of the tick Q of A_2's
clock which W considers to be simultaneous with P (§22.6). We ask:
What is W's opinion about the behaviour of A_2's clock at event Q? (Cf.
§§3.34–6, 13.15 for a similar question about an inertial observer's
opinion.)

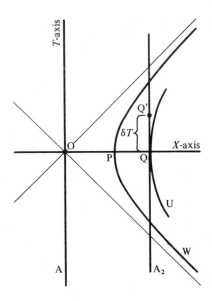

Figure 22.22

Let Q' be a later tick of A_2's clock. Being stationary relative to each
other, A and A_2 agree about the time δT between Q and Q'. So with δt
indicating W's version of this time, (22.27) gives

$$\delta T \sim (1 + fx)\delta t . \tag{22.31}$$

The answer to our question should be embodied in this near-equation,
but it is not easy to see how to cope with the $\sim$ sign.

22.23 The task would have been easier if we'd had an actual equation

$$\delta T = (1 + fx)\delta t . \tag{22.32}$$

In that case, what would be W's opinion of A_2's clock?

Here δT is an actual time recorded by A_2's clock, while δt is what W,
observing from afar, thinks that time should be. Take the case where W

is accelerating *towards* A_2. Then Q is on the positive side of W, x is positive, and $1 + fx$ is greater than 1. So (22.32) says that the time recorded by A_2's clock is greater than W thinks it should be, and therefore W thinks that A_2's clock is running *fast*. **Prove for yourself that** if W is accelerating *away from* A_2, he thinks the clock is running *slow*.

For brevity, let's speak of the factor by which W thinks A_2's clock is running fast or slow (in any particular set of circumstances) as the *clock rate factor* (what A_2's clock records divided by what W thinks it should have recorded). We have proved that if (22.32) were true, the clock rate factor would be $1 + fx$.

22.24 However the relationship is not (22.32) but (22.31), which we could write as

$$\delta T = \bar{r}\delta t ,$$ (22.33)

where

$$\bar{r} \sim 1 + fx .$$ (22.34)

By the same argument as in §22.23, $\bar{r}$ is the clock rate factor *averaged over the whole period* δT from Q to Q'. (That's why I've called it $\bar{r}$ – read as 'r bar'; the bar is a standard symbol for an average.) And the size of this averaged clock rate factor depends on the size of δT – that's why we can't have $=$ in (22.34). So it seems to W that the clock's behaviour is continually changing. But (22.34) tells us that by taking δT small enough we can make $\bar{r}$ as nearly equal to $1 + fx$ as we wish (definitions of §§20.20 and 21.42; and as in §21.13 we don't have to worry about *proportionate* differences).

22.25 However, we are concerned with W's opinion of how A_2's clock is behaving *at* event Q – not just averaged over the period Q to Q'. And that brings us up against a problem like the one we met in §§17.13–14 (which please reread). We can't measure the clock rate factor *at* an event – we need a *time-interval* over which to measure it. But $\bar{r}$, the clock rate factor averaged from Q to Q', can be taken as an approximation to the clock rate factor *at* Q; and the smaller δT, the better the approximation. So, just as in §17.13, we are led to a *definition:*

> The clock rate factor *at* event Q is defined to be the quantity that we can calculate *as accurately as we wish* by taking it to be the clock rate factor averaged over the time-interval δT from Q to Q' and making δT small enough.

22.26 If r stands for the clock rate factor *at* W, this definition translates into $r \sim \bar{r}$ (check that against §§20.20, 21.13 and 21.42). And using §21.10 to combine this with (22.34), we get

$$r \sim 1 + fx \, . \tag{22.35}$$

How would you interpret that statement?

22.27 Did you say '*r* nears $1 + fx$' – in other words, *r* and $1 + fx$ can be made as nearly equal as we wish by taking δT small enough? If so, **criticise yourself.**

It seems an odd assertion, since making δT smaller – i.e., bringing Q' nearer to Q – can't have any effect on *r* (the clock rate factor *at* Q) nor on $1 + fx$. **Think again.** And if that gets you nowhere, **study §§17.15– 16 again and apply similar thinking to the present problem.**

I hope you reached the conclusion that

$$r = 1 + fx \tag{22.36}$$

– that we are entitled to change the $\sim$ of (22.35) into $=$. In other words, we deduce that the clock rate factor *at* event Q is *exactly* $1 + fx$ after all – that A_2's clock, in W's opinion, is running fast or slow by that factor.

22.28 We can now make a better and neater job of proving that point than we could in Chapter 17. We have two assertions, both known to be true:

(i) the ratio $(1 + fx)/r$ can be made as near to 1 as we wish by taking δT small enough – that's what $r \sim 1 + fx$ says (§§20.20, 21.42); and

(ii) changing the size of δT doesn't have the slightest effect on this ratio (since it doesn't affect *r* or $1 + fx$ separately).

There is only one possible way in which both can be true. **Can you spot it?**

The only possibility is that the ratio $(1 + fx)/r$ is *always* as near to 1 as we could possibly wish. In other words, it *is* 1 – which leads straight to the *equation* $r = 1 + fx$.

22.29 This step that transforms (22.35) into (22.36), changing $\sim$ to $=$, is one of those subtle bits of reasoning that often cause trouble and doubt. You'll have to worry at it for yourself. But a few comments may help.

You'll surely agree that if changing the size of δT has no effect on the size of *r* or $1 + fx$, then making δT smaller cannot bring their ratio any nearer to 1 than it was before. So it can't make the ratio as near to 1 as we wish *unless it already was 1 all the time.*

On the other hand, if $(1 + fx)/r$ is equal to 1, there's no trouble. Statement (i) means in essence: 'You say how close to 1 you want the ratio to be, and I'll tell you how small you must make δT to bring this about.' And no matter what degree of closeness you demand, I'll simply say 'Take δT to be any size you like, and you'll find your demand has been satisfied.' And I'll be right!

You probably feel that this is a bit fishy, because the '~' assigns to the increment the task of making the two quantities as nearly equal as we wish, whereas here the job gets done without the increment's help. But complaining about this is like refusing to pay the fire brigade on a day when there are no fires to put out.

The simple fact is that if r and $1 + fx$ are unequal, you can't make (i) and (ii) both true; and if they are equal, you can. So they must be equal.

22.30m The interesting thing in the chain of deductions in §§22.24–9 is that our input consists of the *near*-equations $\bar{r} \sim 1 + fx$ and $r \sim \bar{r}$, and yet we managed to deduce an *exact* equation $r = 1 + fx$. It is surely important to know precisely when we can do this – in just what circumstances a *near-equation* like (22.35) can be changed to an *equation* (22.36). The case we've dealt with is an application of a proposition which, because of its importance within the as-accurately-as-we-wish technique, I'm going to call the

> *First Fundamental Theorem*
> If two variable numbers or quantities, u and v, are connected by the near-equation $u \sim v$, and if the sizes of u and v are unaffected by the size of the relevant increment (or other specified quantity, §20.20), then $u = v$.

Can you prove it?
We've done so already. If in §22.28 you change r and $1 + fx$ into u and v, and read δT as 'the relevant increment (or other specified quantity)', you'll find you have a proof. **Please check the details.** The commentary of §22.29 can be similarly translated.

2.31m The condition that the values of u and v must be unaffected by the size of the increment is absolutely vital. Without it, the conclusion that $u = v$ is *false*. Suppose, for example, that the relevant increment is δu, and that v is $u + \delta u$. You can easily **check from the definition** that $u \sim u + \delta u$. Yet it is obviously not true that $u = u + \delta u$.

22.32 Since §22.22 we have been concerned with W's opinion of the clock of *inertial* observer A_2. He thinks that at event Q it is running fast or slow by a factor of exactly $1 + fx$. But what would he think about the clock of a *non*-inertial observer U who is momentarily stationary relative to him at event Q (Figure 22.22)?

Instantaneously at Q, such a clock agrees with A_2's, and so W's opinion is unchanged. A strict argument for this point would closely follow the line of thought we developed in Chapter 20, where we were concerned with stating precisely, and then testing, this thesis that an accelerated clock momentarily agrees with an inertial clock that is moving with it. For once I'll omit the details.

So we're not confined to inertial clocks. We can generalise the conclusion we've already reached for A_2's clock, and so make the very important prediction that

> If W is accelerating *towards* a distant clock (which is momentarily stationary relative to him), he thinks it is running fast; if he is accelerating *away from* it, he thinks it is running slow – and by a factor of $1 + fx$ in either case (fx being positive for approach and negative for recession).

22.33 You'll notice some resemblance between this conclusion and the dilation of time in its §3.36 form. But the differences are more important than the resemblances. **List them.**

I'll mention the vital features of our new result, and you can contrast them with time dilation. It (1) depends on acceleration – which (2) is an absolute – and (3) occurs even when clock and observer are relatively stationary; (4) the clock may seem to be going fast *or* slow, (5) depending on the direction of the acceleration, and (6) by an amount which varies with the distance between clock and observer. And (7) the relation between U and W is not symmetrical.

22.34 At last we can clear up some remaining doubts about the clock paradox or space-twin paradox (§§4.14–20, 14.22–4 and much of Chapter 20, especially §§20.28–9). In the journey depicted in Figure 14.22 (slot!) A thinks that D's clock is always running slow (except momentarily at turn-round) – which agrees with our conclusion that D experiences less time than A between parting and reunion. **See if you can now work it out in terms of D's opinion of A's clock.**

During the inertial portions of D's journey he will think that A's clock is running slow. And we seem to be on the verge of a contradiction. **But what does D think during his three periods of acceleration?**

At take-off and touch-down (near P and Q) he is accelerating away from A's clock. So he thinks it is running even slower. But the distance, x of §22.32, is small and so the additional effect is negligible. **But what about the turn-round acceleration near O?**

Here W is accelerating *towards* A's clock and so thinks that it is running fast – and by a large amount, since it's a long way away. Working out the quantitative details is a nasty job (we have to cope with accelerations starting and ending). But in fact D thinks that A's clock gains more during turn-round than it loses during all the rest of the journey – which again agrees with more time elapsing for A than for D.

The space travel file is now closed!

22.35 If you reconsider from the traveller's point of view the experiments of §§20.6–9 and 20.26, you will see that they act as verifications of our §22.32 prediction that a clock towards which an observer is accelerating seems to him to be running fast (obviously to get good verifications we ought to work out the quantitative details). And in the deductive process leading to that prediction, the only new assumption is the one in §22.3 about how W's versions of time-over-there and distance are defined in terms of his radar co-ordinates. Thus we have some confirmation that this assumption is a good one.

22.36 While the ideas of §§22.22–31 are fresh in our minds, let's get a rather similar job done, ready for when we need it. It concerns velocity. And we'll take it very generally (though still in the 1D universe) – the velocity, which may be varying in any way you like, of something we'll call B (observer, light signal, body, what you will) relative to an observer who can be inertial or have constant or variable acceleration. All quantities will be this observer's versions. Suppose the distance of B at time t is x, and its distance at time $t + \delta t$ is $x + \delta x$ – so that it moves distance δx in time δt. **Can we now say anything about its velocity?**

Only about its *average* velocity between t and $t + \delta t$ – which we'll denote by $\bar{v}$ (cf. $\bar{r}$ of §22.24). Using the familiar relation that distance travelled is equal to average velocity multiplied by time taken, we have

$$\delta x = \bar{v}\delta t . \qquad (22.37)$$

22.37 But we want to know B's velocity *at* time t – call it v. **Any suggestions?**

Just as in §§17.13 and 22.25, we are led to define B's velocity *at* time t as the quantity that we can calculate as accurately as we wish by taking it to be $\bar{v}$ and making δt small enough. Or in symbols (cf. §22.26) $v \sim \bar{v}$. Now combine this with (22.37) – using §§21.10–11 – and we find that

If v is B's velocity at time t, then

$$\delta x \sim v\delta t . \qquad (22.38)$$

22.38 Suppose next that in dealing with the motion of something or other – still called B – we are able to show that

$$\delta x \sim u\delta t . \qquad (22.39)$$

Comparing this with (22.38) can we infer that u is B's velocity?

It's not quite straightforward. Using §21.10, we can combine the two to get $u\delta t \sim v\delta t$, whence $u \sim v$ (§21.11). And we seem to have proved only that u *nears* B's velocity. **What do you say?**

You'll have thought of the First Fundamental Theorem (§22.30). To be able to apply it, we must add the further condition that the size of u does not depend on the size of δt. Clearly v, the velocity *at* time t, is not affected by the size of δt. So the Theorem now applies and we have proved that

> If x stands for B's distance at time t, and if $\delta x \sim u\delta t$, where the size of u is not affected by the size of δt, then u is B's velocity.

23

The Principle of Equivalence

23.1 Suppose that W – the observer of the previous chapter moving with constant acceleration – watches any inertial observers that may be around. **How will their motion appear to him?**

Think, for example, about several cars moving at various steady speeds along a straight road, and about what their motion would look like when observed from another car that has constant acceleration. I think you'll conclude that the distance between W and any inertial observer E is changing in a constantly accelerated manner, and that therefore

> If W regards himself as stationary, he must think that all inertial observers are moving with constant acceleration f (in the opposite direction to that of his own constant acceleration).

(A possible worry: maybe it won't work out quite like that – because E and W don't agree about times and distances. Actually if you do the calculations, based on our Chapter 22 near-equations, you discover that the statement is only strictly true if E is near to W and moving slowly relatively to him – both conditions to be interpreted in an as-accurately-as-we-wish sense. But that's good enough for us, since we'll only want to use this proposition in nearby, slow-speed conditions.)

23.2 Maybe you protested that W knows (by the particle test) that he is accelerated, and so he can't regard himself as stationary. But **do you know of any circumstances in which an observer's test particle runs away and yet he insists that he's at rest?**

Your test particle accelerates rapidly away from you (cf. §5.4). Yet you have no difficulty in thinking of yourself as stationary. In fact, despite all the mental gymnastics we've been through, you probably still have difficulty in thinking of yourself as *anything but* stationary. I do!

So when W observes his test particle moving away, he has the choice of agreeing that he is accelerated or of asserting that he is stationary – just like you – with gravity making his test particle fall away.

23.3 The experiences of W, moving with constant acceleration in his gravity-free spacetime are exactly like your experiences sitting in your armchair, with gravity acting on you and everything around you. *Your* theory about gravity would account for what *he* observes, and *his* theory about constant acceleration would account for everything *you* observe (cf. §17.20, where we broached this idea briefly).

As W sees it, all inertial bodies (those not acted on by forces) move with constant acceleration – the *same* constant acceleration in the *same* direction. As you see it from your chair, all inertial bodies fall – which is only another way of saying that they all move with the *same* constant acceleration in the *same* direction ('downwards'). There *are* things that don't share in this general tendency to accelerate downwards – the vase on that table, for example. But forces are required to prevent them from falling – the table exerts an upward force on the vase. Friend W also observes things that don't fall – things that maintain fixed positions relative to him. And again forces are required – to keep things accelerating (as an outside inertial observer would see it) at the same rate as W.

You are conscious now of a force that is preventing you from falling – the seat of your chair pressing upwards. Similarly W feels a force acting on him. He could say 'That's the force that makes me accelerate.' But he could equally well adopt your point of view and say 'That's the force that saves me from sharing in this general tendency of things to accelerate (fall) towards the rear of my rocket – the force that keeps me stationary in my rocket.'

23.4 What I've been saying is only true if we stick to a small region of space in which we can take gravity to be constant in strength and direction – your room, for example. Experiments on a larger scale would reveal obvious differences between your experience and W's (think, for example, about the *directions* in which free-falling objects accelerate). But till further notice we'll confine ourselves to small-scale-constant-gravity conditions. And in that case it's beginning to look as if your experiences and W's are indistinguishable – as if the effects of constant gravity are identical with those of constant acceleration of the observer.

23.5 To avoid being distracted by irrelevant long-distance observations, let's consider an observer confined to a small totally enclosed room. And let's consider four different sets of conditions, as in Figure 23.5. In the first two cases the small room is the cabin of a lift, which happens to be stationary in case (i), so that the observer shares the normal experiences of anybody standing on the Earth's surface. In (ii) the cable has broken, so that the room and its contents are falling freely. In the other two cases the room has been housed in a rocket so far from stars and galaxies that we can forget about gravity. The rocket engines are working in (iii), but idle in (iv).

23.6 Case (iv) is the one you ought to feel at home with. True, you've never met it in practice, but we spent our first nineteen chapters theorising about it. The observer is inertial, and there is no gravity; so his is the world described by the Special Theory of Relativity (§5.5). His test particle stays with him. Or if he gives it a shove, it travels (relative to him) in a straight line at constant speed (§5.6). No force is needed to

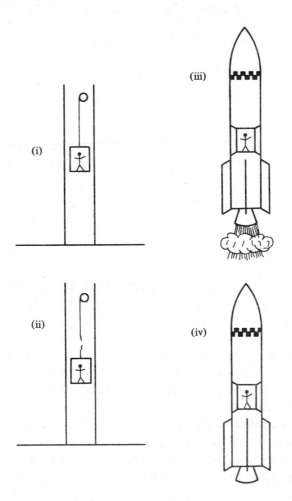

Figure 23.5

keep him in place – he is weightless. **Now compare the experiences of the case (ii) observer.**

They are the same. His test particle falls alongside him. So as he sees it, it stays with him – he is inertial. Or if he projects it sideways, instead of just releasing it, its motion relative to the Earth's surface will be a combination of (*a*) constant speed sideways and (*b*) constant acceleration downwards. But he shares motion (*b*) with the particle. So as far as he is concerned, it only has motion (*a*) – in a straight line at constant speed, just as in case (iv). Also the floor of the lift is falling at the same rate as his feet and therefore exerts no force on them – he too is weightless.

In summary, provided we stick to a small room and a short drop, the experiences of cases (ii) and (iv) are identical.

23.7 Similarly we can **show that the experiences of observers (i) and (iii) are indistinguishable.**

Observer (i)'s test particle falls to the floor. In case (iii) an inertial outsider would see the floor accelerating to catch up with the particle. But the chap in the rocket would just see the particle accelerating 'downwards' to hit the floor.

If (i) projects his particle horizontally, its motion will be that same compound of (*a*) and (*b*) that we had in §23.6. If (iii) projects his particle perpendicularly to the direction of rocket thrust, then according to an inertial outsider the particle continues to move at constant speed in that direction – motion (*a*). But the observer's own motion will give *him* the impression that the particle also has motion (*b*). And so its resultant motion, according to him, is the same combination of (*a*) and (*b*) as it was according to (i).

Observer (i) is conscious that he has weight – that the floor exerts an upward force on his feet, and that if this force ceased to act (? dry rot), he would fall through the floor. Similarly observer (iii) left to himself would move at constant speed. And with the rocket accelerating, that could only mean that he would 'fall' through the floor, were it not for the force that this exerts on his feet.

I'm suggesting that *every* observation or experiment would give the same result in case (i) as in case (iii).

23.8 This last conclusion plays a central role in the further development of our theory. We state it formally under the title

The Principle of Equivalence
There is no means of distinguishing between the effects of constant gravity and those of a constant acceleration of the observer.

So the Principle states that if an observer sees his test particle run away, there is no experiment that would allow him to decide whether this arises from the effects of gravitation or from his own acceleration.

No experiment! But should we not restrict our conclusion to mechanical experiments? – since we've considered no others. Well Einstein once asked us to shun a similar restriction (§§5.9–11); and the result was the great achievement of his Special Theory. Later on, when he was seeking to understand gravity, he asked the world to reject any 'mechanical' restriction on the Principle of Equivalence. Shall we follow him again in the hopes of another good harvest?

23.9   In reading §23.7 you perhaps felt that cases (i) and (iii) must be distinguishable because all the effects in the former are caused by the *force* of gravity, while no analogous force is involved in the latter. But are you sure that gravitation *is* a force? **Is this force of gravity a hard fact of observation, or is it just a hypothesis introduced to explain why things fall in the way that they do?**

Newton's theory of gravity as a force is so firmly built into our culture that we've come to feel that it's obvious that things fall because a force is pulling them down. Yet if you examine your experience, you'll see that you have no direct evidence of this force. Try an experiment on yourself – a small-scale one, I suggest, like jumping off a chair. You don't feel a force pulling you down.

On the other hand when you sit down again, *not* falling, you *do* feel a force – the upward pressure of the chair on your posterior. That observation would seem, if anything, to favour the case (ii) point of view – a rocket's thrust, transmitted by floor and chair, is *pushing* you upwards, giving you an upward acceleration. Maybe the force of gravity is an illusion or an invention!

23.10   This alleged force of gravity has some very odd features. For every other force, we know a method of stopping it acting. We can cut tow ropes, demagnetise magnets, and so on. But no matter what we do, we cannot cancel or vary this force (if so it be) of gravity. Again, the fact that all bodies (in the same place) fall with the same acceleration implies that this gravitational force must be proportional to the body's mass (§18.12) – and this has been verified to one part in a million million. That makes gravity very odd indeed, since in no other case is the strength of the force necessarily related to the mass of what it's acting on. On the other hand, the falling-with-the-same-acceleration behaviour is an obvious corollary of an explanation on the lines of case (iii). If the force needed to explain falling has to be so different from other forces, ought we not to feel suspicious about its reality?

So the simplest interpretation of what we observe would be to say that *we* are accelerated. Then we don't need a force of gravity. But if we insist on maintaining that we are stationary, we have to *invent* this distinctly odd force to explain what we observe about things falling.

23.11 Such illusory or invented forces arise in other connections. In a car speeding round a circular track, any unrestrained object accelerates outwards (so it seems) – away from the centre of the circle. Sitting in the car, you are aware that forces are needed to prevent you from doing likewise. It seems as if everything in the car is subjected to a force driving it away from the centre – a 'centrifugal force', as it's commonly called – whose strength is again proportional to the mass of whatever it's acting on.

But you know that this force is illusory. A trackside observer sees that if an object is not forcefully constrained to follow the car, it simply continues going straight ahead, and the car moves away from it. He needs no force to explain why things tend to move outwards relative to the car. For he sees that really the car turns aside, while unattached objects go straight on.

It is you in the car who are accelerated (§20.5) – towards the centre. But you habitually use yourself as reference system, thinking of yourself as stationary. And then to explain the behaviour of unrestrained objects you have to *invent* centrifugal force.

Could the force of gravity be similarly illusory? Obviously there will be complications when we consider things falling in both England and New Zealand – just as there are complications if the man in the car tries to explain the motion of things in another car on the opposite side of the track. But it looks as if there may be *some* sense in saying that the force of gravity is an illusion that arises because we deny being accelerated when we really are.

23.12 Let's summarise. In constant gravity conditions, the free-falling observer – (ii) of §23.5 – has the same experience as an inertial observer in the absence of gravity. And so his world will be described by the Special Theory of Relativity (§5.5) – by Chapters 1–19 above. And a supported observer (i) is related to this free-falling observer (ii) in exactly the same way as, in gravity-free conditions, a constantly accelerated observer (iii) is related to inertial observer (iv) – that is, in the same way as accelerated observer W of Chapter 22 is related to A. We know how to cope with A's experience – that's the Special Theory in which we have become expert. We also know (Chapter 22) how to translate A's experience into W's. And the same piece of translation should allow us to deduce the experience of the chap standing on Earth – where gravity is at work – from that of the free-falling observer. It looks as if we have the beginnings of a theory of gravitation.

23.13 The Principle of Equivalence is, of course, just another hypothesis, an assumption to be tested as usual by comparing predictions with experiment.

If you apply the thinking of §23.7, paragraph 3, to a light signal sent out sideways – starting with case (iii) and using the Principle of Equivalence to deduce what would happen in (i) – **you will easily reach the prediction that** a light ray shining across the direction of gravity will be bent downwards. But alas, a quick calculation shows that if the experiment is on a small enough scale to let us regard gravity as constant, the amount of bending would be far too small to detect. Disappointing!

However, there is another prediction, easily derived by applying the Principle of Equivalence to something we worked out in Chapter 22, which *can* be tested experimentally. **Can you spot it?**

23.14 **Use the Principle of Equivalence to interpret the conclusion of §22.32 in gravitational terms.**

If a supported observer (one who is restrained from falling) observes a clock *above* him, he should get the same effect as W does when accelerating *towards* the clock. So we deduce (please check the details) that

> In constant gravity conditions a clock at height x above a supported observer will appear to him to be running fast by a factor of $1 + fx$, where f is the acceleration of a free-falling body according to this supported observer

– with a similar 'running slow' statement when the clock is below (x negative). This phenomenon is known as the 'gravitational shift'.

23.15 To calculate the expected effect for a clock a little way above the Earth's surface, we have only to change the acceleration f into g (§§17.20–1). To ease the arithmetic, take the year as time unit, as in §17.21, so that the distance unit is a light year (§1.8) and $g = 1 \cdot 03$. For a clock 22·5 metres above the observer we have

$$gx = \frac{1 \cdot 03 \times 22 \cdot 5}{1000 \times 9 \cdot 5 \text{ million million}},$$

and so it should seem to be running fast by roughly one part in 400 million million. The Mössbauer effect (§20.26) is sensitive enough to test this prediction.

23.16 Using that height of 22·5 metres, the test was carried out at Harvard

University by Pound and Rebka in 1960 and repeated with refinements by Pound and Snider in 1964. It is convenient to express the results as follows.

Suppose $1 + p$ is the factor by which we theoretically expect the clock to seem to be running fast (i.e., $p = gx$), and suppose the experiment actually gives a factor of $1 + q$. The ratio q/p will indicate how well the experimental result agrees with the prediction; $q/p = 1$ would mean perfect agreement. The 1964 experiment gave

$$\frac{q}{p} = 0\cdot9990 \pm 0\cdot0076 \ . \tag{23.1}$$

(If you don't know what the $\pm 0\cdot0076$ means, see §23.17.) Clearly the prediction has been confirmed to quite high accuracy.

23.17 Experiments inevitably involve errors. So we expect only *approximate* agreement with the theoretical predictions. But *how good* must the agreement be if we are to claim a verification? Two sorts of error are involved.

Systematic errors are ones which produce a definite bias in the results. They can arise from such things as using a measuring rod that is a fraction too long, or forgetting to allow for the effect of temperature on its length, or something more sophisticated. Experimenters simply take all the precautions they can think of to eliminate such errors. And they may attempt to estimate an upper limit for any remaining systematic errors in their work – Pound and Snider put it at $0\cdot010$. (But there *might* be other systematic errors arising from causes hitherto unsuspected.)

However, even if *systematic* error is entirely eliminated, no measurement is ever exact. The actual reading given by the measuring instrument will sometimes be a little too high, sometimes a little too low. Such errors are *random* in character, and so nothing definite can be said about the accuracy of any particular measurement.

A quantity like q will be calculated from several (perhaps many) measurements, each involving its own random errors. The total error in q will arise from a combination of all these component errors – sometimes adding together, sometimes partially cancelling. And since each component error is random, one can get information about the total error by using the Theory of Probability.

The total error will sometimes tend to make the result too small, sometimes too large. So obviously a better answer can be got by repeating the experiment several times and taking the average of the results. The $0\cdot9990$ of (23.1) is such an average.

What is not so obvious, but does follow from Probability Theory (provided certain conditions about the nature of the errors are satisfied), is that the size of the error in this averaged result can be estimated by considering how widely spread are the individual results from which the average has been calculated.

The ' ± 0.0076 ' of (23.1) – known as the *standard error* – expresses this estimated size of error. The sign $\pm$ is a combination of $+$ and $-$, which we read as 'plus or minus' (because we are considering the chances that the true result is actually 0.9990 plus or minus some specified size of error). And in essence (23.1) says that there is a

68 per cent chance that the true value of q/p differs from 0.9990 by not more than 0.0076 either way – i.e., that it lies between 0.9914 and 1.0066;

and a 95 per cent chance that it differs from 0.9990 by not more than 2×0.0076 either way;

and a 99.7 per cent chance that it differs from 0.9990 by not more than 3×0.0076 either way.

So (23.1) gives us two pieces of information:

(i) Since 0.0076 is quite small, it tells us that the experiment was a rather precise one, with random errors kept well under control.

(ii) The fact that the theoretically expected value of $q/p = 1$ lies well inside the 95 per cent probability range, 0.9838 to 1.0142, makes it highly unlikely that it was some lucky fluke that gave an experimental result so close to 1. So this experimental outcome really does constitute a strong verification of our prediction that q/p should be 1 – a verification that the clock really did appear to run slow by the predicted amount.

23.18 The clocks-around-the-world experiment of §20.9 gives another verification of the §23.15 prediction, since the gravitational effect on the flying clocks was comparable in size to the effect arising from their motion. Further verifications were done with clocks aboard a Soviet *Cosmos* satellite (1973) and aircraft above Chesapeake Bay (1975).

So we have strong evidence in favour of the new assumption that we've called the Principle of Equivalence (§23.8). And that gives good reason to hope that the method suggested in §23.12 might lead to a theory of gravity.

However, the Principle of Equivalence applies only in regions of spacetime so small that we can take gravity to be constant (cf. §23.4), whereas a gravitational theory won't be much use unless it can cope with regions like that of the Sun and its planets where gravity varies in both strength and direction. I'm afraid we have a long way to go before we can deal with that. Still we've made a start.

24

The metric

24.1 From now on the quantity interval will play an important role. So please do any necessary revision of Chapter 12 (and maybe 11) and §§13.1–8 and 13.21–4. Our discussion there was confined to observers and events on the 1D universe (§5.19). But now I want to show that interval, as defined in §12.15, is still invariant – the same according to all inertial observers – even when we're working in all three dimensions of space. For simplicity I'll only deal with the case of time-like interval (§13.22), so that (12.5) of §12.15 applies. And until I say otherwise 'observer' means *inertial* observer.

24.2 First take the special case where observers A and B are moving relative to each other *on* the 1D universe, and consider the interval between O (the event of A and B passing) and an event Q which is *not on* the 1D universe (forgive the strange language of a 'universe' with events outside it!).

Figure 24.2 is not a spacetime diagram, but an ordinary 'snapshot' showing event Q happening and the situation on the 1D universe KL at the very moment when, *in A's opinion,* Q is happening. And R is an event *on* the 1D universe, which is simultaneous with Q according to A, the direction RQ being perpendicular to KL. Using an extension of §2.22, **can you prove that B will agree that Q and R are simultaneous?**

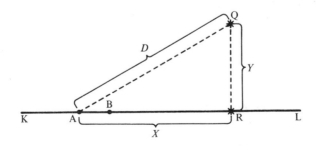

Figure 24.2

Consider two more observers, C and D stationary relative to A and B respectively, who pass each other at event R. We can trust our slow-speed experience when it tells us that relatively stationary observers agree about simultaneity. So C agrees with A's view that Q and R are simultaneous. Then §2.22 says that D also agrees, as does relatively stationary B – which proves the point. **Using §2.27, prove that A and B agree about the distance between R and Q.**

Give C and D rods which, as they pass, stretch from R to Q. These are perpendicular to the direction of relative motion. We then have the situation of the last sentence of Chapter 2, so that C and D agree about the lengths of these rods. And using them as distance measurers, C and D will agree about the distance between R and Q. Finally, another appeal to what we know about *relatively stationary* observers shows that A and B will also agree. Let Y stand for this distance.

24.3 But A and B will not agree about the distances of Q and R from O. The distance between O and R according to A is the same as the distance of R from A according to A (cf. §12.5). Call it X, and let D stand for the distance from O to Q according to A. By Pythagoras' Theorem (§3.15),

$$D^2 = X^2 + Y^2.\tag{24.1}$$

Similarly if x and d stand for B's versions of the distances O to R and O to Q, then

$$d^2 = x^2 + Y^2.\tag{24.2}$$

24.4 Let the time between events O and Q be T according to A and t according to B. Since they agree that Q and R are simultaneous, these are also their times for event R.

With R on the 1D universe, A and B agree about the interval between O and R (§12.15). In fact, from (12.7),

$$T^2 - X^2 = t^2 - x^2.$$

Subtracting Y^2 from both sides gives

$$T^2 - X^2 - Y^2 = t^2 - x^2 - Y^2.$$

And since subtracting two quantities successively has the same effect as subtracting their sum, this can be rewritten as

$$T^2 - (X^2 + Y^2) = t^2 - (x^2 + Y^2),$$

whence, using (24.1) and (24.2),

$$T^2 - D^2 = t^2 - d^2.$$

In other words, A and B do agree about the quantity called interval as defined in §12.15.

24.5 We can easily extend this to the general case where the observers, A_1 and B_1 say, do not move on the same straight line and are not present at either event. We just consider observers A and B, stationary relative to A_1 and B_1 respectively, and both present at O. Since either of these moves in a straight line relative to the other (§5.6) and since they are together at O, they must move on *one* straight line – which we take as KL.

Relatively stationary observers agree about interval, since they agree about time and distance separately. So A_1 and B_1 agree about interval with A and B, who agree with each other (§24.4). Thus

> The invariance of interval, as defined in §12.15, extends to inertial observers and events in ordinary spacetime with its three spatial dimensions.

Now let's get back to our 1D universe, knowing that we can handle the 3D case when we need it.

24.6 Remind yourself of the close analogy between distance and interval (§12.19 and perhaps §11.17–21). They are the invariants in the geometries of the plane and spacetime, respectively (§§11.20, 12.15). And we have said that a geometry is completely characterised by its invariant (§§12.21–2).

Up till now 'spacetime' for us has meant the spacetime of the Special Theory of Relativity – a spacetime free from gravity (§5.5). But from now on we shall be increasingly interested in gravity. And one of Einstein's most profound insights led him to suggest that the spacetime of a universe with gravity would be different from the gravityless spacetime of Special Relativity. Or in the sense we've just been discussing, its geometry would be different. **What will characterise this new geometry?** A different relation between invariant and co-ordinates. And usually, we shall find, this relation will have to be expressed by a near-equation, instead of the simple equation – such as (12.7) – that does for Special Relativity. To prepare for this, let's restate the latter in terms of the 'δ' notation (§21.39–40).

24.7m We said in §21.40 that we must think of δt or δx or the like as a *single symbol* for the increment in the quantity concerned. From this there follows the *most important convention* that

> δt^2 means 'the square of the increment δt'

(and similarly for δx^2, δy^2, what you will). It does *not* mean 'the increment in t^2'.

24.8 Using this convention, let's express the ordinary Special Relativity formula for interval in 'δ' language. Consider event Q (Figure 24.8), which happens at time T and distance X according to A; also event Q', which happens at time $T + \delta T$ and distance $X + \delta X$. Then δT is the time between Q and Q', and δX is the distance between them. And so if we use δs to stand for the interval between Q and Q', (12.5) tells us that

$$\delta s^2 = \delta T^2 - \delta X^2. \tag{24.3}$$

(We'll stick to the time-like case, §13.22, from now on.)

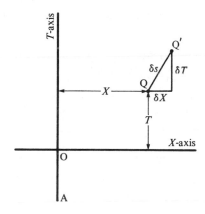

Figure 24.8

24.9m The interpretation of (24.3) in terms of increments needs a little care. An increment has to be an increment *in* some quantity – the amount by which that quantity increases (§21.38). Obviously δT and δX are the increments in the quantities T and X – the amounts by which these increase when you go from Q to Q'. But there is no obvious quantity s for δs to be an increment in.

   However, in all the cases that matter for us we shall find, when we come to them, that there *is* a suitable quantity s for δs to be an increment in. And in the meantime there will be no harm in using the δs form as a sort of 'courtesy title' for the interval between Q and Q'.

24.10 The equation or near-equation which shows how to calculate the invariant from the increments in the co-ordinates is called the *metric*. Thus a geometry is completely characterised by its metric (§§24.6, 12.21–2).

24.11 So (24.3) is the metric for the spacetime of Special Relativity on the 1D universe. Similarly in the geometry of the plane the invariant is distance, for which we can again use the symbol δs. Then, using co-ordinates T and X of §§11.10–11, **prove that the metric is**

$$\delta s^2 = \delta T^2 + \delta X^2. \tag{24.4}$$

Figure 24.8 will still serve, provided you interpret T and X as the 'thatway' and 'cross' of §11.10. And applying Pythagoras' Theorem (§3.15 and cf. §11.19) gives (24.4).

24.12 Equation (24.3) gives the metric for a no-gravity spacetime in terms of the co-ordinates of an inertial observer. We can now work out the metric for the same spacetime, but in terms of the co-ordinates of W, the observer with constant acceleration. We have only to use (22.27) and (22.28) to translate A's measurements into W's. So substituting for δT and δX in (24.3) gives

$$\delta s^2 \sim (1 + fx)^2 \delta t^2 - (1 + fx)^2 \delta x^2. \tag{24.5}$$

We've used §17.25 to write the square of $(1 + fx)\delta t$ as $(1 + fx)^2 \delta t^2$.

24.13M A useful approximation states that if y is small compared with $1 -$ so small that y^2 is too tiny to matter – then

$$(1 + y)^2 \approx 1 + 2y \tag{24.6}$$

(see §20.17 for the difference between $\approx$ and $\sim$). To prove this, redraw Figure 13.12, but making lengths AE $= 1$ and EB $= y$. Then

ABCD $\approx$ AEKH + EBFK + KGDH

(the negligible error arising because the area y^2 of KFCG is missing). Check (on the lines of §13.12) that this translates into (24.6). Changing y to fx gives

$$(1 + fx)^2 \approx 1 + 2fx . \tag{24.7}$$

24.14 We have already agreed (§22.20) that we shall only consider situations in which fx is very small compared with 1. So it will be good enough for us if we use (24.7) to change $(1 + fx)^2$ to $1 + 2fx$ in (24.5), thus obtaining

$$\delta s^2 \sim (1 + 2fx)\delta t^2 - (1 + 2fx)\delta x^2 \tag{24.8}$$

as the metric of a gravity-free spacetime in terms of W's co-ordinates. (Cf. also the last sentence of §22.20.)

This is more typical than previous cases of the sort of metric we shall meet in future. It has $\sim$ instead of $=$. And on the right it has the squares of the increments *each multiplied by an expression involving one of the co-ordinates.*

24.15 Let's be clear about what (24.8) says. We have an event Q which happens at distance x from W; and another event Q' which happens time δt after Q and distance δx farther away (all according to W). Then we're asserting that the interval δs between Q and Q' is given by (24.8). The quantities x, δt and δx are all measured or calculated by W.

But the interval δs is the invariant that all observers agree about. We proved long ago that all *inertial* observers agree about δs; and the calculation of §§24.12–14 ensures that W also agrees.

24.16m We must also examine the meaning of '$\sim$' in this context. Please re-read §§20.20 and 21.42. The latter copes with near-equations involving *two* increments. But (24.8) has three.

When we look back at how we arrived at (24.8), we see that it came from near-equations (22.27) and (22.28). And (putting it crudely) these state that certain equations can be made as nearly true as we wish by taking δt (in the first of them) and δx (in the second) small enough. So (24.8) clearly means (again putting it crudely) that the equation got by changing $\sim$ to $=$ can be made as nearly true as we wish by making *both* δt and δx small enough. You'll have no difficulty in dealing similarly with future near-equations involving more than two increments. If you want a formal statement, here it is:

> When a near-equation involves the increments of several quantities, it means that by making small enough (but not zero) all the increments whose size can be independently varied, the proportionate difference between the two sides can be made as small as we wish, or (as alternative form) the ratio of the sides can be made as near to 1 as we wish.

And in this context 'increments' includes quantities like δs which have been given that 'courtesy title' (§24.9).

24.17 To obtain (24.8) we used (22.27) and (22.28), which are true *only if* $T = 0$ – i.e., if Q lies on A's distance axis, as in §22.16. But no matter when Q occurs, we can choose as A a suitable inertial observer according to whom $T = 0$ (see start of §22.16). So (24.8) is true at *all* times.

24.18 The reasoning that led us to (24.8) involved two approximations that depend on fx being small compared with 1 – first (22.27) and (22.28) as approximations to (22.22) and (22.23); and later (24.8) as approximation to (24.5). So this form of the metric will only hold good if fx is so small compared with 1 that we can forget about the errors introduced by these approximations. We must check that this is so whenever we use (24.8) or anything derived from it.

(And we are spared embarrassing questions about what happens when $fx = -\frac{1}{2}$. This is far too large a value for our approximations to be any use. There's no problem when we use the exact form of §22.21.)

24.19 We have proved, then, that (24.8) is the metric for a gravity-free spacetime as viewed by accelerated observer W – i.e., expressing δs in terms of W's co-ordinates. **But can you think of another interpretation?**

By the Principle of Equivalence (§23.8, and see also §23.12) we can equally well regard (24.8) as the metric of a spacetime in which there is gravity of constant strength, as viewed by observer W who is supported against the general tendency to fall. This is the interpretation that will most interest us in future.

Note how the relation between δs and the increments in the co-ordinates now depends on distance, x, from the observer – on height above or depth below him. Contrast the Special Relativity metric (24.3), in which the relation between δs and the co-ordinate increments is everywhere the same.

I'm sorry to have to insert three articles of necessary detail before we begin to explore what can be learnt from studying (24.8) as the constant gravity metric.

24.20M It's about time we justified the assertion of §12.12 that $(-a)^2 = a^2$. The basis of the justification is that negative quantities must have such properties that the ordinary rules of arithmetic work out correctly for them. For example, (16.10) must be true for negative quantities as well as positive ones. So by (16.10)

$$b^2 - (-a)^2 = \{b - (-a)\}\{b + (-a)\}$$

§§15.23–4
$$= (b + a)(b - a)$$

(16.10)
$$= b^2 - a^2,$$

which requires $(-a)^2 = a^2$. So a square, as we said in §12.12, cannot be negative.

24.21M Thus the square root (§13.4) of a^2 can be *either a or −a*, since squaring both gives a^2. Any number or quantity has *two* square roots. They have the same absolute value (§13.20), but one is positive and one is negative.

24.22 It's an obvious deduction from §22.38 that if x stands for the distance of something at time t, and if $\delta x^2 \sim U\delta t^2$, where the size of U is not affected by the size of δt, then U is the square of the velocity.

24.23 According to §12.23 the complete theory of a gravity-free spacetime as it appears to inertial observers can be deduced from the Special Relativity metric. Could we hope that the new metric (24.8) would similarly allow us to deduce how things behave in constant gravity conditions?

Pursuing that line will simply mean approaching familiar phenomena by a different (more difficult!) route. But on the way you'll learn new tricks and exercise your skills in familiar surroundings before setting off into the unknown.

24.24 We can easily cope with the behaviour of light in constant gravity conditions on the 1D universe – light signals travelling straight 'up' or 'down'. If the signal goes from event Q to event Q', then the interval δs between Q and Q' is zero. (We proved that for inertial observers in §13.24; and in §24.12 we arranged that interval shall be the same for W as for inertial observers.) In fact we can now *take* δs = 0 *as the condition that characterises the motion of a light signal*. So we should be able to answer our problem by putting δs = 0 in (24.8).

.25m The result of doing so would be

$$(1 + 2fx)\delta t^2 - (1 + 2fx)\delta x^2 \sim 0 .\qquad(24.9)$$

Have a look at the definitions in §§20.20 and 21.42 and **see if you're satisfied with that last statement.**

Call the left-hand side Y for brevity. Then the difference between the sides is $Y - 0 = Y$. And the proportionate difference is either Y/Y or $Y/0$ (according to which side you divide by). But $Y/Y = 1$, and so *can't* be made as small as we wish by taking the increments small enough. And $Y/0$ – dividing a non-zero number by zero – is nonsense. **Check for yourself** that (24.9) also fails to conform to the ratio form of the definition. So (24.9) just does not make sense as a near-equation.

It is actually possible, in certain conditions, to have genuine near-equations with zero on one side.† But we shall not meet any. So we'll simply take action to avoid ever writing anything that *looks* like a near-equation but has zero on one side. **How can we do that in the present case?**

24.26 Add $(1 + 2fx)\delta x^2$ to both sides of (24.8) – and we'll examine the validity of that step in §24.28 – obtaining

$$\delta s^2 + (1 + 2fx)\delta x^2 \sim (1 + 2fx)\delta t^2\qquad(24.10)$$

before putting δs = 0 to get the near-equation describing the light's motion. So this will be

$$(1 + 2fx)\delta x^2 \sim (1 + 2fx)\delta t^2,$$

whence (§21.11) $\delta x^2 \sim \delta t^2$. **Draw a conclusion.**

Article 24.22 (with $U = 1$) tells us that the square of the velocity of light is 1.‡ And so (§24.21) the velocity of light is either 1 or −1. The

---

† For instance if the size of Y is not affected by the size of the relevant increment(s), $Y \sim 0$ turns out to be O.K.; and then §22.30 leads to $Y = 0$. What makes this possible is that the ratio 0/0 can take various values according to the circumstances.

‡ Since this is constant, we could deduce that actually $\delta x = \delta t$. Hence the left-hand side of (24.9) is zero. And so in *this* case the near-equation with zero on one side *would* have been valid. But the considerations of §24.25 will apply without reservation in the rest of our work.

latter merely refers to a signal going in the negative direction – downwards. Thus in constant gravity conditions a stationary (supported) observer finds that light going straight up or down has constant speed – the same speed as is measured by a free-falling (inertial) observer. (See §23.12 if the relation between supported and free-falling observers is not clear.)

24.27 But of course it *had* to work out like that. The way we defined 'distance according to W' in §22.3 arranges that the speed of light according to W will be 1 (cf. §§7.10–13). All we've done is to exhibit in one very simple case how the behaviour of something can be deduced from a knowledge of the metric.

Unfortunately this is a deceptively simple example – there's much more to be done before we can deduce anything else. But at least it has suggested the lines on which we shall have to work. Putting it very roughly, to solve a gravitational problem we must

> (1) Find the solution in the conditions of the Special Theory of Relativity (in this example the solution is: speed of light = constant = 1);
> (2) Express this solution in terms of the metric only (in this case, $\delta s = 0$); and
> (3) Assume that the solution expressed in this form will still be valid in terms of whatever metric applies to the more complicated gravitational situation.

These misty ideas will acquire firm solidity as we go on.

24.28m Adding the same thing to both sides of an *equation* is always an allowable operation (§9.12). But adding the same amount to both sides of a *near*-equation does not necessarily give another near-equation. For example, you can **use the definitions (§§20.20, 21.42) to prove that** if K is a constant, then $\delta x + K \sim 2\delta x + K$; but what you get by adding $-K$ to both sides, namely $\delta x \sim 2\delta x$, is clearly false (the ratio of the sides is 2). So we are required to justify the addition that took us from (24.8) to (24.10).

The difference between the sides is the same in both cases: $\delta s^2 - (1 + 2fx)\delta t^2 + (1 + 2fx)\delta x^2$. So the *proportionate* difference for (24.8) is this quantity divided by δs^2; and for (24.10) it is the same quantity divided by $\delta s^2 + (1 + 2fx)\delta x^2$, which is larger. Thus the proportionate difference is less for (24.10) than for (24.8).

Now the meaning of (24.8) is that we can make its proportionate difference as small as we wish by taking δt and δx small enough. And so the same is clearly true for the smaller proportionate difference of (24.10).

25

Introducing geodesics

25.1 Let's consider how we might apply the procedure vaguely described in §24.27 to investigate the motion of a free-falling (inertial) body under the influence of gravity. Step (1) requires us to say **what characterises inertial motion in the conditions of Special Relativity.**

By §5.6 it is motion in a straight line at constant speed (relative to some inertial observer whose co-ordinate system we are using).

25.2 For step (2) of §24.27 we need to express that answer in terms of the metric. Let's make a start by considering the 'straight line' part of the answer in isolation; and let's think in classical slow-speed terms, so that we can ignore the partial interchangeability of space and time. **Then the invariant is . . . ?**

Distance, of course (§§11.20–1). Now can you express the straight line idea in terms of distance – **can you give, in terms of distance, a definition that will enable you to pick out straight lines from among all other curves?**

The straight line between two points is the shortest of all curves joining them. Let's speak of the *distance measured along the curve,* rather than the length of the curve. Then this recipe for finding straight lines becomes

> Among all the curves joining two points, the straight line is the one along which the distance between the points has a minimum value.

The last four words are just a more technical (but often convenient) way of saying 'has least size' or 'is shortest' (§21.2, footnote).

25.3 To make this recipe work we must, of course, know how to measure distance along a curve. We can't use a flexible tape measure (how do you know it keeps constant length?). Distance for us is defined by the radar method (§7.13). We must start from that (or maybe its more complicated alternative, the rigid measuring rod, §7.17). And these measure straight line distance only.

If we divide the curve into portions like PQ, QR and RS of Figure 25.3, then we can say that the sum of the straight line distances P to Q, Q to R, R to S, etc., gives an *approximation* to distance measured along the curve. And obviously we can make this approximation as good as we wish by taking the portions small enough. To put that properly we'll need again to use near-equations.

Figure 25.3

25.4 Let's concentrate on one portion PQ (Figure 25.4). If *s* stands for distance measured along the curve from some agreed starting point O, the distance from P to Q along the curve will be an increment in *s* – denoted by δs. And we'll use s_i to stand for the straight line distance from P to Q. **Why do I call it s_i rather than δs_i?**

Because this length is not an increment in anything. More precisely, it is not an increment in any variable quantity that is relevant to the problem in hand (though it *is*, for example, an increment in length measured along the line KL – which is *not* relevant). And in various matters that will soon arise this distinction is vital. Let's examine it more carefully.

25.5m If you think of a moving point that starts from O and travels steadily along the curve through P, Q, R, . . . , then the distance measured along the curve from O to this moving point steadily increases. As it goes from P to Q, s increases by the distance PQ measured along the curve. So this last *is* an increment in s. And so also is distance QR measured along the curve – it is the amount by which s increases while the moving point goes from Q to R. And again distance PR along the curve is an increment. Here we notice one vital property of increments: two of them added end to end make another increment – a larger one.

But the straight line distances PQ and QR do *not* add up to make straight line distance PR. So there is no continuously growing quantity of which these three are increments.

25.6 Still referring to Figure 25.4, we can now assert that

$$\delta s \sim s_i . \tag{25.1}$$

For this merely says (cf. §25.3) that the distance δs along the curve is approximately equal to the straight line distance s_i, and that by taking

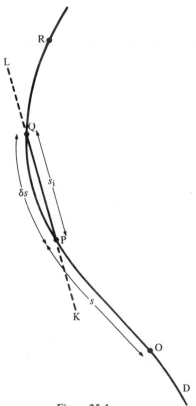

Figure 25.4

Q close enough to P we can make the approximation as good as we wish (as defined more precisely in §§20.20, 21.42).

In an actual problem we might know some way of calculating s_i for each portion. That would give an approximation to the corresponding δs. Carrying out this calculation for all the portions and then adding would give an approximation to the total distance. And we could make this as near as we wish to the true value by taking the portions small enough. That's our as-accurately-as-we-wish recipe for calculating distance measured along the curve. The process is very like that of §§20.22–5. I suggest you revise those articles, and then start again from §25.3.

25.7 When we speak, without further qualification, of the distance between two points, we mean the *straight line* distance between them. Similarly, coming back to spacetime, when we speak without qualification about the interval between two events, we mean the interval measured directly – the interval as given by the metric (24.3) in terms of some inertial observer's co-ordinates. But we can also speak of the interval between two events *measured along the world line* of some accelerated observer who is present at both – the spacetime analogy of distance measured along a curve. A very similar recipe will do for calculating it.

Figure 25.4 will serve again if we interpret the curve OPQR as the world line of accelerated observer D. Then s_i will stand for the interval from P to Q measured directly. Notice that s_i is again not an increment. But if s stands for interval measured along the world line, starting from event O, then the interval from P to Q along the world line *is* an increment, δs, in s.

Then s_i is an approximation to δs, which we can make as good as we wish by taking Q close enough to P. In other words, (25.1) holds good with these new meanings for s_i and δs. So we could calculate approximately the total interval along the world line – say from O to R – by dividing it up into portions, taking δs for each portion to be equal to s_i, and finally adding all the increments together. The method would be just like that of §§20.22–5. And the argument of §20.25, slightly adapted, shows that we can make the approximation as good as we wish by using small enough portions.

25.8 The basic thing that we shall do over and over again with the metric is to use it to calculate interval measured along a world line (or signal line). Then δs *is* a true increment and our worries of §24.9 are at an end.

But (24.3) refers to interval measured direct. The δs of that equation is the quantity that we called s_i in §25.7. So really we should have written (24.3) as $s_i{}^2 = \delta T^2 - \delta X^2$, and then (25.1) would have given

$$\delta s^2 \sim \delta T^2 - \delta X^2.\tag{25.2}$$

So this, rather than (24.3), is the preferred form for the Special Relativity metric. All our other metrics will be near-equations anyway, and so this question will not arise again.

The form (25.2) will also serve for the metric in a complete space-time with three dimensions of space plus time, provided we interpret δX as the *distance* between P and Q – not just as the increment in some X co-ordinate. This distance can itself be expressed, when necessary, in terms of the increments in the space co-ordinates.

25.9 Still interpreting the curve of Figure 25.4 as D's world line, let δt stand for the time from P to Q by D's clock. **Can you guess the relation between δs and δt?**

If necessary, reread §13.22 and **try again.**

If the observer were inertial, we should have $\delta s = \delta t$. So we might guess that this relation would hold good in all cases. Let us try to prove that.

25.10 Let t_i be the time from event P to event Q as measured by an inertial observer E, who is present at both. Figure 25.10 reproduces Figure 25.4, with times marked instead of intervals and line KL renamed as E's world line. Notice that δt is an increment (**in what quantity?**), but t_i is not. **Then what is the relation between s_i and t_i?**

By §13.22

$$s_i = t_i \, .\tag{25.3}$$

And what is the relation between δt and t_i?

The Additional Assumption (§20.19, with δt instead of t) gives

$$\delta t \sim t_i \, .\tag{25.4}$$

What emerges when we combine (25.1), (25.3) and (25.4)?

If we use (25.3) to put s_i instead of t_i in (25.4), we get $s_i \sim \delta t$. Then, using §21.10, the combination of this with (25.1) gives

$$\delta s \sim \delta t \, .\tag{25.5}$$

and not $\delta s = \delta t$ as we were hoping. What a disappointment!

25.11 But we've done better than you'd think. Please check up on the meaning of near-equations (§§20.20, 21.42). And then have a look through §§21.20–2 – where you'll translate $k_1 - k_0$ and $ft_1 - ft_0$ into δk and $\delta(ft)$, of course. In summary (25.5) states that the proportionate error involved in saying that δs is actually equal to δt can be made as small as we wish by taking δt small enough. And §21.22 exemplifies a method of investigating the size of this error: by dividing the step into smaller ones – subdividing the increment δt in the present case. Instead of going from P to Q in one increment δt, we could use two successive increments of $\frac{1}{2}\delta t$ (Figure 25.11).

Let δs_1 stand for the increment in s in the first of these steps – the amount by which s increases when t does its first $\frac{1}{2}\delta t$ increase. And similarly let δs_2 be the increment in s in the second half-step. Then applying (25.5) to each gives $\delta s_1 \sim \frac{1}{2}\delta t$ and $\delta s_2 \sim \frac{1}{2}\delta t$. So it is approximately true that

$$\delta s_1 = \frac{1}{2}\delta t \text{ and } \delta s_2 = \frac{1}{2}\delta t \ . \tag{25.6}$$

But $\delta s = \delta s_1 + \delta s_2$ (this is the vital adding-end-to-end property of increments that we noted in §25.5). And substituting from (25.6) leads again to the result that, approximately, $\delta s = \delta t$. We get precisely the same result for our approximation whether we do the calculation in

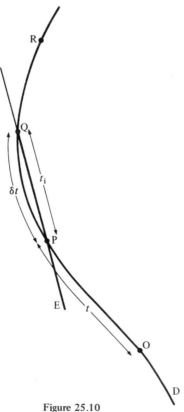

Figure 25.10

two half-steps or one whole step. And you can easily check that no matter how much we subdivide our step into smaller ones, we shall always reach the same conclusion: namely that the supposedly improved approximation to the truth is *still* $\delta s = \delta t$.

Interesting! We can always get an approximately true equation by changing $\sim$ to $=$. By taking the increments small enough we can arrange for this approximation to be as good as we wish – that's what '$\sim$' says. Yet when we *do* take smaller steps, the approximately true equation that comes out is always the same, namely $\delta s = \delta t$. **What conclusion would you draw?**

The only possibility is that the statement $\delta s = \delta t$ is, after all, an *exact* equation. Then *of course* we can arrange, by taking the steps small enough, for it to be as accurate as we wish – since it is already as accurate as anybody could possibly wish. If this argument worries you, read over §§22.26–30, which involved a rather similar kind of reasoning, and think again. I'm sure you'll be convinced that though we started with the *near*-equation $\delta s \sim \delta t$, we can deduce from this the *exact* equation $\delta s = \delta t$.

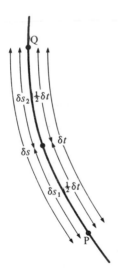

Figure 25.11

5.12m The First Fundamental Theorem (§22.30) also stated that in certain conditions we can upgrade $\sim$ to $=$. In that case the quantities on either side of $\sim$ had to have the special property that their size is not affected by the size of the increments. And (25.5) also has a special feature, though a different one: the quantities connected by $\sim$ are *the incre-ments only*. Contrast, for example, (21.20) – which brings in the quantity k as well as δt and δk. Will it always work out like this when *only the increments* are involved? This suggests the:

25.13m
Second Fundamental Theorem
If variable quantities y and z are related in such a way that the near-equation

$$\delta y \sim \delta z \qquad\qquad (25.7)$$

connecting their increments is always true (i.e., true for increments starting from any value of y and the corresponding value of z), then actually
(i) $\delta y = \delta z$;
(ii) $y - y_0 = z - z_0$,
where y_0 and z_0 are one pair of corresponding values of y and z; and
(iii) $y = z + a$,
where a is a constant quantity.

And *please note well:* all this follows because (25.7) involves *only* the increments. In a case like (21.20) we already know that we can't change $\sim$ to $=$.

25.14m **Can you prove (i) of §25.13?**
I asked a similar question in §22.30. If I caught you out then, **don't let it happen again.**

Article 25.11 makes no reference to what s and t (or their increments) actually represent. They could stand for *any* two quantities. So if you change δs and δt to δy and δz, you've got the required proof.

Conclusion (ii) of §25.13 is simply conclusion (i) restated with the increments written in the more primitive way we used in Chapter 21 before we developed the 'δ' notation. Finally, if we add y_0 to both sides of (ii), we get $y = z + y_0 - z_0$. And renaming $y_0 - z_0$ as a gives form (iii). So the Second Fundamental Theorem is really rather easy to prove.

25.15m **Can you see why I was at pains in §§25.4–5, 25.7 and 25.10 to emphasise that s_i and t_i are not increments?**
If t_i really were an increment, then the Second Fundamental Theorem applied to (25.4) would have given $\delta t = t_i$ – contradicting all we've discovered about the non-existence of Public Time, the experiences of the space-travelling twin and the like. Actually the argument that we used to prove the Second Fundamental Theorem (§25.11) would break down in this case, because it depends on the increments adding end to

end (§25.5). The quantities corresponding to t_i for the single step and the two half-steps would not have this property (the Three Clocks again!).

25.16 We can now go back to complete our work on the relation between time by an observer's clock and interval measured along his world line (§§25.9–11). If s and t are measured from the same zero (O in Figures 25.4 and 25.10), then a pair of corresponding values are $s_0 = 0$ and $t_0 = 0$. So applying conclusion (ii) of the Second Fundamental Theorem to (25.5), we deduce that $s = t$. In other words,

> Interval measured along an observer's world line is equal to time as measured by his clock.

(And so the instrument for measuring time-like interval is a clock.)

25.17 There's a wheels-within-wheels structure about what we're doing. So let's recall that our main problem at present is to find a way of describing inertial motion in a Special Relativity spacetime in terms of the metric (§24.10). We coped with the purely spatial aspects of that problem in §25.2 – but only by way of illustration. Now let's tackle the total problem in spacetime. Actually the answer is staring you in the face – provided you bring to mind the relevant point from our earlier studies. **So what feature expressed in terms of interval, distinguishes inertial observers from accelerated ones?**

Take the conclusion we established in §20.28 (in its first form) and combine it with the conclusion of §25.16, **and you deduce that**

> Among all the observers who are present at two events, the inertial observer is the one along whose world line the interval between the events has a maximum value.

(where the last four words just mean 'is biggest', cf. §25.2).

25.18 Compare that with §25.2. Just as that earlier conclusion is a recipe for *picking out the straight line* from among all the curves joining two points, so also our latest result is a recipe for *picking out the inertial observer* from among all the observers who are present at two particular events. The two prescriptions are very much alike, except that one calls for the invariant to be minimum, the other for it to be maximum.

From (25.2) it follows that we can only get greatest interval by having least distance. And so the 'straight line' aspect of inertial motion is automatically included when we describe it in maximum-interval terms.

25.19 The invariance of interval and the proposition about time being greatest for the inertial observer both hold good in a normal spacetime of three spatial dimensions plus time (§§24.5, 25.8, 20.28). And so the same applies to the conclusion of §25.17 about interval being greatest along an inertial world line. (For the meaning of 'world line' in such a three-plus-one-dimensional spacetime, see §20.30.)

On the other hand, we've established this property of inertial motion *only* for the conditions in which the Special Theory of Relativity applies – no gravity and using the co-ordinate system of some inertial observer, so that the metric is (25.2). Nevertheless it *is* expressed in terms of the metric only, and so §24.27 encourages us to hope that it will still hold good even when gravity presents us with a more complicated geometry characterised by a more complicated metric. If that proves to be true, we should have a recipe for studying inertial motion – free fall – under the action of gravity.

25.20 To illustrate this idea, let's consider the possibility of doing a similar generalisation for the 'least distance' specification of a straight line in space. To make it simpler, let's work in only two dimensions. If our two-dimensional space is an ordinary plane, the metric is (24.4). And this is the case for which the 'least distance' recipe of §25.2 is known to work.

But the surface of a sphere is also a two-dimensional space. Please try *not* to think about the three-dimensional space it's embedded in. In the language of §12.21 the geometry of a spherical surface means the rules for working with measurements on that surface without ever referring to anything not on the surface. More imaginatively, it is the geometry that would be worked out by creatures who live on the sphere's surface and who have no consciousness of anything not on it.

On this surface you can't set up rectangular co-ordinates like those of Figure 24.8. If in doubt, try drawing them on an orange – you'll find, among other troubles, that the *T*- and *X*-axes meet twice. But other sorts of co-ordinate system are possible – for example the latitude and longitude used in geography. One degree of latitude corresponds to 110 kilometres along a meridian. But the conversion factor from degrees of *longitude* to kilometres of distance depends on the latitude – one degree is 110 kilometres at the equator, 70 kilometres at the latitude of London. Thus the metric is more complicated than (24.4). In suitable units it can be expressed as

$$\delta s^2 \sim \delta(\text{latitude})^2 + (\text{conversion factor}) \times \delta(\text{longitude})^2, \quad (25.8)$$

where the size of the conversion factor depends on the latitude.

25.21 There are no straight lines on a sphere. So what happens if we apply the 'shortest distance' criterion there? Mathematically that's a tough question. So let's answer it experimentally. Stick two pins in an orange and stretch an elastic band tightly between them. The band now marks the shortest curve joining the feet of the pins.

This shortest curve turns out to be part of what is called a *great circle*. Looked at from outside, a great circle is the curve marked out on the surface by a plane passing through the centre of the orange. Sticking to the surface itself, it is what its name says – the largest circle you can have on the sphere. Or, more relevant for present purposes, it is the path of an ant crawling on the outside of the orange and always going straight forward. So it is the *straightest possible* path.

There are no straight lines on the surface of a sphere. But the 'shortest distance' recipe gives the straightest curve that does exist there. The method begins to look promising.

25.22 Though we resorted to experiment, it would be possible, of course, to reach that conclusion by calculation. Imagine Figure 25.4 drawn on the sphere. Knowing the co-ordinates at the end points, P and Q, of a portion, we could use the metric (25.8) to calculate the increment δs in distance (and since this metric already incorporates a $\sim$ sign, we don't need to bother about the distinction between δs and s_i). Then the process of §§25.3–6 would enable us to calculate the distance along the curve from O to R. And we could compare the distances along various curves and prove that the shortest is indeed a great circle.

25.23 Similarly, if we have a spacetime with a metric more complicated than (25.2), we could use the method of §25.7 to calculate the interval measured along any world line between two events, and then pick out the particular world line for which this interval has maximum value. And the hope expressed in §25.19 is that this would still turn out to be the world line of an inertial observer – that is, a free-falling observer.

This, of course, is no more than an intelligent guess (and we're ready for the facts to prove us wrong). We've found that when the metric of a *space* is more complicated than (24.4), the 'least distance' recipe produces a natural and useful generalisation of the straight line idea. By analogy we're hoping that the 'greatest interval' recipe will equally naturally cover free fall no matter what the spacetime metric may be. And actually we can do a little more immediately towards making this suggestion plausible.

25.24 Consider again the creatures we imagined in §25.20, whose consciousness is confined to things happening on the surface of a sphere. This surface would be all of space for them;† and their spacetime would be the spherical surface plus time – call it S-type spacetime for short. Its metric would state that $\delta s^2 \sim \delta t^2$ *minus* the right-hand side of (25.8). What would inertial motion be like in S-type spacetime?

Maximum interval implies minimum distance (the argument for that is exactly as in §25.18, even though the formula for distance is more complicated). So (leaving aside the question of speed) our suggested maximum-interval recipe (§§25.19, 25.23) says that the spatial path of anything moving inertially in S-type spacetime will be one along which distance is minimum. And that, we know, is a great circle (§§25.21–2). Does experiment confirm this?

This inertial motion in S-type spacetime would be the motion of a body which, as you and I would describe it from outside, is constrained to move on the spherical surface but is otherwise not affected by forces. You can make a fair approximation to that sort of motion by tying something heavy to the end of a piece of string, the other end of which you hold in your hand, and whirling it round so fast that the effects of gravity on its motion are of minor importance. **What path does it follow?**

Ideally (for it's probably not really going fast enough) it *is* a great circle – as predicted. So the maximum-interval recipe does give the right answer for the purely spatial aspects of inertial motion in an S-type spacetime. This is admittedly a rather trivial example; but so far as it goes, it does support the suggestion that the 'world line along which interval is greatest' recipe for inertial motion will work out correctly even when the metric is more complicated than (25.2).

25.25   In that case the more complicated metric arose because we selected a particular two-dimensional space, carved out (as it were) from our ordinary three-dimensional space. But maybe such analogies make it not unreasonable to pursue our hope that the maximum-interval recipe will still work when the complication of the metric arises from the action of gravity – as with (24.8) interpreted as in §24.19, and more significant examples to follow.

25.26m   The prescription for a straight line requires a minimum value for the invariant; that for an inertial world line demands a maximum. It's convenient to embrace the two in the phrase 'extreme value' – which is to be regarded simply as an abbreviation for 'either a maximum or a minimum value'.

---

† Assuming they move slowly enough to ignore the interchangeability of space and time (Chapter 11).

So the recipes for straight lines and inertial observers (§§25.2, 25.17) could be rewritten so that both end ' . . . has an extreme value'.

25.27m These curves along which the invariant has an extreme value are called *geodesics*. If we agree that in a spacetime context 'point' means 'event' and 'curve' means 'world line or signal line' (the invariant being distance or interval as appropriate), then we can give the definition:

> The geodesic between two points is that curve, out of all the curves joining them, along which the invariant has an extreme value.†

Let's take it for the present that this extreme value is not zero. Then the curve is called an *ordinary* geodesic. So we've met ordinary geodesics in §§25.2, 25.17 and 25.21.

In most contexts it's best not to bother about whether the extreme value is a minimum or a maximum – though we know, when we need it, that it's the former for space and the latter for spacetime.

25.28 Our hope about the maximum-interval recipe as a means of characterising inertial or free-falling motion in all conditions can now be restated in the language of geodesics. Let's put it formally as the

> *Geodesic Assumption*
> Even when there is gravity, making the metric more complicated than (25.2), the world line of an inertial (free-falling) body is an ordinary geodesic.

A glance back at §24.27 will show you that we now have a clear programme for studying free fall under gravity. But to put it into practice we need to know how you decide which curves are geodesics when you are told the metric that characterises the gravitational conditions that are being investigated. And that needs a chapter to itself.

---

† We shall never have more than one geodesic between two points. But that *can* occur in other contexts, and then the definition needs broadening.

26

How to find ordinary geodesics

26.1M This will be a tough chapter. But at least the preliminaries are simple. Using (16.10), we have

$$(x - b)^2 - b^2 = (x - b - b)(x - b + b) = (x - 2b)x.$$

Or, interchanging the sides and using §16.12,

$$x^2 - 2bx = (x - b)^2 - b^2. \tag{26.1}$$

Being derived from identities, this is also an identity (§16.9) – its truth does not depend on what x and b stand for.

26.2M We now ask: for what values of x does the expression $x^2 - 2bx$ have an extreme value (b being a constant)? Well the term $(x - b)^2$ in (26.1) cannot be negative (§24.20). So its minimum value is zero, and this occurs when $x = b$. Since $x^2 - 2bx$ differs from $(x - b)^2$ only by the constant b^2, it follows that this expression is also minimum when $x = b$.

On the other hand, we can clearly make this quantity as big as we wish by taking x large enough. So its only extreme value is this minimum.

26.3M Now consider the more general expression $mx^2 - 2nx$, where m and n are constants. Using §16.12, we can put this in the form

$$m\left(x^2 - 2\frac{n}{m}x \right).$$

And then §26.2 (with $b = n/m$) tells us that

$$mx^2 - 2nx \text{ has an extreme value when } x = n/m.$$

(Whether this is minimum or maximum depends on whether m is positive or negative.)

26.4M The problem we're now going to tackle – how to find geodesics when you're given the metric – is, I regret to say, the toughest and trickiest bit of mathematics we've met yet, maybe even the worst in the whole book. Not that it's conceptually difficult or particularly subtle. But it consists of one close-knit piece of reasoning all the way from here to §26.23. You're probably not used to such sustained arguments. And there's a limit to what I can do to help. For example, it wouldn't be much use giving you a series of verbal paraphrases of the various mathematical statements, as I used to do. That would only make the long train of reasoning even longer and therefore harder to grasp as a whole. On the contrary, I must shorten the reasoning as much as possible, and that means making more thorough use of mathematical language's remarkable powers of abbreviation.

So you will have to carry yet farther the attitude of being ready to give much more thought per sentence than you would in most types of reading (§16.23). And of course when you can't easily see the meaning of a mathematical statement, you know how to expand its abbreviations till it becomes clear. You'll seldom need – as you did in the early days – to translate it totally into words; you'll find a suitable compromise. For example, when you come to (26.9) – glance ahead now; never mind what σ_P and σ_Q stand for, but K is a *constant* number – you might well say to yourself 'I can see that $(1/K + 1)$ is also a constant. So this tells me that the square of $\sigma_P + \sigma_Q$ is proportional (§15.3) to $K\sigma_P^2 + \sigma_Q^2$.' And (26.12) might provoke the first thought of 'That's fearsomely complicated,' quickly followed by 'Ah, but I can see it says that this $K\sigma_P^2 + \sigma_Q^2$ can be expressed as the sum of a lot of bits, each of which is reasonably simple' – and then you'd do as much detail about each bit as you need.

To start with, however, I hope you won't bother much with details. Just concentrate on getting a grasp of the *general shape* of this long argument. To help, I've put it in a rather special form. The original problem is too hard for us. So we find a way of simplifying it. But still it's too hard, and so we simplify it again. . . and again. . . and again. After the fourth simplification we've got a problem we *can* solve. And then step by step we put back the complications we previously left out. I think you'll have no serious difficulty in following this general strategy.

The big thing is to avoid getting bogged down in detail. And in three places where there is a danger of dull detail distracting you from the essentials, I have simply said 'it can be proved that. . .', and given a reference to a later article where, in due course, you can check up. So, with the emphasis always on the broad plan, not the detail, will you now carry on to §26.27, where further advice awaits you.

26.5 Let's consider a space or spacetime with three co-ordinates that we'll call x, y and z. These may be distances, or time and distance, or any other measurements that would specify a point or an event. Leaving them thus undefined will give us more flexibility in future.

When we apply our results, there will usually be two space co-ordinates and one of time. But for the present you may find it easier to think of *x*, *y* and *z* as three spatial co-ordinates – even visualise them as distances from the floor and two walls of your room. But if you do visualise like this, try to imagine that some strange magic has distorted the geometry – so that, for example, Pythagoras' Theorem needs modifying.

26.6 The invariant will be denoted as usual by *s* – which we can interpret as distance or interval as required. And (again for the sake of flexibility) we'll express the metric in the form

$$\delta s^2 \sim \alpha \delta x^2 + \beta \delta y^2 + \gamma \delta z^2, \qquad (26.2)$$

where† α, β and γ stand for quantities whose value (size) may depend on the value of *x*, but *not* on the values of *y* and *z*.

That last restriction is vital. Without it, the following proof would not work. But it enormously simplifies the job. And all the metrics we shall meet will have this property that α, β and γ vary with *only one* of the co-ordinates – (24.8) is an example.

It is convenient to sweep the $\sim$ into a corner by putting (26.2) in the form‡

$$\delta s \sim \sigma , \qquad (26.3)$$

where

$$\sigma^2 = \alpha \delta x^2 + \beta \delta y^2 + \gamma \delta z^2. \qquad (26.4)$$

Why have I called it σ, not $\delta\sigma$?

To remind us that this is not an increment (§§25.4–5). Otherwise we might misguidedly apply the Second Fundamental Theorem (cf. §25.15).

26.7 We seek a recipe for specifying the geodesic (§25.27) between two points (or events) O and S (Figure 26.7) – for picking out the geodesic from among all the curves joining O and S. In other words, we want a rule for fixing the co-ordinates of every point on the geodesic. But in that form the problem is much too hard. So we start our process of successive simplifications.

26.8 *First simplification.* We assume the problem already solved so far as the *x* co-ordinate is concerned. That is, we assume that every point already has the correct *x*, and we have only to alter the *y* and *z* co-ordinates of each point till they give an extreme value to the invariant measured along the curve from O to S.

---

† These are alpha, beta and gamma – the Greek a, b and c.

‡ This σ is sigma, the Greek s.

Since changing y and z has no effect on α, β or γ (§26.6), we can now consider these as constants at each point of the curve (though they will vary from point to point). That makes things easier, but the problem is still too hard for us; and so we start preparing for the second simplification.

26.9 Part of our difficulty arises because we know δs only through a *near-equation*, (26.2). An obvious idea would be to substitute σ for δs. But that will only be good enough if we work with small enough portions of the curve. And so we divide the whole curve from O to S into portions (marked by dots in Figure 26.7). If KL is a typical portion, δx, δy and δz are the increments in x, y and z in passing from K to L; and the corresponding quantities δs and σ are given by (26.2) and (26.4).

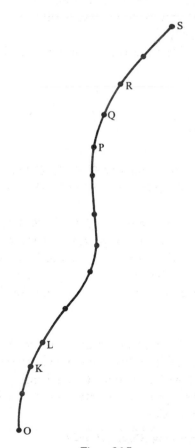

Figure 26.7

The enlarged diagram of the portion KL (Figure 26.9) must be used only as a conventional mnemonic. Increment δx is omitted, since we are assuming that part of the problem solved. The δy and δz increments are represented *conventionally* as horizontal and vertical. The increment δs in the invariant measured along the curve is appropriately marked. And the broken line reminds us that σ does not refer to the invariant at all, but only to its near relation as defined in (26.4).

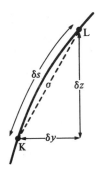

Figure 26.9

The total invariant s measured along the curve from O to S would be got by adding all the increments δs together. But that's too hard for us. So we settle temporarily for taking the σ (instead of the δs) for each of the portions into which the curve has been divided, and adding them all together. Let† Σ stand for the sum of all these quantities σ. By (26.3) the σ for each portion is an approximation to the corresponding δs; and so, on summing all the portions, Σ is an approximation to s. This leads to the

26.10 *Second simplification.* We search for the conditions that give an extreme value to Σ (instead of s).

But dealing with the whole curve at once is too complicated. So we consider a simpler problem concerned with three points. That's the

26.11 *Third simplification.* Out of the portions into which the curve has been divided, we select two consecutive typical portions, PQ and QR (Figure 26.7). We assume that our doubly simplified problem has already been solved for points P and R. And we aim to find a third correct point, Q, in between.

That is to say, we must specify the co-ordinates of Q in such a way that σ from P to Q *plus* σ from Q to R has an extreme value.

† This is the capital form of σ.

Since we're assuming (§26.8) that we know the x co-ordinate in each case, it follows (§26.6) that we could calculate the values of α, β and γ for each point; so we give them names. We'll use the subscript $_P$ to distinguish the values of α, β and γ *at* P, and of δx, δy, δz and σ *starting from* P. And subscript $_Q$ will refer to similar quantities at Q or starting from Q. More explicitly α_P, β_P and γ_P stand for the values which α, β and γ take at P; and α_Q, β_Q and γ_Q stand for their values at Q. And δx_P, δy_P, δz_P, and σ_P stand for the values of δx, δy, δz and σ between P and Q; and δx_Q, δy_Q, δz_Q and σ_Q for their values between Q and R. Figure 26.11, showing an enlargement of this bit of Figure 26.7, will again serve as a mnemonic. Despite its complexity, the notation has a simple logic. Then by (26.4) we have

$$\sigma_P^2 = \alpha_P \delta x_P^2 + \beta_P \delta y_P^2 + \gamma_P \delta z_P^2, \tag{26.5}$$

and

$$\sigma_Q^2 = \alpha_Q \delta x_Q^2 + \beta_Q \delta y_Q^2 + \gamma_Q \delta z_Q^2. \tag{26.6}$$

The problem as it now stands is to find values of the increments which give an extreme value to $\sigma_P + \sigma_Q$.

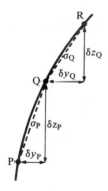

Figure 26.11

26.12 When we use (26.5) and (26.6) to express this in terms of the co-ordinates, the problem gets restated as: Find relations among the increments that will give an extreme value to

$$\sqrt{(\alpha_P \delta x_P^2 + \beta_P \delta y_P^2 + \gamma_P \delta z_P^2)}$$
$$+ \sqrt{(\alpha_Q \delta x_Q^2 + \beta_Q \delta y_Q^2 + \gamma_Q \delta z_Q^2)} . \tag{26.7}$$

The complexities of (26.7) would make even an experienced mathematician shudder. He'd tackle it if he had to. But first he'd try to find a way round it. And that would lead him to our

26.13 *Fourth simplification.* There are an infinite number of correct positions of Q – in Figure 26.11 *any* position on the curve between P and R would do. If we think of Q as moving along the curve from P to R, then clearly σ_P is growing steadily larger and σ_Q steadily smaller. And so the ratio σ_Q/σ_P can be any positive number. Thus if we choose a particular value of K for this ratio, we simply specify how far along the curve between P and R the new point Q is to be. So we say that we are going to look for the particular position of Q that makes

$$\frac{\sigma_Q}{\sigma_P} = K . \tag{26.8}$$

26.14 Using (26.8), we can prove that

$$(\sigma_P + \sigma_Q)^2 = \left(\frac{1}{K} + 1\right)\left(K\sigma_P^2 + \sigma_Q^2\right). \tag{26.9}$$

(The proof is in §26.29, but ignore it for the time being.) It follows that if $\sigma_P + \sigma_Q$ has an extreme value, then so has $K\sigma_P^2 + \sigma_Q^2$; and conversely, if $K\sigma_P^2 + \sigma_Q^2$ has an extreme value, so has $\sigma_P + \sigma_Q$. (The truth of the latter depends on the fact that $\sigma_P + \sigma_Q$ cannot change from positive to negative or vice versa, since we measure one way along the curve.) So instead of the nasty problem stated in §26.12, we are merely required to *discover what values of the increments in the co-ordinates give an extreme value to $K\sigma_P^2 + \sigma_Q^2$.* And that, you'll be glad to know is a problem that we *can* solve.

26.15 Since P and R are fixed (third simplification), the increments in y and z between these two points are constant quantities – call them b and c. Then since increments add end to end (§25.5),

$$\delta y_P + \delta y_Q = b , \tag{26.10}$$

and

$$\delta z_P + \delta z_Q = c . \tag{26.11}$$

Then it can be proved (see §26.30) that (take a deep breath!)

$$\begin{aligned} K\sigma_P^2 + \sigma_Q^2 = {}&K\alpha_P\delta x_P^2 + \alpha_Q\delta x_Q^2 \\ &+ (K\beta_P + \beta_Q)\delta y_P^2 - 2\beta_Q b\delta y_P + \beta_Q b^2 \\ &+ (K\gamma_P + \gamma_Q)\delta z_P^2 - 2\gamma_Q c\delta z_P + \gamma_Q c^2 . \end{aligned} \tag{26.12}$$

And so we have to find the values of the increments that give an extreme value to the quantity on the right of this equation.

26.16 Now that's not as complicated as it looks. Since we are assuming the x co-ordinate already correct, δx_P and δx_Q are constant quantities; and α_P, β_P, γ_P, α_Q, β_Q and γ_Q are also constant (§26.8). So the right-hand side of (26.12) contains only four variable terms, which we can group as

$$(K\beta_P + \beta_Q)\delta y_P^2 - 2\beta_Q b\delta y_P \tag{26.13}$$

and

$$(K\gamma_P + \gamma_Q)\delta z_P^2 - 2\gamma_Q c\delta z_P . \tag{26.14}$$

Then all we have to do is to find the value of δy_P which gives an extreme value to (26.13) and the value of δz_P which gives an extreme value to (26.14). **And you know how to do that.**

26.17 Applying the results of §26.3, we find that (26.13) has an extreme value if

$$\delta y_P = \frac{\beta_Q b}{K\beta_P + \beta_Q} , \tag{26.15}$$

which leads (see §26.31) to

$$K\beta_P \delta y_P = \beta_Q \delta y_Q . \tag{26.16}$$

Similarly (26.14) has an extreme value if

$$K\gamma_P \delta z_P = \gamma_Q \delta z_Q . \tag{26.17}$$

This is the solution to our quadruply simplified problem. If δy_P, δz_P, δy_Q and δz_Q are chosen so as to make equations (26.16) and (26.17) true, then the quantity $K\sigma_P^2 + \sigma_Q^2$ will have an extreme value; and (by §26.14) so also will $\sigma_P + \sigma_Q$.

So now we set about reversing the simplifications.

26.18 *Undoing the fourth simplification.* We get rid of K (and so permit Q to be *any* point of the curve between P and R) by using (26.8). Applied to (26.16) that gives

$$\frac{\sigma_Q}{\sigma_P}\beta_P \delta y_P = \beta_Q \delta y_Q ,$$

whence

$$\frac{\beta_P \delta y_P}{\sigma_P} = \frac{\beta_Q \delta y_Q}{\sigma_Q} , \tag{26.18}$$

and similarly

$$\frac{\gamma_P \delta z_P}{\sigma_P} = \frac{\gamma_Q \delta z_Q}{\sigma_Q} . \tag{26.19}$$

26.19 *Undoing the third simplification* – that is, making the transition from the portions PQ and QR to the curve as a whole. Equations (26.18) and (26.19) say that the quantities $\beta\delta y/\sigma$ and $\gamma\delta z/\sigma$ are to have the *same value* for the portion QR as for PQ. As these are two typical portions, we could apply the argument over and over again to show that (under the same conditions) these two quantities, calculated for any of the portions into which the curve has been divided, must *always* have the same value. In other words

$$\frac{\beta\delta y}{\sigma} = B \quad \text{and} \quad \frac{\gamma\delta z}{\sigma} = C , \tag{26.20}$$

where B and C stand for constant quantities. Or in more convenient form,

$$\beta\delta y = B\sigma \quad \text{and} \quad \gamma\delta z = C\sigma . \tag{26.21}$$

If these two equations are true for *every one* of the portions into which the curve is divided, then the quantity Σ (§26.9) has an extreme value – which solves the problem as it stood after the second simplification.

26.20 *Undoing the second simplification* – making the transition from Σ and σ to s and δs. Using (26.3), we can change (26.21) to

$$\beta\delta y \sim B\delta s \quad \text{and} \quad \gamma\delta z \sim C\delta s . \tag{26.22}$$

26.21 Though these latest near-equations no longer contain σ, nevertheless they still refer to the problem of finding an extreme value for Σ, not for s. Passing on from Σ to s involves a type of argument that we've met before (§§20.22–5 and 25.6–7).

From (26.3) we know that $\delta s \sim \sigma$ for each portion separately. So taking all the portions together, we get $s \sim \Sigma$. In other words, by dividing the curve into a large enough number of small enough portions we can arrange for Σ to be as good an approximation to s as we wish.†

That last statement holds good in regard to *any* curve joining O and S – whether its shape is or is not such as to give an extreme value to Σ. And so it implies that by dividing the curve into small enough portions we can arrange that the conditions which give an extreme value to Σ shall be as near as we wish to the conditions which give s an extreme value. But the conditions that give Σ an extreme value are simply that the near-equations (26.22) shall be true for all portions of the curve. And these already embody the notion of taking small enough portions. So they will also be the conditions which give s an extreme value.

Our conclusion is that for the curve to be a geodesic, the near-equations (26.22) must be true for *any* of the portions into which it is divided. Put otherwise, they must be true when δy, δz and δs are the increments (like those shown in Figure 26.9) starting from *any* point of the curve.

26.22 *Undoing the first simplification.* We can't get away any longer with the pretence that we have already solved the problem so far as concerns the x co-ordinate. We must show how to find the correct value of x at every point – or what amounts to the same thing, find a near-equation for δx to supplement those for δy and δz that we have already.

This time we can't use the method that worked for the y and z

† A complete justification of this assertion would require an argument like that of §20.25, which I forbear to repeat.

co-ordinates. For the deduction of (26.15) depends on $K\beta_P + \beta_Q$ and $\beta_Q b$ being constant (like m and n of §26.3), which is no longer true when we allow x to vary.

However, all we need is another near-equation relating δx to δs; and if it also involves δy and δz, that doesn't matter, since we could use (26.22) to get rid of these. **There's a very obvious candidate.**

The third near-equation is provided by the metric itself, (26.2).

26.23 So our conclusion is that for a curve to be a geodesic, the near-equations (26.2) and (26.22) must be true for the increments δx, δy, δz and δs starting from any point of the curve.

Even if you are still far from confident that you've followed that long train of deduction, don't try to study it further yet. Instead, concentrate on understanding what the conclusion *means;* and then we'll do a very simple example of how it is *used,* before you get down to any necessary further study.

I can't spare the space to go over the meaning in detail. Just a few words of guidance. First be clear what the increments δx, δy, δz and δs are – Figure 26.9 will help – and what α, β and γ mean (§26.6). And then make sure you understand what the three near-equations say, using (as required) §§20.20 (and perhaps passages leading up to it), 21.42 and 24.16. When you're concerned with getting a general impression (rather than stating things precisely) it is good enough to think of (say) the first item in (26.22) as saying that we can make the *equation* $\beta\delta y = B\delta s$ as nearly true as we wish by taking δy small enough. In (26.2) it is δx, δy and δz that have to be made small enough.

You will see that at rock bottom these three near-equations simply say something about how – if the curve is to be a geodesic – the quantities δx, δy, δz and δs must change, in relation to one another, as we shift K and L (Figure 26.9) to different points of the curve.

When you feel reasonably confident about the meaning of this conclusion, move on to the example that follows.

26.24 Let us find the geodesics in the spacetime of the Special Theory of Relativity on the 1D universe. Having only two co-ordinates to cope with, we suppress the δz term in (26.2), getting

$$\delta s^2 \sim \alpha\delta x^2 + \beta\delta y^2.$$

If we now change x to X and y to T, and put $\alpha = -1$ and $\beta = 1$, we have

$$\delta s^2 \sim \delta T^2 - \delta X^2, \tag{26.23}$$

which is the Special Relativity metric (25.2).

In this simple case α and β are both constant. But to provide a faint foreshadowing of things to come, let us behave as if they varied with X – so that the equivalent of (26.22) can be applied only to δT. The first of these near-equations, with y changed to T and $\beta = 1$, gives $\delta T \sim B\delta s$, which we can write as

$$\delta s \sim D\delta T ,\qquad\qquad (26.24)$$

where $D = 1/B$ is also a constant. So for a geodesic the increments must be related by (26.23) and (26.24).

We are not interested in δs, which we therefore eliminate by using (26.24) to substitute $D\delta T$ for it in (26.23) getting

$$D^2\delta T^2 \sim \delta T^2 - \delta X^2,$$

whence (§§3.19, 9.12, 16.12)

$$\delta X^2 \sim (1 - D^2)\delta T^2.\qquad\qquad (26.25)$$

By the Geodesic Assumption (§25.28), this last should describe the motion of an inertial observer – he should move in such a way that (according to another inertial observer who is doing the measuring) his distance increases by δX in time δT, where these quantities are related by (26.25). **Can you deduce anything about his speed?**

26.25 Article 24.22 applied to (26.25) tells us that the square of his speed is equal to $1 - D^2$ – which is also constant. So the speed itself is constant. And we have checked that the Geodesic Assumption, along with our newly discovered prescription for finding geodesics, leads to the correct conclusion that an inertial observer moves at constant speed relative to another inertial observer. **Can you infer any limitation on the possible relative speed of these two?**

26.26 The square, D^2, must be positive (§24.20). Thus the square of the speed (being $1 - D^2$) is less than 1; and so is the speed itself. Our geodesic method has also given us the correct result that the speed of one inertial observer relative to another is necessarily less than the speed of light.

We've been using an enormously powerful steam-hammer to crack a very ordinary nut and discover that the kernel is familiar. But perhaps this abuse of power has helped you to understand how the geodesic steam-hammer works. Later we'll apply it to tasks that would be well beyond the powers of ordinary nutcrackers.

26.27M I can't know how well you've coped with all that. Maybe you're so clever that you can go straight on – Good Luck to you! Maybe you still don't even appreciate the general strategy. If so, you'll have to go over it again (and again?). Making your own brief summary of each article is often helpful; and then you can place it within the general 'simplify, solve and de-simplify' strategy. Eventually, when the general plan *is* clear, you have to make sure of the details. And here the advice of §26.4, paragraph 2, becomes particularly relevant. Finally you'll want to check up on those it-can-be-proved points in §§26.29–31.

6.28M Repeated application of §16.12 yields another useful identity:
$$(a + b)(a + b) = a(a + b) + b(a + b)$$
$$= a^2 + ab + ba + b^2 ,$$
and on tidying up,
$$(a + b)^2 = a^2 + 2ab + b^2. \tag{26.26}$$
Again, by (15.8),
$$(a - b)^2 = \{a + (-b)\}^2$$
$$= a^2 + 2a(-b) + (-b)^2 .$$
And so, using §24.20,
$$(a - b)^2 = a^2 - 2ab + b^2. \tag{26.27}$$

26.29 Proving (26.9). From (26.8) we have
$$\sigma_Q = K\sigma_P$$
and
$$\sigma_P = \frac{1}{K}\sigma_Q .$$
So
$$\sigma_P\sigma_Q = \sigma_P K\sigma_P = K\sigma_P^2,$$
and also
$$\sigma_P\sigma_Q = \frac{1}{K}\sigma_Q^2 .$$
Adding corresponding sides of these two,
$$2\sigma_P\sigma_Q = K\sigma_P^2 + \frac{1}{K}\sigma_Q^2. \tag{26.28}$$
Then by (26.26)
$$(\sigma_P + \sigma_Q)^2 = \sigma_P^2 + 2\sigma_P\sigma_Q + \sigma_Q^2$$
(26.28)
$$= \sigma_P^2 + K\sigma_P^2 + \frac{1}{K}\sigma_Q^2 + \sigma_Q^2$$
§16.12 (in reverse)
$$= (1 + K)\sigma_P^2 + \left(\frac{1}{K} + 1\right)\sigma_Q^2$$

§16.12, and since K times $1/K$ is 1,

$$= \left(\frac{1}{K} + 1\right)K\sigma_P^2 + \left(\frac{1}{K} + 1\right)\sigma_Q^2$$

§16.12

$$= \left(\frac{1}{K} + 1\right)\left(K\sigma_P^2 + \sigma_Q^2\right).$$

26.30 Proving (26.12). From (26.10),

$$\delta y_Q = b - \delta y_P, \tag{26.29}$$

whence, using (26.27), $\delta y_Q^2 = b^2 - 2b\delta y_P + \delta y_P^2 = \delta y_P^2 - 2b\delta y_P + b^2$. Similarly, $\delta z_Q^2 = \delta z_P^2 - 2c\delta z_P + c^2$. Substituting these in (26.6), we have

$$\sigma_Q^2 = \alpha_Q\delta x_Q^2 + \beta_Q(\delta y_P^2 - 2b\delta y_P + b^2) + \gamma_Q(\delta z_P^2 - 2c\delta z_P + c^2)$$

§16.12

$$= \alpha_Q\delta x_Q^2 + \beta_Q\delta y_P^2 - 2\beta_Q b\delta y_P + \beta_Q b^2 + \gamma_Q\delta z_P^2 - 2\gamma_Q c\delta z_P + \gamma_Q c^2.$$

Adding each side of this to K times the corresponding side of (26.5) gives (26.12).

26.31 To prove (26.16). Multiplying both sides of (26.15) by $K\beta_P + \beta_Q$, using §16.12, and subtracting $\beta_Q\delta y_P$ from both sides (write that out in stages if you need to) gives

$$K\beta_P\delta y_P = \beta_Q b - \beta_Q\delta y_P$$

§16.12

$$= \beta_Q(b - \delta y_P)$$

(26.29)

$$= \beta_Q\delta y_Q.$$

26.32 The first of the (26.22) pair, after dividing both sides by B, can be rewritten as $\delta s \sim B'\beta\delta y$, where $B' = 1/B$ is another constant. Its mate can similarly be put in the form $\delta s \sim C'\gamma\delta z$, where $C' = 1/C$.

26.33 For future reference, let's summarise:

> We are concerned with a space or spacetime in which the metric is
> $$\delta s^2 \sim \alpha\delta x^2 + \beta\delta y^2 + \gamma\delta z^2, \tag{26.30}$$
> where α, β and γ stand for quantities whose values may depend on the value of x, but not on that of y or z. Let δx, δy and δz stand for the increments in the co-ordinates in passing from one point of a curve (or world line) to another.
> Then this curve (world line) will be a geodesic if the incre-

ments are related by (26.30) and by

$$\beta\delta y \sim B\delta s \quad or \quad \delta s \sim B'\beta\delta y \,, \tag{26.31}$$

together with

$$\gamma\delta z \sim C\delta s \quad or \quad \delta s \sim C'\gamma\delta z \,, \tag{26.32}$$

where B, C, B' and C' are constants.

The choice of alternatives is a matter of convenience. But altogether we need *one* near-equation from (26.31), and *one* from (26.32), along with (26.30).

There are, of course, many geodesics – many lines in a plane, many great circles on a sphere, many inertial observers. And the arbitrary constants B and C merely enable us to decide *which* of them we're talking about.

We've had one example of how to use these near-equations to find geodesics. Let's get more practice while learning about a new co-ordinate system.

5.34M I'm sure you know the common system of measuring angles – with a right angle divided into 90 degrees (90°), each subdivided into 60 minutes (60'), further subdivided into 60 seconds (60"). This is used in most practical circumstances. But for theoretical purposes an alternative system is preferable.

Draw a circle with its centre at the apex O of the angle (Figure 26.34) meeting the arms at P and S. Measure the length of the *arc* PS – not the straight line distance (the 'chord') from P to S, but the distance along the circle. The ratio of this length to the radius OP is used to measure the size of the angle – in units called *radians*. So we have the relation

$$\text{size of angle POS in radians} = \frac{\text{length of arc PS}}{\text{length of radius OP}} \,, \tag{26.33}$$

from which it follows that

$$\begin{aligned}\text{length of arc PS} = \ &(\text{length of radius OP})\\ &\times (\text{size of angle POS in radians}).\end{aligned} \tag{26.34}$$

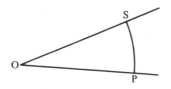

Figure 26.34

Except in a few practical calculations, angles in all that follows are in radians. Don't confuse θ and φ used for the sizes of angles with the radar times of earlier chapters.

26.35M We introduce a new co-ordinate system for specifying positions in a plane (an alternative to the familiar system of §§11.11 and 24.11). We start from a fixed point O, called the *pole,* and a fixed line OL – Figure 26.35. Then the *polar co-ordinates* of point P are

(i) the length, r, of the line joining O to P; and

(ii) the size, θ, of the angle between the line OL and the line OP.

The length r is called the *radius vector.* It is always taken to be positive (since a negative r would be equivalent to adding 180° to θ). And the angle θ (positive if you get from OL to OP by swinging anticlockwise; otherwise negative) is the *vectorial angle.*

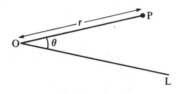

Figure 26.35

26.36 Now let's find the metric for the geometry of a plane expressed in terms of polar coordinates. Consider two points: P with polar co-ordinates r and θ, and Q with polar co-ordinates $r + \delta r$ and $\theta + \delta\theta$. Thus θ is the size of the angle LOP (Figure 26.36) and $\delta\theta$ that of POQ. Lengths OP and OQ are r and $r + \delta r$, respectively. Draw an arc of the circle whose centre is O and radius is OP, and let it meet OQ at S. Then length OS = length OP = r, and so length SQ = δr.

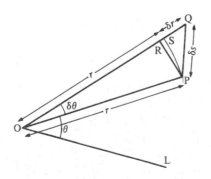

Figure 26.36

We want to express δs, the distance P to Q, in terms of δr and $\delta\theta$. Let PR be the perpendicular from P to OQ. By Pythagoras' Theorem

$$\delta s^2 = (\text{length RQ})^2 + (\text{length PR})^2 . \qquad (26.35)$$

As we make $\delta\theta$ smaller, we bring S closer to R, and so

$$\text{length RQ} \sim \delta r . \qquad (26.36)$$

At the same time the arc PS of the circle comes closer to coinciding with the line PR, so that

$$\text{length of straight line PR} \sim \text{length of arc PS.}$$

And by (26.34) the latter is equal to $r\delta\theta$, so that

$$\text{length PR} \sim r\delta\theta . \qquad (26.37)$$

Then substituting these in (26.35) gives

$$\delta s^2 \sim \delta r^2 + r^2 \delta\theta^2 \qquad (26.38)$$

as the metric we are seeking.

26.37M Just as a^2 means $a \times a$, so also a^3 is used as an abbreviation for $a \times a \times a$, and a^4 for $a \times a \times a \times a$, and so on.

Notice that $(r^2)^2 = r^2 \times r^2 = r \times r \times r \times r = r^4$. And so (cf. §17.25 for first step)

$$\left(\frac{r^2}{B}\delta\theta\right)^2 = \frac{(r^2)^2}{B^2}\delta\theta^2 = \frac{r^4}{B^2}\delta\theta^2 .$$

Also

$$r^4 \times \frac{1}{r^2} = \frac{r^4}{r^2} = r^2 .$$

We'll use these soon.

26.38 Using the proposition of §26.33, helped by the example in §26.24, I wonder if you can work out **what are the near-equations characterising geodesics when the metric is (26.38).**

If you failed, see how well you can do when prompted. We again suppress the last term of (26.30). In §26.33 it is laid down that the values of α and β may depend on co-ordinate x *only*. So **which co-ordinate in (26.38) plays the role corresponding to x?**

Well r appears in (26.38), but θ (as contrasted with $\delta\theta$) does not. So r plays the role of x. Thus to turn (26.30) into (26.38) we have to write r in place of x, and θ in place of y. **What takes the place of α?**

In (26.30) α is associated with δx^2, which has become δr^2 – which is multiplied by a mere 1 in (26.38). So $\alpha = 1$. **And what about β?**

Since δy^2 has become $\delta\theta^2$, which is multiplied by r^2 in (26.38), we have $\beta = r^2$. **So what does the first alternative of (26.31) become?**

Changing β to r^2 and y to θ, as agreed, this gives

$$r^2\delta\theta \sim B\delta s \ .$$ (26.39)

Together with the metric (26.38), this defines our geodesic. But we are not interested in δs. **How do we get rid of it?**

From (26.39) we have

$$\delta s \sim \frac{r^2}{B}\delta\theta \ .$$ (26.40)

Substituting in (26.38) gives

$$\frac{r^4}{B^2}\delta\theta^2 \sim \delta r^2 + r^2\delta\theta^2 \ ,$$

(see §26.37 if in difficulty), whence

$$\delta r^2 \sim \frac{r^4}{B^2}\delta\theta^2 - r^2\delta\theta^2 \ ,$$

or more neatly

$$\delta r^2 \sim r^4\left(\frac{1}{B^2} - \frac{1}{r^2}\right)\delta\theta^2$$ (26.41)

(§26.37 – and I can't remind you *every* time about §16.12).

So this last should be the near-equation connecting the increments in r and θ as we pass from one point of a geodesic to another.

26.39 Now a geodesic in a plane should be a straight line (§25.2) no matter what co-ordinates we use. To boost our confidence we need an independent confirmation that (26.41) really does give a straight line. You could, of course, do a numerical check by calculating a table of r against θ on the lines of our Chapter 21 calculation for k and ft. For simplicity take $B = 1$. Your starting value of r must then be *greater* than 1 (to avoid δr^2 or $\delta\theta^2$ being negative, §24.20). As your choice of OL is arbitrary, you may as well start from θ = 0. Choose how big your $\delta\theta$ will be, remembering that you're working in radians (and later you can check whether your steps are small enough by the method of §21.22). Then (26.41) enables you to calculate the corresponding δr, and hence the value of r at the start of the second step. And so on. Finally you can draw the resulting curve, using ruler and protractor, and verify that it is nearly a straight line.

I've always *said* all that to my classes, but I've been too busy (? lazy) to work it out. Thankfully, just before I finalised this article, my Nottingham student Stan Zobel (age 69) did it. He chose $r = 1\cdot2$ at the start and $\delta\theta = 0\cdot02$ radians. And he found that – even without using the method of §§21.19–20 to enhance accuracy – by the time he'd got to θ = 0·2 radians (11°27′33″) his r differed from the value required for an exact straight line by only 0·56 per cent. **I think you should undertake a similar exercise.**

26.40 Numerical exercises are helpful. But really we need a *general* proof – not involving geodesics, of course – that (26.41) describes correctly the relation connecting the increments in r and θ between two points of a straight line. So take P and Q to be any two points on a line MPQ (Figure 26.40). We use h for the perpendicular distance, OM, from O to this line. The rest of the diagram is the same as Figure 26.36.

Triangles PQR and OQM have the same angle at Q and right angles at R and M. So they are the same shape, and (cf. §9.13) there is a scale factor such that

$$\text{length PQ} = \text{length OQ} \times \text{scale factor}$$

and

$$\text{length PR} = \text{length OM} \times \text{scale factor}.$$

Dividing corresponding sides of these equations gives

$$\frac{\text{length PQ}}{\text{length PR}} = \frac{\text{length OQ}}{\text{length OM}} . \qquad (26.42)$$

As we make the increments smaller, Q gets closer to P and so length OQ $\sim r$. Also lengths PQ $= \delta s$, OM $= h$ and, by (26.37), PR $\sim r\delta\theta$. Substituting in (26.42) we have

$$\frac{\delta s}{r\delta\theta} \sim \frac{r}{h}, \quad \text{whence} \quad \delta s \sim \frac{r^2}{h}\delta\theta .$$

If we put $B = h$, this is the same as (26.40). So we have derived this last by methods that have nothing to do with geodesics. And that was already true of the other near-equation – the metric (26.38) – that we used in the calculations of §26.38. So we now have our independent

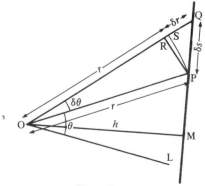

Figure 26.40

confirmation that (26.41) is correct. Yet again the geodesic method has given the right answer!

Having discovered that $B = h$, we can rewrite the near-equation (26.41) connecting increments between points of a line as

$$\delta r^2 \sim r^4\left(\frac{1}{h^2} - \frac{1}{r^2}\right)\delta\theta^2 , \qquad (26.43)$$

where h is the perpendicular distance from O to the line.

26.41 One more exercise will bring us very close to the sort of thing we'll be doing when we start using geodesics in earnest. Let's apply the method to investigate inertial motion in a plane, using polar co-ordinates – still with no gravity, of course. If δs now refers to interval, we have $\delta s^2 \sim \delta t^2 -$ (increment in distance)^2. And (26.38) gives (increment in distance)$^2 \sim \delta r^2 + r^2\delta\theta^2$. Hence

$$\delta s^2 \sim \delta t^2 - \delta r^2 - r^2\delta\theta^2 \qquad (26.44)$$

is the metric for the problem in hand. **Work out the near-equations for the geodesics.**

26.42 Once more r plays the role that x does in §26.33 – as the co-ordinate involved in α, β and γ. So we can identify r, θ and t with x, y and z of §26.33. And so $\alpha = -1$, $\beta = -r^2$ and $\gamma = 1$.

This time it's easier if we use the second alternative in (26.31) and the first in (26.32), getting

$$\delta s \sim Br^2\delta\theta \qquad (26.45)$$

and

$$\delta t \sim C\delta s, \qquad (26.46)$$

where B (written instead of $-B'$) and C are constant. (If you find this hard, write out the near-equations in their §26.33 form, and then translate into r, θ, t terms, using the 'dictionary' provided in the previous paragraph.) So (26.44)–(26.46) describe a geodesic, which (by §25.28) should be the world line of an inertial observer.

26.43 Let's assume (as will often be the case in 'real life' problems later on) that we're interested only in the path this observer moves along, not in his speed. So we must eliminate δt as well as δs from the near-equations. (I shall leave you to worry out some bits of algebra – they're all points we've already covered.)

Using (26.45) to substitute $Br^2\delta\theta$ for δs in (26.46), we get $\delta t \sim BCr^2\delta\theta$. Then we can use this last and (26.45) to substitute for δt and δs in (26.44), obtaining

$$B^2r^4\delta\theta^2 \sim B^2C^2r^4\delta\theta^2 - \delta r^2 - r^2\delta\theta^2,$$

whence $\delta r^2 \sim (B^2C^2r^4 - B^2r^4 - r^2)\delta\theta^2$. If we agree to write $B^2C^2 - B^2$ as $1/h^2$, a little further manipulation leads to

$$\delta r^2 \sim r^4 \left(\frac{1}{h^2} - \frac{1}{r^2} \right) \delta\theta^2 .$$

So we've deduced, as comparison with (26.43) shows, that an inertial observer in the absence of gravity moves in a straight line (relative to the other inertial observer whose measurements are our co-ordinates). Another long road round to something we already know. But also another confirmation that the geodesic approach gives results.

26.44 Let's see whether this geodesic method gives the expected answer for free fall in constant gravity conditions. If you skipped §§17.24 and 17.27 on first reading, please tackle them now – you'll find them easy.

The metric this time is (24.8) – see §24.19. The co-ordinate involved in α and β is x, which must therefore be identified with the x of (26.30) (whose δz term we again suppress). And the δy of (26.30) becomes the δt of (24.8). So $\alpha = -(1 + 2fx)$ and $\beta = 1 + 2fx$. Then (26.31) in its second form gives $\delta s \sim B(1 + 2fx)\delta t$, where B is a constant. Substituting this value for δs in the metric (24.8) makes it

$$B^2(1 + 2fx)^2\delta t^2 \sim (1 + 2fx)\delta t^2 - (1 + 2fx)\delta x^2,$$

whence (on dividing both sides by $1 + 2fx$, etc.)

$$\delta x^2 \sim \{1 - B^2(1 + 2fx)\}\delta t^2. \qquad (26.47)$$

According to the Geodesic Assumption (§25.28) this near-equation should describe the motion of a free-falling body. At some instant it is at distance x (from the supported observer whose measurements we are using, §24.19). In time δt after that it moves a further distance δx, where δx and δt are related by (26.47). **What about its speed?**

26.45 Using v for its speed, §24.22 says that

$$v^2 = 1 - B^2(1 + 2fx) . \qquad (26.48)$$

The size of the constant B merely decides the value of x at which $v = 0$ – that is, the height from which the object starts falling (or starts falling back if projected upwards). And the simplest case is where $B = 1$, giving $v^2 = -2fx$ – so that it starts falling from $x = 0$. The negative sign indicates acceleration in a negative (downwards) direction. So if we measure distance downwards, we can change this to

$$v^2 = 2fx . \qquad (26.49)$$

Does that ring a bell?

It's the same as (17.16), which describes the relation between (slow) speed and distance-travelled (starting from rest) for motion *with con-*

stant acceleration. So we have verified that applying the Geodesic Assumption to the constant gravity metric does lead to the correct
answer – that any free-falling body moves with constant acceleration.

By the way, if you are not content with the case where the faller
starts from zero height (right beside the supported observer), you may
care to prove that if $v = 0$ at height $x = x_0$, then $v^2 \approx -2f(x - x_0)$
– leading to the same conclusion. This deduction depends on fx_0 and fx
being very small compared with 1 (§24.18) and uses the approximations $(1 + 2fx)/(1 + 2fx_0) \approx (1 + 2fx)(1 - 2fx_0) \approx 1 + 2f(x - x_0)$.

26.46 That partially answers the question we posed in §24.23. Given the
constant gravity metric, we *can* – by the geodesic method – deduce the
behaviour of free-falling objects. So now we've checked that the Geodesic Assumption gives the right answer in two cases: no gravity and
constant gravity. Maybe it's reasonable to hope it will still work when
we come to gravity that varies from place to place.

The suggestion is that once you know the metric – which characterises the geometry of spacetime (§§12.21–2, 24.6) – then, by using the
machinery of geodesics, you can automatically work out how free-
falling or inertial observers will move; or more briefly, inertial motion
is a mere matter of geometry, irrespective of whether there is gravity
or not. If that proves true, it is surely a profound revolution in outlook. For it is being suggested that *natural motion* – unconstrained,
inertial motion – *is determined by geometrical considerations alone.*
That's the outlook of Einstein's General Theory of Relativity, towards
which we are plodding.

26.47 Even in this simple example of free fall in constant gravity conditions,
we can begin to see the contrast between the Newtonian and Einsteinian approaches to the problem of gravity. Altering his language to
facilitate comparison, Newton's view would be:

(N1) The geometry of spacetime is Euclid's geometry with time added on;

(N2) When no force is acting on it, the world line of a body is a
geodesic – which in view of (N1) means motion in a straight line at
constant speed. Unfortunately things are found to fall – accelerate
downwards – even when no observable force is acting (cf. §§23.9–11).
To explain this Newton adds

(N3) But there *is* a force which acts on everything – the force of
gravity. To explain why all bodies have the *same* acceleration, he pos-
tulates that the force is proportional to the mass of the body it is acting
on (§23.10).

Einstein's contrasting approach may be summarised as:

(E1) The geometry of spacetime is characterised (in the case we're considering) by the metric (24.8); and

(E2) When no force is acting on it, the world line of a body is a geodesic.

Then, as we deduced in §§26.44–5, it follows automatically *without the need to postulate a force of gravity* that any body that's left to itself will accelerate 'downwards' with the same constant acceleration.

You'll notice that Einstein's theory uses one less assumption than Newton's. Assumptions (N1) and (E1) make statements about the geometry of spacetime; and (N2) and (E2) are the same – anything left to itself follows a geodesic. From (E1) and (E2) alone, Einstein's theory deduces correctly how unsupported bodies behave, whereas Newton's theory gives the wrong answer till he adds (N3).

Since it involves one less assumption, Einstein's theory is *simpler in principle* than Newton's (even if the technical working out is harder).

27

Inverse square law gravity

27.1 Constant gravity exists only as an approximation that does well enough in small regions of space. And on this small scale the only case where Einstein's theory gives a detectably different prediction from Newton's is that of §§23.14–16. For further tests we have to consider vast spaces in which gravity obviously varies from place to place. And to cope with these more complex conditions we shall have to modify two of our most basic assumptions.

27.2 Our most fundamental assumption of all – Einstein's Principle of Relativity (§5.11) – involves the explicit condition 'In the absence of gravity . . .'. In *constant* gravity conditions **would it still be true that all inertial observers are equivalent?**

Yes – since the inertial observers in this case are exactly like those of the no-gravity case, but *observed by* an accelerated observer (§§23.6, 23.12).

27.3 If, on the other hand, gravity varies from place to place, then this Principle is no longer true. For simplicity imagine that the Earth is the only massive body in the Universe. Then (ii) and (iv) of §23.5 are inertial observers, one at the Earth's surface, the other very very far away in space. Our demonstration (§23.6) that these have identical experiences only worked because we confined our observers to small rooms. If we let them watch things moving inertially a few thousand miles away, **will their experiences be the same?**

No. As (ii) sees it, such things will have accelerations of different sizes in different directions (depending on position relative to the Earth's centre). But (iv) will detect no deviation from motion in a straight line at constant speed. And indeed if these observers can measure accurately enough, then even inside their little rooms their experiences will be different. For when (ii) releases test particles at arm's length on either side of him, they will drift slowly towards him (since he and they are converging on the Earth's centre), whereas A's similar particles would keep their distance.

Of course we could set them to do experiments confined to a cubic millimetre of space and a nanosecond of time. And then they wouldn't observe this difference in test particle behaviour – *unless* they could measure still more accurately . . . And following out that line of argument you will see that the smaller the scale of the experiment in space and time, the more precise would the observers' measurements have to be if they are to detect any difference between their results. Or putting it another way, by insisting on sufficiently small-scale experiments, we can make inertial observers as nearly equivalent as we like. **Can you think how that could be said by means of a single near-equation?**

27.4 We've seen (§§12.23, 24.8, 24.10) that the whole of the Special Theory of Relativity, *including* the equivalence of inertial observers, can be summed up by saying that the metric is (24.3). So we've only got to change = to ~, and take our metric to be

$$\delta s^2 \sim \delta T^2 - \delta X^2 \tag{27.1}$$

(which we've had to do anyway to cope with interval measured along a world line, §25.8).

If δT and δX refer to the measurements of whatever inertial (free-falling) observer we are concerned with at the moment, then (27.1) says that we can make inertial observers as near to being equivalent as we wish by simply taking δT and δX small enough – that is, by working in a small enough region of spacetime.

So in future near-equation (27.1) will take the place of our former statement that all inertial observers are equivalent.

27.5 Then again the Principle of Equivalence, as stated in §23.8, refers very definitely to *constant* gravity. And the argument of §23.7 only works when we confine our observers to small rooms. If we allow the case (i) observer to take account of things falling at the antipodes, his experience is nothing like that of case (iii).

But once more it's the scale of the experiment that counts. Inside the little rooms the deviation from constant gravity will not be detectable, and so no departure from the Principle of Equivalence would be observable – *unless* these chaps can measure very very accurately, in which case (i) will notice that test particles released at arm's length on either side of him fall along converging paths, while (iii)'s particles move strictly in parallel. So again we might insist on experiments within a cubic millimetre and a nanosecond . . . And continuing that line of argument **you can prove that** even with variable gravity we can make the Principle of Equivalence as nearly true as we wish by confining observations to a small enough region of spacetime. This again can be put in the form of near-equations, which will arise naturally as we proceed.

27.6 The realm of the Sun and the inner planets is the chief testing ground for General Relativity. It is almost the only variable-gravity region where we can measure accurately enough to test whatever predictions the theory may make.

The planets move round the Sun in ellipses – but ellipses that are very near to being circles. So in the next few articles, let's make things easy by assuming that the planetary orbits *are* circles, with the Sun as their common centre.

27.7 Is a planet's motion inertial? Its own gravity interferes with the particle test (except perhaps in a laboratory at the planet's centre!). In the closely analogous case of an orbiting cosmonaut (§5.4) his own and his vehicle's gravitational influence is so small that the test can be done – and it indicates inertial motion, as you've probably seen on TV. In any case, there are no signs of anything exerting significant forces on a planet – no strings attached, only very weak magnetic fields, and so on. We can surely agree that its motion is inertial.

Yet instead of keeping to a straight line, the planet orbits round the Sun. So, starting from his first two assumptions (§26.47) Newton was driven to postulate the existence of a force of gravity dragging it towards the Sun. Without accepting Newton's *force* of gravity, let's agree with him that a planet is continually *falling* towards the Sun.

27.8 But how can it be falling towards the Sun when it always stays about the same distance away? Here's an explanation that stems from Newton himself. Instead of Sun and planet, consider the Earth and a cannon ball fired horizontally from the top of a mountain (and assume there's no air resistance).

If the shot leaves the gun with modest speed, it will drop pretty steeply along path 1 (Figure 27.8), hitting the ground not far from the mountain. With a bigger muzzle velocity its path might be the one marked 2. Its greater speed has carried it further before hitting the ground. A very much faster shot would actually travel *away* from the Earth – path 4.

With a suitable muzzle velocity, somewhere between the last two, we can arrange for the ball to travel right round the Earth – on circular path 3 – and hit the gunner in the back. However, he nimbly removes himself and the gun to a safer place. And so the ball starts off on another identical trip – and another, and . . . It is now in orbit.

If the cannon ball is falling on flights 1 and 2, it must also be falling on flight 3. It stays in orbit *because* it is falling – and because its speed is adjusted to give the right relation between rate of fall and forward progress. Similarly for planet and Sun.

27.9 Falling is accelerated motion. The planet is accelerated towards the Sun. It moves (in the circular orbit case) at *constant speed,* and yet it's accelerated.

If that still worries you, first reread §20.5, pondering particularly on the last three sentences, and then think again about the gunnery of §27.8. If the gunner drops his ball down a vertical shaft, it is obviously accelerated towards the Earth's centre. If he shoots it along path 1 or 2, you'd agree that it still has acceleration, though now combined with the transverse motion he gave it. When he projects it at the right speed to give circular orbit 3, it will have constant speed. But surely if you agree that it's accelerated towards the Earth's centre in the previous cases, you are not going to deny a similar acceleration on path 3.

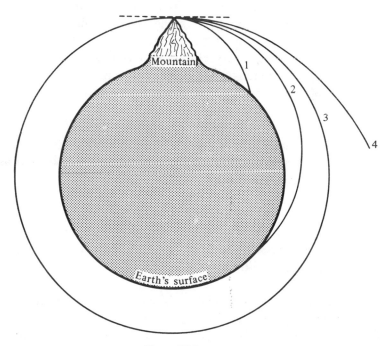

Figure 27.8

In vertical fall the ball's speed keeps increasing, while its direction of motion stays fixed. On paths 1 and 2 both speed and direction are changing. And on path 3, while its speed stays constant, its direction of motion keeps on changing. So with velocity defined in terms of speed *and* direction, all these cases involve change of velocity – and that is acceleration.

27.10 We need to know how big this acceleration is. Consider a body P moving at constant speed v round the circumference of a circle (centre O) whose radius is r (Figure 27.10). If it were not accelerated towards the centre, it would move in a straight line along the tangent PQ. In time δt it would go distance $v\delta t$ from P to Q. However, the inwards acceleration ensures that after time δt the body actually arrives at R on the circumference of the circle. So its circular motion from P to R is a combination of (1) the straight motion from P to Q at constant speed, and (2) the accelerated motion from Q to R.

We could repeat the manoeuvre starting from R – at constant speed along the tangent for time δt, and falling straight inwards to get back on the circle again – and repeat it as often as we wish. The result would be a saw-tooth approximation to circular motion – indicated by broken lines in Figure 27.10. And by taking δt small enough, we could arrange for this combination of motions (1) and (2) to be as good an approximation as we wish to the actual motion in a circle. And taking δt small enough will also arrange for arc length PR to be as nearly equal as we wish to length PQ $= v\delta t$ – thus giving the correct speed v along the circular path.

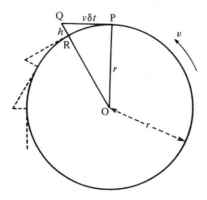

Figure 27.10

27.11 Let f stand for the size of the body's acceleration, and h for the distance QR which it falls in time δt. Then (17.11) gives

$$h \sim \tfrac{1}{2}f\delta t^2. \tag{27.2}$$

(At §17.24, we didn't know about near-equations; but you can easily check that it really should be $\sim$ instead of $\approx$.)

Tangent PQ is perpendicular to radius OP. So Pythagoras' Theorem gives $(r + h)^2 = r^2 + v^2\delta t^2$. Using identity (26.26), this becomes $r^2 + 2rh + h^2 = r^2 + v^2\delta t^2$, whence $2rh + h^2 = v^2\delta t^2$. Substituting for h from (27.2), we get $fr\delta t^2 + \tfrac{1}{4}f^2\delta t^4 \sim v^2\delta t^2$ (§§24.7, 26.37 for the meaning of δt^4), which leads to

$$fr + \tfrac{1}{4}f^2\delta t^2 \sim v^2. \tag{27.3}$$

(I shall often be forced to compress like that. If it seems difficult, you'll find it gets easier when you set it out with each equation on a line of its own.)

By making δt smaller, we can make $\tfrac{1}{4}f^2\delta t^2$ as insignificant as we wish compared with fr and v^2. So it *looks* as if

$$fr \sim v^2. \tag{27.4}$$

How would you prove that properly?

Go back to the definition (§§20.20, 21.42).

The ratio of the sides $= v^2/fr$

(27.3)

$$\sim \frac{fr + \tfrac{1}{4}f^2\delta t^2}{fr} = 1 + \tfrac{1}{4}\frac{f}{r}\delta t^2.$$

I'm sure you can finish the proof. **Can you deduce something firmer from (27.4)?**

The size of δt does not affect the size of fr nor of v^2. So the First Fundamental Theorem (§22.30) allows us to change (27.4) into a true equation,

$$fr = v^2. \tag{27.5}$$

–which tells us how the acceleration is related to the speed and the size of the circle.

27.12 That applies to *any* case of constant speed motion in a circle. But we are aiming to discover how a planet's sunward acceleration depends on its distance from the Sun. For that purpose we need a further piece of information about how planets *actually do* behave – how speed is related to size of orbit. This new information is provided by Kepler's Third Law (1619), which states that

If r is the radius† of a planet's orbit and T is the time that it takes to go once round, then r^3 is proportional (§15.3) to T^2. I.e.,

$$r^3 = AT^2,\tag{27.6}$$

where A is a constant.

This is not a theoretical deduction, but an empirical discovery made by collating the facts of observation.

The distance round the orbit is $2\pi r$ (where π – 'pi', the Greek p – is the famous number, approximately 3·1416, giving the ratio of a circle's circumference to its diameter). As this distance is covered in time T at speed v, we have $T = 2\pi r/v$. And substituting in (27.6), we get $r^3 = 4\pi^2 A r^2/v^2$, which reduces to $v^2 r = 4\pi^2 A$. But $4\pi^2 A$ is a constant, to which we can give the new name of λ ('lambda', the Greek l), so that the last equation becomes

$$v^2 r = \lambda .\tag{27.7}$$

This is still an empirical relation between the planet's speed and the radius of its orbit. Then using (27.5) to substitute fr for v^2 gives $fr^2 = \lambda$, or

$$f = \frac{\lambda}{r^2},\tag{27.8}$$

which tells us how the planet's acceleration is related to its distance from the Sun. To reach that result we had to assume circular orbits. But Newton, using a more elaborate analysis, was able to deduce the same conclusion from the elliptical planetary motion that is actually observed.

Thus the planet's acceleration is proportional to $1/r^2$ – a quantity that is traditionally called the 'inverse square' of r. When gravitational acceleration varies in this way, we speak of *inverse square law gravity*.

This inverse square law relationship is found to apply whenever the gravitational influence of a massive body causes accelerations in much less massive bodies around it. The value of λ varies from one case to another and acts as a measure of the gravitational power with which the central body acts on its surroundings. It will be simpler if I normally talk in terms of the Sun and its planets. But to make our conclusions apply (say) to the Earth's gravitational influence on its natural and artificial satellites, you have only to change the value of λ.

27.13 Newton interpreted these accelerations in terms of a *force* of gravity – any two particles of matter are attracted towards each other by a force that is proportional to the product of their masses and the inverse

† For an elliptical orbit the relevant quantity is half of the ellipse's longest axis.

square of their distance apart. Following Einstein, we'll say nothing about force. And nothing in this book will need generalisations about 'any two particles'. All we need is a statement about the accelerations of bodies in the neighbourhood of a very massive body like the Sun.

The mass of a planet is small compared with that of the Sun. But subject to that restriction, we can take planets as typical of free-falling bodies, and so draw the conclusion that

> If a free-falling (inertial) body, having small mass compared with the Sun's, is at distance r from the Sun, then it falls towards the Sun with acceleration whose size f is given by (27.8), where λ is a constant quantity.

And obviously this is a description of free-fall as it appears to a supported observer – one who is constrained by forces that keep him from falling. There is no such animal of course. But we are used to describing the Solar System from the point of view of a hypothetical supported observer – geocentric descriptions have long been out of fashion!

27.14 The quantity λ first occurs in (27.7) – which says (since v in natural units is a fractional pure number, §8.39) that λ is a fraction of a distance, and is therefore itself a distance or length. Twice this length (which later proves to be more important) is known as the *Schwarzschild radius*.

In Newton's formulation, the attractive power of the central body is expressed in terms of mass; and therefore it is usual to denote it by GM or by m, even in Relativity. But the unorthodox λ – with its constant reminder that we are talking about a length – will suit us better.

27.15 The Earth's orbital speed v is roughly a ten-thousandth of the speed of light (§1.8), while the radius of the orbit is about 150 million kilometres. So by (27.7) λ, in the case of the Sun, is about a hundred-millionth of 150 million kilometres – or about 1·5 kilometres. A more accurate calculation gives 1·477 kilometres. Thus λ turns out to be remarkably small on a Solar System scale.

Now all the differences between Newton's and Einstein's theories depend on the number got by dividing λ by whatever distance you're concerned with. So you can see why these differences are exceedingly minute.

27.16 Now let's try to calculate the metric (§24.10) for a spacetime with inverse square law gravity. But to simplify matters at the start let's confine ourselves again to a 1D universe – a line through the centre of the Sun.

The method is in principle similar to that of §§24.12, 24.14 and 24.19 – finding a relation between the co-ordinates of supported and free-falling observers and using this to translate the Special Relativity metric into the one we want. But the change from constant to varying gravity complicates the calculation.

27.17 We'll work out the metric in terms of the co-ordinates of a supported observer, E, situated at a very great distance from the Sun, S. Being supported, E stays at the same distance from the Sun. Figure 27.17 schematically uses verticals to represent 'at the same distance from E according to E'. And so E's and S's world lines are shown vertical. And the horizontal line marked $t = 0$ represents events happening at zero time according to E. All distances in what follows will be *according to* E, but they will be *measured from* S.

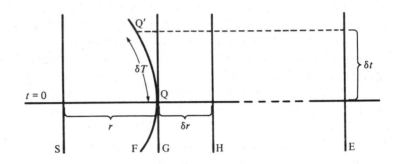

Figure 27.17

Let G and H be two more supported observers – stationary, therefore, relative to S and E, and having vertical world lines. And let† r and $r + \delta r$ stand for the distances of G and H from the Sun according to E.

27.18 Let F be a free-falling observer who at zero time – event Q – is at the same place as G and momentarily stationary relative to all the other observers. As E sees it, F is falling towards S with an acceleration whose size, f, depends on his distance from the Sun.

27.19 Now let's temporarily pretend that gravity between G and H has *constant* strength f (and we'll have to put that right later). **Then what does G think about H's clock?**

It is at distance δr above him, and so he thinks it is running fast by a factor of $1 + f \delta r$ (§23.14).

---

† We use r, rather than x, because this quantity will eventually become the radius vector in polar co-ordinates (§26.35) with the Sun as pole.

27.20 In view of E's great distance, we can't pretend that gravity is constant between G and E. But, without worrying about quantitative detail, **what would E think, in general terms, about G's clock?**

Variable gravity will not alter the general nature of the phenomenon described in §23.14. So G's clock (being below E – nearer the Sun) will seem to E to be running slow by an amount that depends in some way on position. But the detailed formula of §23.14 cannot apply when gravity varies. So let's assume that E thinks G's clock is running slow by a factor of $1 + z$, where z is a negative number whose size depends on r. We have to discover how z is related to r.

27.21 This factor $1 + z$ will play a role similar to the $1 + fx$ of §23.14. And in discovering the relation of z to r, we shall make use of the §23.14 result. The latter was derived from the conclusion of §22.32, which arose from an interpretation of near-equation (22.27). And this last is only valid if fx is very small compared with 1. So the deductions that follow will only be valid if z is also much smaller than 1.

27.22 **If we combine the contents of §§27.19 and 27.20, what emerges concerning** E's opinion of H's clock?

With E, G and H stationary relative to each other, there are no disagreements about simultaneity. And so we calculate the clock rate factor (§22.23) between E and H by simply multiplying together the clock rate factors between E and G and between G and H (the procedure that would apply when comparing clocks in daily life). So E considers that H's clock is running slow by a factor of $(1 + z)(1 + f\delta r)$ $= 1 + z + f\delta r + zf\delta r$ (§16.12).

But ever since §22.20 we've been restricted to conditions where $f\delta r$ (equivalent to the former fx) is a small fraction of 1. And the same goes for z (§27.21). So $zf\delta r$ is a small fraction of a small fraction, and is therefore so small that we can forget about it. Thus the factor by which E thinks H's clock is running slow can be taken to be

$$1 + z + f\delta r . \tag{27.9}$$

27.23 But we can also **express this factor in another way.**

If G's clock at distance r from the Sun seems to E to be running slow by a factor of $1 + z$ (§27.20), then H's clock at distance $r + \delta r$ will seem to him to be running slow by a factor of $1 + z + \delta z$, where δz is the increment in z that corresponds to increment δr in r. And comparing these two ways of writing the same factor, **we see that**

$$\delta z \sim f\delta r . \tag{27.10}$$

Why have I written $\sim$, not $=$?

In calculating (27.9) we assumed gravity to be constant between G and H (§27.19). This is not true, but can be made as nearly so as we wish by taking δr small enough. The $\sim$ sign expresses this idea. Putting it in is equivalent to modifying the Principle of Equivalence to deal with variable gravity, as described in §27.5.

27.24 The reasoning that led to (27.10) would apply for gravity whose strength is related in *any* way to the distance from the Sun. We now specify inverse square law gravity by saying that the actual relation is given by (27.8). And substituting this in (27.10) gives

$$\delta z \sim \frac{\lambda \delta r}{r^2} . \tag{27.11}$$

But what we want is an equation involving z and r, not just their increments. The process of deriving this from (27.11) requires a little more mathematical preparation.

27.25m If x increases by δx, then $1/x$ decreases – acquires a negative increment $\delta(1/x)$. Let's find a near-equation expressing $\delta(1/x)$ in terms of x and δx.

Consider first the increment $\delta(xy)$ in the product of two quantities x and y. If x increases to $x + \delta x$ and y increases to $y + \delta y$, then xy increases to $(x + \delta x)(y + \delta y) = xy + x\delta y + y\delta x + \delta x \delta y$ (§16.12). And $\delta(xy)$ is simply the amount by which this exceeds xy. Thus

$$\delta(xy) = x\delta y + y\delta x + \delta x \delta y . \tag{27.12}$$

27.26m Now let

$$y = 1/x . \tag{27.13}$$

Then xy is fixed in size (being $= 1$), so that $\delta(xy) = 0$, and (27.12) becomes $x\delta y + y\delta x + \delta x \delta y = 0$, whence $x\delta y = -y\delta x - \delta x \delta y$. Notice that as δx and δy are made smaller, $\delta x \delta y$ will decrease in size much more rapidly than the other terms. This suggests that

$$x\delta y \sim -y\delta x. \tag{27.14}$$

Check that this is correct (by the familiar method exemplified in §27.11).

Substituting (27.13) in (27.14) and dividing both sides by x gives

$$\delta\left(\frac{1}{x}\right) \sim -\frac{\delta x}{x^2} \tag{27.15}$$

– which is the result we wanted. A negative increment is a decrease. So this says that when you *increase* x by amount δx, then $1/x$ *decreases* by approximately $\delta x/x^2$; and by taking δx small enough we can make the approximation as good as we wish.

27.27m **Work out near-equations for** $\delta(a/x)$ **and** $\delta(-\lambda/r)$, where a and λ are constants. By (22.8) and (27.15)

$$\delta\left(\frac{a}{x}\right) = a\delta\left(\frac{1}{x}\right) \sim -\frac{a\delta x}{x^2}. \tag{27.16}$$

Changing a to $-\lambda$ and x to r gives

$$\delta\left(-\frac{\lambda}{r}\right) = -\frac{(-\lambda)\delta r}{r^2},$$

and so (§15.24)

$$\delta\left(-\frac{\lambda}{r}\right) \sim \frac{\lambda\delta r}{r^2}. \tag{27.17}$$

27.28 **Does that suggest anything in relation to our main problem?**
Combining (27.11) and (27.17) gives $\delta z \sim \delta(-\lambda/r)$. **And the next step?**
This near-equation satisfies the conditions for the Second Fundamental Theorem (§25.13) – *increments only* on both sides. **What do you get if you use the Theorem in form (iii)?**
You have to change y and z of §25.13 to z and $-\lambda/r$, respectively, getting

$$z = -\frac{\lambda}{r} + a,$$

where a is some constant number. And so (§27.20) E thinks that G's clock is running slow by a factor of

$$1 - \frac{\lambda}{r} + a. \tag{27.18}$$

If we can find the size of a, we shall know this clock rate factor completely.

27.29 Let's make the opening sentence of §27.17 more precise by adding that E's distance, R, from the Sun is to be so great that λ/R can be taken to be indistinguishable from zero – and no matter what accuracy you insist on, we can, of course, make R large enough to satisfy that condition. (If you want to say 'make R infinite', by all means do so; I am merely avoiding that dangerously deceptive concept.)
Now consider the case in which G coincides with E. Then $\lambda/r = \lambda/R$ will also be indistinguishable from zero; and the clock rate factor reduces to $1 + a$. But this factor must be 1 – for E thinks his own clock is going just right! And so $a = 0$. Furthermore, since a is a constant, it must be zero for *all* values of r, not just for $r = R$. And so we have proved that E thinks G's clock is going slow by a factor of $1 - \lambda/r$.

27.30 At the Earth's distance from the Sun, gravity is so weak ($\lambda/r \approx$ 0·000 000 01) that we can take our own time measurements to be the same as E's (even though we are free-falling). Putting G on the Sun's surface, we obtain the prediction that an atomic clock (§20.9) at the surface of the Sun would seem to us to be running slow by a factor of $1 - \lambda/r$, where r is now the Sun's radius, 696 000 kilometres. In other words, the frequency of light from the Sun should be reduced by that factor. This prediction has been verified to an accuracy of about 5 per cent – which is very good, considering the difficulties of the observation. Similar predictions for the light from very dense stars have also been verified.

These tests show that the Principle of Equivalence, modified as in §27.5, holds good in inverse square law conditions, and that the calculations of the present chapter so far are sound. They do *not* (as careless formulations sometimes imply) provide evidence for the wider aspects of the General Theory of Relativity.

27.31 At event O, observers F and G are momentarily moving together. **What will E think about the behaviour of F's clock at that moment?**

In no-gravity conditions an accelerated clock keeps time with an inertial clock that is momentarily moving with it – that's what we proved in Chapter 20 (though we had to state it more precisely, §§20.10–19). Using the Principle of Equivalence, that translates into the statement that when there *is* gravity, a supported clock keeps time with a free-falling clock that is momentarily stationary beside it – G's clock keeps time with F's. So in E's opinion, F's clock behaves, at event Q, just like G's – he thinks it is running slow by a factor of $1 - \lambda/r$.

27.32 Let Q' be an event happening to F some time after event Q (Figure 27.17). Let δT stand for the time from Q to Q' by F's clock, and let δt stand for the time between Q and Q' according to E. Note that δT is an actual time by F's clock, whilst δt is merely what E thinks it should be. **Can you suggest a near-equation connecting δT and δt which states that E thinks F's clock is running slow by a factor of $1 - \lambda/r$?**

The answer is

$$\delta T \sim \left(1 - \frac{\lambda}{r}\right)\delta t . \tag{27.19}$$

It's $\sim$, not $=$, since F's clock only agrees *momentarily* with G's. If it were $=$, the interpretation would be obvious – the actual time δT by F's clock is only $1 - \lambda/r$ of what E thinks it should be. A strict demonstration, taking account of the $\sim$, would follow closely the lines of

§§22.22–8. You may care to follow it out – changing A_2 to F and W to E, reading $1 - \lambda/r$ instead of $1 + fx$, and using the First Fundamental Theorem to shorten the job.

27.33 Near-equation (27.19) relates E's version of the time increment between Q and Q' to the version of an on-the-spot inertial (free-falling) observer. So we can use it to express the δT^2 term of the Special Relativity metric in terms of E's δt^2 – cf. the role of (22.27) in §24.12. Squaring both sides and using approximation (24.6) with $y = -\lambda/r$, gives

$$\delta T^2 \sim \left(1 - \frac{2\lambda}{r}\right)\delta t^2 .\tag{27.20}$$

27.34 To find our metric by the method of §24.12, we should also need to discover a similar relation between δX (the increment in F's distance measurement between Q and Q') and δr. Alas, I do not know of any simple way of doing so at this stage. So for the present we have to admit ignorance and assume that

$$\delta X^2 \sim \alpha \delta r^2,\tag{27.21}$$

where α stands for some unknown expression involving r. Then we must carry on with our researches, hoping that we shall eventually find a way of discovering the correct form for α.

Presumably α differs only slightly from 1 at values of r that concern us. The larger we make r, the nearer must it get to 1. It may be some close relative of $1 - 2\lambda/r$.

27.35 Expressed in terms of a free-falling observer's co-ordinates, the metric is (27.1). Substituting for δT^2 and δX^2 from (27.20) and (27.21) gives

$$\delta s^2 \sim \left(1 - \frac{2\lambda}{r}\right)\delta t^2 - \alpha \delta r^2\tag{27.22}$$

as the metric for inverse square law gravity, expressed in terms of E's co-ordinates.

27.36 In working out this metric we assumed $t = 0$. But an argument on the same lines as §24.17 shows that it holds good at all times.

Our reasoning used approximations that are only valid if λ/r is very small compared with 1. In the Solar System testing ground this condition is certainly satisfied for all space outside the body of the Sun – the greatest possible value of λ/r is 0·000 002 at the Sun's surface (§§27.15, 27.30).

27.37 Given the metric (27.22), **how could we make deductions about the motion of light?**

The motion of a light signal is characterised by having $\delta s = 0$ (§24.24). So we add $\alpha \delta r^2$ to both sides of (27.22) (to avoid the trap mentioned in §24.25; and cf. §24.28) and then put $\delta s = 0$ and divide both sides by α, getting

$$\delta r^2 \sim \frac{1}{\alpha}\left(1 - \frac{2\lambda}{r}\right)\delta t^2 . \tag{27.23}$$

This tells us the distance, δr, that a signal will move in time δt, if it is already distance r from the Sun.

27.38 **So what can we say about the speed of light?**

By §24.22 we deduce that at distance r from the Sun's centre

$$(\text{speed of light})^2 = \frac{1}{\alpha}\left(1 - \frac{2\lambda}{r}\right) . \tag{27.24}$$

At great distances this will be almost $1-$as might be expected, since gravity there is negligible, giving a Special Relativity situation. But (27.24) says that in these inverse-square-law-gravity conditions the speed of light (according to E) depends on distance from the centre (unless it turns out that $\alpha = 1 - 2\lambda/r$).

27.39 However, this deduction is based on the assumption that the t and r of (27.22) are actually the time and distance that observer E would measure (or some Earth-bound observer might attribute to E by 'correcting' his own measurements). And it's by no means obvious that such an assumption is justified. In §§22.4–5 we expressed doubts about the quantities that we took to be time and distance according to an observer with constant acceleration. Further developments since then, especially in the present chapter, can only increase our scepticism.

This does not, of course, diminish the value of t and r as *co-ordinates* specifying events–it merely warns us to be cautious in interpreting their physical meaning. We'll return to this question later. Meanwhile, note the last paragraph of §22.5.

27.40 In §§27.40–4 you'll probably find it helpful to do a step by step comparison with §§26.44–5 (and you may want to revise from §26.33 on). **How can we begin to study the motion of a free-falling body when the metric is (27.22)?**

According to §25.28 we have to find the ordinary geodesics, using the recipe of §26.33. **See how much of that you can do.**

For reasons like those of §26.44, paragraph 2, we must suppress the δz term of (26.30) and identify its x, y, α and β with r, t, $-\alpha$ and $1 - 2\lambda/r$ of (27.22). Then (26.31) gives $\delta s \sim B(1 - 2\lambda/r)\delta t$, where B is a constant. Substituting in (27.22) gives

$$B^2\left(1 - \frac{2\lambda}{r}\right)^2 \delta t^2 \sim \left(1 - \frac{2\lambda}{r}\right)\delta t^2 - \alpha\delta r^2 .$$

Add $\alpha\delta r^2$ to, and subtract $B^2(1 - 2\lambda/r)^2\delta t^2$ from, both sides; and then apply §16.12 to the terms involving δt^2 and divide both sides by α. The outcome is

$$\delta r^2 \sim \frac{1}{\alpha}\left(1 - \frac{2\lambda}{r}\right)\left\{1 - B^2\left(1 - \frac{2\lambda}{r}\right)\right\}\delta t^2 . \qquad (27.25)$$

According to §25.28, this describes the motion of a free-falling observer – call him F. If at a certain moment he is distance r from the Sun, (27.25) tells us what further distance δr he will move in time δt.

27.41 **So what about F's speed, u?**

Interpreting (27.25) by means of §24.22, we get

$$u^2 = \frac{1}{\alpha}\left(1 - \frac{2\lambda}{r}\right)\left\{1 - B^2\left(1 - \frac{2\lambda}{r}\right)\right\} . \qquad (27.26)$$

Since u varies with distance, it follows that F has an acceleration (according to E). This is the acceleration that we describe as gravitation towards S. To explain it Newton had to invent his force of gravity (§§26.47, 27.7). But in our approach to the subject no force is involved. The acceleration is merely a consequence of the fact that the 'geometry' of this spacetime is the one described by the metric (27.22).

The idea that gravitational accelerations are consequences, not of forces, but of spacetime geometry is central to the General Theory of Relativity. It is also the point of departure for our next chapter – a much less mathematical one than usual. And if you're suffering from mathematical indigestion, you're welcome to skip the rest of the present chapter – which is interesting but not essential. But glance at the last sentence of §27.45 and the comment in §27.46.

To investigate F's acceleration in more detail, we need a near-equation connecting speed with distance travelled, as a preliminary to which a bit more mathematics is required.

27.42m The increment in x^2 is written $\delta(x^2)$ – contrast δx^2 of §24.7. Putting $y = x$ in (27.12) gives $\delta(x^2) = 2x\delta x + \delta x^2$, and so

$$\delta(x^2) \sim 2x\delta x . \qquad (27.27)$$

Check the last step against the definition (§§20.20, 21.42; cf. §§27.11, 27.26).

27.43 Consider a body moving with acceleration f – which can vary in any way, subject to the restriction that its speed, v, always stays small

compared with the speed of light. Then we can use the classical definition of acceleration as rate of change of speed, so that

$$f \sim \frac{\delta v}{\delta t} \tag{27.28}$$

(where the $\sim$ deals with *varying* acceleration just as it dealt with varying speed in §§22.36–8). And from (22.38),

$$v \sim \frac{\delta r}{\delta t} \ . \tag{27.29}$$

Then using successively (27.27), (27.29), a minor rearrangement at
the $=$, and (27.28), we have

$$\delta(v^2) \sim 2v\delta v \sim 2\frac{\delta r}{\delta t}\delta v = 2\frac{\delta v}{\delta t}\delta r \sim 2f\delta r \ .$$

I.e.,

$$\delta(v^2) \sim 2f\delta r \ . \tag{27.30}$$

This is a generalisation of (17.16) to cope with varying acceleration.
It tells us (as accurately as we wish) how much v^2 increases by, when
the body moves distance δr.

27.44 Working out the size of F's acceleration from (27.26) is rather complicated. It's much simpler if, instead of u, we consider the speed expressed as a fraction of the *local* speed of light (which will be only
slightly different). Using v to denote this version of the speed, it follows from (27.24) and (27.26) that

$$v^2 = 1 - B^2(1 - 2\lambda/r) \ . \tag{27.31}$$

The constant B serves to fix the distance at which F finishes rising, is
momentarily stationary ($v = 0$) and starts falling – event Q of Figure
27.17). In fact, if $v = 0$ at distance $r = r_0$, then (27.31) gives

$$1 - B^2(1 - 2\lambda/r_0) = 0,$$

whence

$$B^2 = \frac{1}{1 - \dfrac{2\lambda}{r_0}} \ . \tag{27.32}$$

27.45 Successively applying §§22.9, 16.12 and 22.10 to (27.31) gives

$$\delta(v^2) = \delta\left\{1 - B^2\left(1 - \frac{2\lambda}{r}\right)\right\} = \delta\left(1 - B^2 + \frac{2B^2\lambda}{r}\right) = \delta\left(\frac{2B^2\lambda}{r}\right) \ .$$

Then (27.16) leads to

$$\delta(v^2) \sim -\frac{2B^2\lambda}{r^2}\delta r \ .$$

And so (27.30) tells us that F's acceleration is $f = -B^2\lambda/r^2$ (where the negative sign indicates acceleration *towards* S). We have to substitute for B from (27.32). But since we are restricted to slow speeds and therefore short falls, we might as well take $r_0 = r$. Then the acceleration at distance r is

$$f = -\frac{\lambda}{\left(1 - \dfrac{2\lambda}{r}\right)r^2}.$$

(27.33)

27.46 So after all the acceleration *doesn't quite* vary according to the inverse square law. But with our restriction to conditions in which $2\lambda/r$ is a small fraction, the deviation from that law will be tiny. In the region outside the Sun it will never exceed one part in 250 000 (§27.36, last paragraph). If we derive the acceleration from u instead of v, the mathematics is more complicated, but we still conclude that the deviation from inverse square law is small.

How is it that we *put in* inverse square law in §27.24, but in the end only *get out* an approximation to it? An interesting question for speculation! I think the short answer is that it arises from our insistence that gravitation must be a matter of geometry only. We pray to the Spirit of Mathematics: 'Please supply a geometry – a metric – that gives inverse square law gravitation.' And the answer comes back: 'Such a geometry does not exist. Here's the nearest thing we've got.'

And after all, why should we be sure that the law really is an inverse square one? Maybe the observations from which Newton and his successors drew their conclusion (§27.12) are not refined enough to detect the 'not quite'.

28

Curved spacetime

28.1 In recent chapters a new view of gravitation has been emerging. What we *observe* is that unsupported, unconstrained bodies accelerate in a direction and at a rate that depends only on their position relative to other bodies. *Newton's explanation* of this universal tendency was that the acceleration is caused by a force (§§26.47, 27.13). In the *new explanation* that is emerging from our recent work, no force is involved. Gravitational acceleration is seen as a consequence of the properties of spacetime (§§26.46–7, 27.41). These spacetime properties are summarised in the metric (§24.10). If the metric is that of Special Relativity, nothing falls – inertial motion is at constant speed in a straight line. If the metric is (24.8), unsupported bodies fall with the same constant acceleration (§26.45). If it is (27.22), they fall (§27.41) with an (almost) inverse square law acceleration (§§27.45–6).

The metric defines the geometrical properties of a spacetime (§§12.21–2, 24.6, 24.10). More vividly, it describes the *shape* of spacetime – since the shape of anything is just a summary of its geometrical properties. And so we can say that in Einstein's theory it's simply the shape of spacetime that makes free-falling bodies move the way they do.

28.2 Speaking of the metric as characterising the 'shape' of a space or a spacetime is not so artificial as it may seem at first. Consider the case where the space concerned is a *surface*. We shall concentrate on its *intrinsic* properties – properties that concern measurements made while sticking strictly to the surface itself: distance covered by a fly crawling on the surface, for instance. You can imagine, if you like, that some cunning physicist has arranged that light signals starting from any point of the surface will travel *in* the surface, following its shape, never leaving it; and that light signals from outside are repelled from it. Then an inhabitant of the surface would know nothing about the rest of space. To him the surface itself would be the whole universe.

We are chiefly concerned with the intrinsic study of the surface that such an inhabitant might carry out. But now and then we'll allow ourselves to learn what we can by looking at it from outside.

28.3 The surface might be a plane. Its metric would then be (24.4), which has a very simple form – the square of the distance increment is equal simply to the *sum of the squares of the increments in the co-ordinates.* And the creatures who live in the plane could discover this by simply doing a lot of intrinsic measurements. Looking at it from outside, *we* can add the information that it is *flat.*

28.4 Now suppose that the surface under consideration is that of a sphere. In §25.20 we found that the metric is now more complicated – the square of one of the co-ordinate increments is multiplied by a factor that depends on position. Of course that was for one particular pair of co-ordinates, latitude and longitude. But it can be proved (though not within the scope of this book) that it's impossible to find *any* co-ordinate system on a sphere that leads to a simple sum-of-the-squares metric like (24.4). No matter how you try, at least one of the squares of the increments will appear in the metric multiplied by a quantity that varies with position.

The creatures to whom the spherical surface is the whole universe could discover this for themselves, using intrinsic measurements only. And looking at it from outside, we can add the information that this time the surface is *curved.*

28.5 Contrast the simple sum-of-squares metric of the flat plane with the more complicated one of the curved sphere. Mathematicians use this distinction to *define* what they mean by saying that a surface is flat or curved.

> A *flat surface* is one on which co-ordinates *can* be chosen in such a way that
> (distance increment)^2 $\sim$ sum of squares of increments in
> co-ordinates.
>
>   A *curved surface* is one in which co-ordinates *can not* be so chosen.

Please note the operative words *can* and *can not.* It's not a question of whether some particular co-ordinate system has or has not got this sum-of-squares property, but of whether a co-ordinates system with this property *can be found.* Polar co-ordinates in a plane, for example, give the more complicated metric (26.38). But obviously choosing different co-ordinates does not turn a plane into a curved surface. The plane is flat because we *can* choose co-ordinates that give the simple metric (24.4).

There is a theory – too tough to do here – by which, if you're given the metric in terms of *any* set of co-ordinates, you can decide whether the surface does or does not satisfy the criterion for flatness.

28.6 'Flat' and 'curved' so defined do not quite correspond to everyday language. On a cylinder, for example, consider co-ordinates set up as follows (Figure 28.6). On a line HK, parallel to the cylinder's axis, choose a point O. Given any point Q, draw the 'line' MQ perpendicular to HK – but this is, of course, actually an arc of a circle following the curve of the cylinder. Then the co-ordinates of Q are the lengths OM and MQ, both measured along the surface.

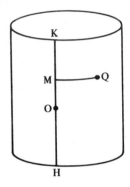

Figure 28.6

A little thought – e.g., about slitting the cylinder lengthwise and flattening it out – will convince you that the metric *does* have the sum-of-squares form. And so a cylinder is flat in the mathematician's sense, though in everyday language it is curved.

28.7 The slitting-and-flattening technique could be used to map the cylinder on a plane *without distortion* – without altering any lengths. By contrast every mapping of a sphere on a plane involves distortion. And this is how the mathematician's view of 'flat' and 'curved' ties in with ordinary usage. If a surface is flat by the definition of §28.5, it may or may not be a plane; but it certainly can be mapped on a plane without distortion. If it is curved, it can't.

Also, when you use the slitting-and-flattening technique, you discover that the geodesics on a cylinder (found, perhaps, by the rubber band method of §25.21) become straight lines. But when you map a sphere on a plane, its geodesics (most of them) are turned into curves.

And the same goes for any other surface that is curved in the mathematical sense. So a mathematician thinks of a cylinder as a flat surface which some busybody has rolled up, but which can be unrolled again. And he contrasts this with a sphere that can never be 'unrolled' onto a plane, except by distorting it.

28.8 The metric is something completely intrinsic to the surface itself. Its form could be discovered by those creatures for whom the surface is the universe. They could set up a co-ordinate system and do the necessary measurements. Then (using the theory mentioned at the end of §28.5) they could decide whether their surface is flat or curved. They would not even have to be capable of conceiving that there is any space apart from their own surface. Thus, with a definition given in purely intrinsic terms, the meaning of saying that a surface is curved does not depend on its being curved *in* a space of three dimensions.

These considerations will, I hope, get you past the difficulties that arise from the way Relativitists talk about 'curved space' and 'curved spacetime'. If you think of 'curved' in terms of being curved as seen from outside – being curved *in* a space of more dimensions – then the idea of a three-dimensional curved space becomes exceedingly difficult to make sense of. And as for curved spacetime. . . ! But if you accept 'curved' as merely saying something about the metric, and therefore about the geometrical properties of the spacetime, the difficulties vanish. 'Curved' becomes a shorthand adjective for a rather long description of the metric.

28.9 To apply the definition to spacetime we need one small alteration – making allowance for the negative terms of the metric. So we say that

> A *flat spacetime* is one in which co-ordinates *can* be chosen so that
> (interval increment)^2 ~ sum or difference of squares of
> co-ordinate increments.
>   A *curved spacetime* is one in which co-ordinates *can not* be so chosen.

And you won't waste any more effort in trying to visualise spacetime curved *in* some sort of super-spacetime!

28.10 So the spacetime of Special Relativity – spacetime without gravitation as it appears to an inertial observer – is, by definition, flat. **What about spacetime with constant gravity?**

The metric in a supported observer's co-ordinates is (24.8), which does not have the sum-or-difference-of-squares form. Nevertheless this spacetime is also flat. For we have only to adopt as co-ordinates the time and distance measurements of a free-falling observer, and that brings us back to the simple flat spacetime type of metric. So, on a par with the cylinder as a flat surface that somebody has rolled up, we can think of constant gravity spacetime as being a flat (no gravity) space-time which somebody has perversely decided to study from the point of view of an accelerated observer.

28.11 But it's very different in the case of inverse square law gravity, where the metric (in E's co-ordinates) is (27.22). It can be proved that there is no possibility of choosing some other pair of co-ordinates that would reduce this metric to the difference-of-squares form.

We can, of course, take the co-ordinates of some free-falling observer, and so get the flat metric in his own immediate neighbourhood (§§27.3–4). But distant free-falling observers will be accelerated relative to him; when he translates their measurements into his own, he will have to incorporate the effects of these accelerations, using equations that involve their distance from him. And so he'll end up with a metric of the curved type.

28.12 A surface need not be curved to the same extent all over. The surface of a fried egg, for example, is strongly curved in the middle, but nearly flat towards the rim. And the spacetime of inverse square law gravitation shows something similar. For as r gets bigger, both α and $1 - 2\lambda/r$ get nearer to 1, so that the metric (27.22) gets nearer to $\delta s^2 \sim \delta t^2 - \delta r^2$. Thus at great distances from the Sun this spacetime is effectively flat. The curvature is only important near the Sun.

28.13 It's difficult to visualise how the curvature of spacetime – the 'shape' of spacetime, as we called it in §28.1 – could produce accelerations in free-falling bodies. It's hard to imagine that the world line of a planet is a geodesic – the curved spacetime analogue of a straight line. Or that the very slight deviations from flatness indicated by the metric (27.22) could make all the difference between straight-line-constant-speed motion and the rather complicated behaviour of a planet in its elliptical motion.

An analogy using curved surfaces may help. (It constitutes one of my many debts to Professor Bondi's television series.) We know that the Earth's surface is curved. But we frequently map it on a plane – one of the easiest ways of doing so being Mercator's projection (§14.12). This, or course, introduces gross distortions.

Suppose an aeroplane flies from London to Vancouver – two places at about the same latitude. Its shortest route – a geodesic – is a great circle (§25.21). It does not keep to the starting and finishing latitude, but swings a long way up north from London and then south again to Vancouver (if you need convincing, use the rubber band method of §25.21).

On the Mercator projection, with parallels of latitude represented by horizontal lines, the plane's shortest route will be as shown in Figure 28.13. The geodesic is represented by a *curve* as we expect when we map a curved surface on a plane (§28.7).

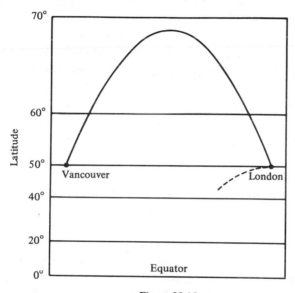

Figure 28.13

28.14 But how would it seem to an extremist Flat Earther who even believes that the Mercator map gives an undistorted scale drawing of the Earth? Many measurements have convinced him that of all routes between London and Vancouver this curve is the most economical in time and fuel. Yet his map informs him that it is not the straight route, the shortest distance.

If he really believes that Mercator tells the unvarnished truth, he must find an explanation for this curious route. Bondi suggested that he might postulate a force pulling the aeroplane towards the equator. So if the pilot had aimed directly at Vancouver, the force (says our friend) would have pulled him into the route shown by the broken line. In order to reach Vancouver he had to aim the appropriate amount towards the north and let the force pull his path into the curve that brings him to the right terminus.

The point is that if the Flat Earther takes a surface that is really curved and insists on regarding it as flat, then the geodesics will appear to him as curves, and his straight lines will not correspond to geodesics. But he assumes that an aeroplane left to itself would follow a straight path. And he has to invent a force to explain why it doesn't.

Pretending that a curved surface is flat turns geodesics into curves. Analogically, if we take a spacetime that is really curved and pretend that it is flat, the geodesics will be changed from the simplest possible world lines into complicated ones. Let's see if these ideas will help us understand the motion of a planet.

28.15 A planet's motion seems complicated. Since it stays in one plane, we could draw its world line in a three-dimensional spacetime diagram. Think of it concretely in terms of the room you're sitting in. We can plot time vertically upwards and represent the plane of the orbit at any instant by a horizontal plane. The Sun's world line is then a vertical pillar. And as a first approximation (assuming a circular orbit) the planet's world line would be the curve called a *helix* – the curve that forms the outside edge of a spiral staircase. But the planet actually moves on an ellipse, with the Sun a bit nearer one end than the other; and its speed depends on its distance from the Sun. So the helix is distorted – elliptical instead of circular, with the Sun's world line placed off centre, and steeper in some parts than in others.

That's very complicated compared to the straight world line that should, according to Newton, represent the motion of a body not acted on by force. **Do a bit of thinking about the description I've just given.**

28.16 Did you notice that in describing the planet's world line I assumed that the spacetime round the Sun could be adequately mapped by the space in your room – by flat space,† in fact. I've taken the curved spacetime of the Sun's neighbourhood and insisted on regarding it as flat. Could that be why the planet's world line seems complicated? Could that be what compels us to invent a force of gravity?

I'm suggesting that if we could think naturally in terms of curved spacetime – and that only means if we could think of the rules of geometry as being rather different from Euclid's – then we should see the planet's world line as that very simple thing, a geodesic. Its motion would look simple and obvious and natural – just as the aircraft's great circle route seems natural to anybody who thinks in terms of a spheri-cal Earth. And just as the Flat Earther's pro-Mercator prejudices made

---

† Any deviation from flatness in your room would be inconceivably small.

the plane's flight seem complicated, so our equally inexcusable predilection for flat spacetime gives the planet's motion an unnecessary appearance of complexity.

28.17 In considering that suggestion we have to remember that the complicated motion of the planets is not something that we *see*. We only see spots of light in the black canopy of the heavens. The spots that we interpret as planets wander among those that we interpret as stars. The theory that planets move at varying speeds in elliptical orbits is what we *infer* from the observational data.

But we habitually do our geometrical theorising in flat space – in terms of the Euclidean geometry that has served us so well in other contexts. So in using the observational data to make inferences about planetary motion, we uncritically assume that flat spacetime is adequate for the job. This leads us, just as surely as it led the chap with the misplaced faith in Mercator, into distortions of everything that happens in spacetime. Hence the apparent complexity of planetary behaviour.

Habit and our cultural heritage make it difficult to think in other than flat and Euclidean terms. And the difficulty is compounded when, in our efforts to cope with unthinkably large distances, we make models or drawings of the Solar System. Being of a size that our imaginations can cope with comfortably, these models and drawings impress themselves on our minds as if *they* were the reality. But their small scale puts them in what is virtually a flat spacetime, which we therefore come to accept as reality.

And I'm suggesting that if we *could* think naturally in terms of the rather more complicated geometry of curved spacetime, then we should see that the world line of a planet is that simplest of curves, a geodesic. The Einstein view would then be much simpler than all the complications of Newton's force of gravity.

28.18 Do keep remembering that 'flat' and 'curved' refer to the definitions of §§28.5 and 28.9 – they merely describe the rules for working with measurements in a space or a spacetime. If that idea has dropped into the background of your thoughts, it might be advisable to reread §§28.13–17 while keeping it well to the fore.

The words 'flat' and 'curved' are helpful when they lead to illuminating analogies like that of the Flat Earther. But often they tend to produce more confusing images than helpful suggestions. When that happens, remind yourself of their definitions in terms of intrinsic geometry.

28.19 Even at the surface of the Sun, $1 - 2\lambda/r$ differs from 1 by only one part in 250 000 (§27.46). So the deviation from flatness even close to the Sun is very small indeed.

Now our measurements of distances in the Solar System are subject to gross uncertainties. So there is no hope of testing this theory by direct measurements of time and distance. But we've all met a tarmac surface that seemed to be flat till we rolled a ball on it. And then the deviation from straight line motion revealed hills and hollows that the eye could not detect. Similarly we might hope that studying how things move in the neighbourhood of the Sun would reveal even this small departure from flatness.

29

The metric around the Sun

29.1 The metric for inverse square law gravitation on a 1D universe leads to no new predictions that we could test. To make further progress we must extend our researches to a plane passing through the Sun's centre (and happily we shall never need to go beyond that).

It will be convenient to use polar co-ordinates (§26.35) with the Sun, S, as pole. So the co-ordinates of event Q (Figure 29.1) are

  t, the time of Q;
  r, the distance from S to Q; and
  θ, the angle between a fixed line SL and the line SQ

– all measurements being made by a very distant supported observer like E of Chapter 27.

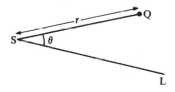

Figure 29.1

29.2 If the spacetime round the Sun were flat – no gravity – the metric would be (26.44). But because of the Sun's gravity, we can expect a more complicated metric, which we can write as

$$\delta s^2 \sim \gamma \delta t^2 - \alpha \delta r^2 - \beta \delta \theta^2. \tag{29.1}$$

where α, β and γ represent expressions involving r (but not t or θ – since the metric must be the same in all directions, and we assume that it is not changing as time goes on). We have to find the forms of these expressions.

If we confine ourselves for a moment to a line through the Sun, then $\delta \theta = 0$ (there is no change of angle), and (29.1) reduces to $\delta s^2 \sim \gamma \delta t^2 - \alpha \delta r^2$. But this brings us back to the case of inverse square law gravity on a line, for which the metric is (27.22). And so we deduce that $\gamma = 1 - 2\lambda/r$.

29.3 A glance back at §26.36 will convince you that the $\delta\theta^2$ term refers to distances measured at right angles to the direction SQ in which gravity is acting. One might reasonably guess that gravity would not affect such transverse distances. And the point can be proved by an argument that I shall merely sketch.

Consider a situation like that of §2.22, but take C to be an inertial observer and D to be an observer who is stationary relative to C at stage (i), but accelerated in the direction perpendicular to PQ – downwards in Figure 2.22. Then the symmetry argument will still show that C and D agree about the simultaneity of P and Q.

Now supply C and D with rods defined as in §2.27. The same argument will work again to establish that C and D agree about their lengths. And using them as measuring rods, C and D will agree about distance measured at right angles to the direction of D's acceleration. Finally, the Principle of Equivalence transforms this into the proposition that a free-falling observer (C) and a supported observer (D) agree about distances perpendicular to the direction of gravitational acceleration. Thus the $\delta\theta^2$ term in our new metric should be the same as in (26.44). And so $\beta = r^2$.

29.4 Substituting for β and γ in (29.1), we see that the metric for inverse square law gravity in a plane is

$$\delta s^2 \sim \left(1 - \frac{2\lambda}{r}\right)\delta t^2 - \alpha\delta r^2 - r^2\delta\theta^2 \, , \tag{29.2}$$

where the form of α is still a mystery to us.

29.5 On previous occasions when we found a new metric, the first thing we did with it was to study the behaviour of light (§§24.24–6, 27.37–8). All we had to do was to put $\delta s = 0$ (**Why?** See §24.24). That gave us a near-equation connecting δx (or δr) and δt. And in a 1D universe, where the only thing to be studied was speed, that single near-equation was enough.

But now that we are concerned with two spatial dimensions, there are two questions: What path does light follow? And how fast does it go? To cope with these we need a second near-equation.

29.6 To work out what this new near-equation is, we follow again the strategy outlined in §24.27. When we applied that method to inertial motion, its straight-line-constant-speed characteristic got translated into the proposition that in the absence of gravitation the world line of an inertial observer is an ordinary geodesic (Chapter 25, culminating in §§25.17–18, 25.27). And then we suggested in §§25.19 and 25.28 that, even when there is gravity, an inertial world line will still be an ordinary geodesic. So far this idea has given promising results. **Can you see how to stretch it a bit further to make it cover the motion of light?**

In no-gravity conditions light also has straight-line-constant-speed motion – but with the additional point that this constant speed is the very special one to which we have given the value 1. So a signal line will again be a geodesic, but with a further equation which generalises for any spacetime the 'speed of light = 1' statement of the no-gravity case. **What is this further equation?**

By the reasoning of §24.24 it is what we get by stating that interval measured along the signal line is zero, i.c., that $\delta s = 0$.

29.7 When a geodesic has this additional property that interval measured along it is zero, it is called a *null geodesic* – contrast the ordinary geodesic of §25.27. And our new suggestion is that

A signal line is a null geodesic,

and that the behaviour of light in any sort of spacetime can be deduced from this statement in the same way as we can deduce the behaviour of free-falling bodies from the Geodesic Assumption (§25.28).

29.8 So we want to work out the near-equations that characterise a null geodesic. Naturally your first thought is to rush back to Chapter 26, where we worked out the near-equations for ordinary geodesics, and simply put $\delta s = 0$ in them. Unfortunately it doesn't work. For when you put $\delta s = 0$ in (26.31), you get $\beta \delta y \sim 0$ – which is forbidden nonsense (§24.25).

We shall have to be more subtle. Much of what we did on ordinary geodesics will still apply. But we'll have to make some modifications, so that we can introduce the condition that $\delta s = 0$ in a way that will not lead to nonsensical statements. I shall simply run through Chapter 26, mentioning which parts still hold good, and describing the alterations that must be made in others. In order to make a connected argument of it, I think you must reread the bits of Chapter 26 as I mention them. Start with §26.5, and then . . .

29.9 The metric is to be (26.2), with the quantities defined as in the first paragraph of §26.6. The argument of Chapter 26 will do unchanged as far as the end of §26.11, where we add the further conditions that at the desired extreme value σ_P and σ_Q will both be zero.

In making the fourth simplification (§26.13), we can't have equation

(26.8) when Q is in its correct position – for σ_P and σ_Q would both be zero, and 'zero divided by zero' is nonsense (or ambiguous). But in the various wrong positions of Q that have to be considered, σ_P and σ_Q will not in general be zero. So we can lay it down that *in our search for the correct position* of Q we shall *consider only those wrong positions* for which (26.8) holds good. Then (26.9) will be true for *all* positions of Q. (The previous proof holds good for wrong positions; and when Q's position is correct, both sides are zero.)

The original reasoning of Chapter 26 now stands firm as far as the end of §26.17. But in undoing the fourth simplification (§26.18) we cannot use (26.8), since σ_P and σ_Q are zero. So to eliminate K we divide each side of (26.16) by the corresponding side of (26.17), getting

$$\frac{\beta_P \delta y_P}{\gamma_P \delta z_P} = \frac{\beta_Q \delta y_Q}{\gamma_Q \delta z_Q} \, ,$$

which replaces (26.18) and (26.19). And by an argument just like that of §26.19, this leads to the equation

$$\beta \delta y = H \gamma \delta z \, , \tag{29.3}$$

where H is a constant number. This takes the place of (26.21). And since we've already got rid of σ, it equally takes the place of (26.22).

In §26.21, the justification for the statement that $s \sim \Sigma$, as briefly described in the footnote, depends on each δs and σ being non-zero – for it uses (cf. §20.25) the notion of 'proportionate error', which becomes meaningless when they are zero. But if they *are* zero, we have $s = \Sigma = 0$, which is stronger than, and includes, $s \sim \Sigma$. So this last holds good whether the curve does or does not give an extreme value to Σ – just as paragraph 3 of §26.21 requires.

The rest of §26.21 stands, except that this time we have to allow for the 'small enough portions' idea by changing = to $\sim$ in (29.3), getting

$$\beta \delta y \sim H \gamma \delta z \, . \tag{29.4}$$

To complete the conditions specifying a *null* geodesic we have only to add the point that interval measured along the curve is zero – which we do by putting $\delta s = 0$ in the metric (after rearranging it to avoid a near-equation with zero on one side, §24.25).

29.10 So with

$$\delta s^2 \sim \alpha \delta x^2 + \beta \delta y^2 + \gamma \delta z^2 \tag{29.5}$$

as metric (the various quantities being defined as in §26.33) our conclusion is that the curve will be a null geodesic if the increments are related by near-equation (29.4), together with a second near-equation obtained by putting $\delta s = 0$ in the (suitably rearranged) metric.

29.11 As a simple exercise (getting any help you need from §§26.41–3) will you **investigate the motion of light in a gravity-free plane using polar co-ordinates; and in particular, find what path a light signal follows across the plane.**

The metric is (26.44). For the same reasons as in §26.42, we can identify r, θ and t with x, y and z of (29.5), so that $\alpha = -1$, $\beta = -r^2$ and $\gamma = 1$. Then (29.4) gives $-r^2\delta\theta \sim h\delta t$ (where h is a constant). And from the re-arranged metric, with $\delta s = 0$, we get $\delta r^2 \sim \delta t^2 - r^2\delta\theta^2$. To find the spatial path we eliminate δt by using the former of these near-equations to substitute $-r^2\delta\theta/h$ for δt in the latter. This gives

$$\delta r^2 \sim r^4\left(\frac{1}{h^2} - \frac{1}{r^2}\right)\delta\theta^2 . \tag{29.6}$$

And comparing with (26.43), we see that the path of a light signal is (as it should be!) a straight line.

29.12 **Now do a similar investigation of the path followed by a light signal under the influence of inverse square law gravity in a plane.**

The metric is (29.2). Again we identify r, θ and t with x, y and z of (29.5), getting $\beta = -r^2$ and $\gamma = 1 - 2\lambda/r$. Then (29.4) gives

$$-r^2\delta\theta \sim H\left(1 - \frac{2\lambda}{r}\right)\delta t , \tag{29.7}$$

which can be rewritten as

$$\delta t \sim \frac{-r^2\delta\theta}{H\left(1 - \frac{2\lambda}{r}\right)} . \tag{29.8}$$

Rearranging the metric and putting $\delta s = 0$ gives $\alpha\delta r^2 \sim (1 - 2\lambda/r)\delta t^2 - r^2\delta\theta^2$. Using (29.8) to eliminate δt from this last, and dividing both sides by α, we have

$$\delta r^2 \sim \frac{1}{\alpha}\left\{\frac{r^4}{H^2\left(1 - \frac{2\lambda}{r}\right)} - r^2\right\}\delta\theta^2 ,$$

which can be rewritten (§§16.12, 26.37) as

$$\delta r^2 \sim \frac{r^4}{\alpha}\left\{\frac{1}{H^2\left(1 - \frac{2\lambda}{r}\right)} - \frac{1}{r^2}\right\}\delta\theta^2 . \tag{29.9}$$

Are you sure you know what that means?

If Q and Q' are two points of the light signal's path (Figure 29.12), then the increments in the polar co-ordinates as we pass from Q to Q' are related by near-equation (29.9). And comparison with (29.6) shows that this time the light's path is *not* a straight line.

Now α must be positive (since the δr^2 term in the metric, being spatial, must be subtracted). And δr^2, $\delta \theta^2$ and r^4 cannot be negative (§24.20). Thus the part of (29.9) inside the curly brackets must be either zero or positive. And so the second of its two terms must be equal to or less than the first – much less except when r is near its minimum.

29.13 So far we've had to put up with our ignorance about α – knowing only that it stands for some expression that involves r. Now I'm going to suggest – rather tentatively – what that expression is. We know from observation that the path of a light ray, even when affected by gravitation in the neighbourhood of the Sun is *very nearly* a straight line – so nearly that nobody suspected anything else until after Einstein made a famous prediction that we'll come to soon.

If it were *exactly* a straight line, then the near-equation for its path would be (29.6) instead of (29.9). So (29.9) should bear a very strong resemblance to (29.6). And the resemblance will be strongest if we take

$$\frac{1}{\alpha} = 1 - \frac{2\lambda}{r}, \tag{29.10}$$

in which case (29.9) becomes

$$\delta r^2 \sim r^4 \left\{ \frac{1}{H^2} - \frac{1}{r^2}\left(1 - \frac{2\lambda}{r}\right) \right\} \delta \theta^2. \tag{29.11}$$

This minimises the effect of the $1 - 2\lambda/r$ factor by attaching it to the smaller of the two terms inside the curly brackets (end of §29.12). It arranges (assuming that H plays a similar role to h) for the nearest thing to (29.6) that is possible when we start from a metric of the type (29.2).

29.14 Now I have to confess that in this long chain of argument by which I have been trying to lead you to Einstein's conclusions, while avoiding the forbidding difficulties of his own method, this guess about α is the weakest link of all. You may very rightly feel suspicious about it. You could also make good *a priori* cases for $\alpha = 1$ (the simplest possible) or $\alpha = 1 - 2\lambda/r$ (the radial speed of light would be constant, §27.38), and perhaps other possibilities.

If we met a situation like this in real research, we should simply go ahead, working out the predictions of the several likely alternatives – till experiment or observation allowed us to decide which (if any) was correct.† Doing that here would be tedious and cumbersome. So I shall

---

† Theoreticians concerned with making comparisons between Einstein's theory and some more recent rivals do, in fact, work on similar lines.

simply follow up the consequences of assuming (29.10) true. And at an appropriate point I'll report on what results we should have got if we had chosen the other assumptions.

  With Einstein's own method – of which I'll give you a vague and simplified description before we part – the correct form follows unambiguously from the assumptions of the theory.

29.15 The larger we make r, the nearer does $1 - 2\lambda/r$ approach to 1, and so the closer does the resemblance between (29.11) and (29.6) become. And even at the Sun's surface it differs from 1 by only one part in 250 000. So we see that (if our guess about α is correct) light in the vicinity of the Sun travels *very very nearly* in a straight line – but not quite.

29.16 Assuming (29.10) true, and using it to substitute $1/(1 - 2\lambda/r)$ for α in (29.2), we can now write the metric for inverse square law gravitation in a plane as

$$\delta s^2 \sim \left(1 - \frac{2\lambda}{r}\right)\delta t^2 - \frac{\delta r^2}{1 - \dfrac{2\lambda}{r}} - r^2\delta\theta^2 . \tag{29.12}$$

  The spacetime defined by this metric is curved, by the definition of §28.9 – though to establish that strictly I should have to deal with the can/cannot point (cf. §28.5).

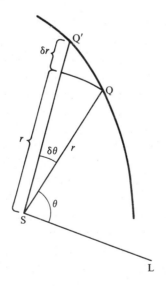

Figure 29.12

29.17 Quite apart from any doubts about α, I've been guilty in §§29.12–15 at least of careless statements, if not of gross deception. **Did you spot the point?**

If not, revise §§28.13–17 and **try again.**

I've been repeating the same false assumption that I made in §28.15 and revealed at the start of §28.16 – taking a curved spacetime and reasoning as if it were flat. Worse indeed! I've been behaving like the Mercator maniac of §28.14. The metric for the *space* (as opposed to the spacetime) around the Sun is got by changing positive terms to negative and vice versa in (29.12) (to obtain a space-like interval, §13.23) and putting $\delta t = 0$ (so that we're talking about events that are separated by space only). So this spatial metric is

$$\delta s^2 \sim \frac{\delta r^2}{1 - \dfrac{2\lambda}{r}} + r^2 \delta\theta^2 \, , \tag{29.13}$$

where δs now refers to distance. Its form shows that even the *space* round the Sun is curved.

So I was being misleading even in §29.1 (and a couple of subsequent occasions) when I spoke of a 'plane' through the Sun's centre. A plane is a surface whose metric in polar co-ordinates is (26.38). And the metric we obtained in the previous paragraph shows that gravity has distorted this plane into a curved surface – curved fairly strongly near the Sun, almost flat at great distances (cf. §28.12). (But *please* don't allow yourself to think of it as a plane with some sort of bulge near the Sun. It's a cross-section of a three-dimensional space which is itself curved. Stick to the definition of §28.5 – it's just a surface in which the rules of geometry disagree, slightly, with Euclid.)

Near-equation (29.6) refers to a straight line *in a plane,* but not in a curved surface like this one. So it was wrong to compare (29.9) or (29.11) with (29.6) and deduce that the path of a light signal is nearly, but not quite, a straight line. We should have said that this is how things *will appear to us* if we behave rather like that Mercator Flat Earther (§§28.13–4) and *assume* (wrongly) that the space around the Sun is flat. *On that assumption,* but *only* on that assumption, (29.6) would correspond to a straight line, and therefore (29.9) or (29.11) would not quite do so.

However, in working out the geometry of the Solar System (and beyond) the only quantities that are actually measured are distances and times on Earth and angles between light rays reaching astronomical observatories. All else is calculated. And astronomers do their calculations on the assumption that they are working in flat space. So the statement that the path of light near the Sun is not quite a straight line describes how things will appear if normal astronomical methods are used. In these conditions, we are saying, the path of light should appear to be slightly bent as it passes the Sun. By *how much?*–that will be the main question for the next chapter.

29.18m But first we must learn a bit more about extreme values (§25.26). We'll prove that

> If two variable quantities, r and θ, are related in such a way that
> $$\delta r \sim U\delta\theta , \qquad\qquad (29.14)$$
> where the size of U depends on the size of r (but not of θ, nor of the increments), then the extreme values of r are among those that make $U = 0$.

This theorem does *not* say that every value of r which makes $U = 0$ is extreme – only that the extreme values (if any) are *among* those that make $U = 0$. Though it applies to quantities of any type, I have adopted the symbols r and θ because we shall only use it in relation to polar co-ordinates.

29.19m To prove this, choose any value you like of θ. Then you could work out the corresponding value of r, from which the value of U could be calculated. If U is not zero, it must be positive or negative. Suppose it to be positive. Then if $\delta\theta$ is positive, (29.14) tells us that δr is also positive – or more strictly, it says that by giving $\delta\theta$ a *small enough* positive value, we can arrange that δr is also positive. In other words, a slight increase in θ gives an increase in r. And if r can be increased, it is not a maximum. Again, if we make $\delta\theta$ negative, (29.14) says that *de*creasing θ by a small enough amount gives a *de*crease in r – which was therefore not a minimum to start with.

So we've proved that a value of r that makes U positive cannot be extreme. **Now you prove the like when U is negative.** Between us, we've covered all the cases, and shown that if U is not zero, then r does not have an extreme value. So one can only occur if $U = 0$.

29.20m We shall actually use this theorem in the form:

> If two quantities, r and θ, are related in such a way that
> $$\delta r^2 \sim V\delta\theta^2, \qquad\qquad (29.15)$$
> where the size of V depends on the size of r (but not of θ, nor of the increments), then the extreme values of r are among those that make $V = 0$.

To prove it, take the square root of both sides of (29.15), getting $\delta r \sim U\delta\theta$, where $U = \sqrt{V}$, and apply the original theorem.

29.21 As an example, consider (29.6), which refers (in flat space) to a straight line. Here

$$V = r^4\left(\frac{1}{h^2} - \frac{1}{r^2}\right) = r^4\left(\frac{1}{h} - \frac{1}{r}\right)\left(\frac{1}{h} + \frac{1}{r}\right),$$

by (16.10). The values of r that make $V = 0$ are obviously 0, h and $-h$ (e.g., if you put $r = h$, the first bracket $= 1/h - 1/h = 0$, and so $V = 0$). If the line does not pass through the pole S, r cannot be zero; and in polar co-ordinates r cannot be negative (§26.35). So we're left with $r = h$ as the only extreme. And this is actually the length of the perpendicular from S to the line (§26.40) – which is well known to be the least value of r.

30

Light and gravity

30.1 The path of a light signal is what we usually call a *light ray*. So if our theory (including the guess about α) is correct, a light ray from a star should appear to be bent (§29.17) by an amount that we can calculate from (29.11). We have to see whether the calculation gives predictions that agree with observation. (You'll find the mathematics only moderately difficult. The advice of §§26.4 and 26.27, suitably amended, may be helpful.)

30.2 Let MLN (Figure 30.2) represent a ray coming in from beyond M, making its closest approach to the Sun at L, and going off in the direction beyond N. And let h stand for its shortest distance from the Sun's centre, S – the distance SL. If r and θ are the polar co-ordinates of any point Q on the ray (taking SL as the line from which θ is measured), and δr and $\delta \theta$ are the increments between Q and another

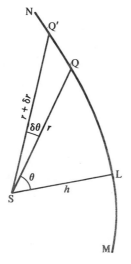

Figure 30.2

point Q' of the ray, then these quantities are related by (29.11). So according to §29.20, the extreme values of r should be among those which make

$$r^4 \left\{ \frac{1}{H^2} - \frac{1}{r^2}\left(1 - \frac{2\lambda}{r}\right)\right\} = 0 \,,$$

or (since r can't be zero), which make

$$\frac{1}{H^2} - \frac{1}{r^2}\left(1 - \frac{2\lambda}{r}\right) = 0 \,.$$

Now h, being the least distance of the ray from S, is an extreme value of r; so putting $r = h$ must make the last equation true – which gives

$$\frac{1}{H^2} = \frac{1}{h^2}\left(1 - \frac{2\lambda}{h}\right) \,. \tag{30.1}$$

And substituting in (29.11), we have

$$\delta r^2 \sim r^4 \left\{ \frac{1}{h^2}\left(1 - \frac{2\lambda}{h}\right) - \frac{1}{r^2}\left(1 - \frac{2\lambda}{r}\right)\right\} \delta\theta^2 \,. \tag{30.2}$$

30.3 Let JLK (Figure 30.3) be the straight line through L perpendicular to SL – so that it is tangent to MLN at L, and its least distance from S is again h. We find the shape of the curve MLN by comparing it with this line.

As before, Q is the point of the light ray whose co-ordinates are (i)

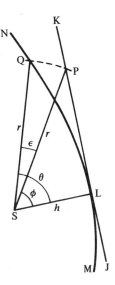

Figure 30.3

the distance r from S to Q, and (ii) the angle θ between SL and SQ. We next take the point P on line JLK which is at the *same* distance r from S. Its polar co-ordinates are (i) length SP, which is r again, and (ii) the angle ϕ between SL and SP.

The relation between δr and $\delta \phi$, got by changing θ to ϕ in (26.43), is

$$\delta r^2 \sim r^4 \left(\frac{1}{h^2} - \frac{1}{r^2} \right) \delta \phi^2 \tag{30.3}$$

– in which, it must be emphasised, h and r are the same as the h and r in (30.2).

A comparison of (30.2) and (30.3) will show how much the line SQ swings round when SP is made to swing through a given angle – and from this we can work out how much MLN is bent.

30.4 The angle PSQ in Figure 30.3 gives a measure of how much the light ray deviates from the straight. We call it† ϵ, so that $\theta - \phi - \epsilon$, from which it follows (§§22.9–10) that

$$\delta \theta - \delta \phi = \delta \epsilon . \tag{30.4}$$

Assume that you know the value of r, and think of ϕ being increased by amount $\delta \phi$. Then by using (30.3), (30.2) and (30.4) in succession we could calculate the corresponding increments δr, and then $\delta \theta$, and finally $\delta \epsilon$. And a process of adding increments, like that of §21.21 would enable us to calculate ϵ itself – which would answer our question about how much the ray is bent.

30.5 Combining (30.2)–(30.4) in order to put the connection between $\delta \phi$ and $\delta \epsilon$ in convenient form involves a tedious bit of manipulation, which I am deferring to §30.19. You can study that in due course. Meanwhile don't let it obscure your appreciation of the broad line of argument. Simply believe me for the moment that from these three near-equations we can deduce that

$$\delta \left(\frac{h}{\lambda} \epsilon \right) \sim U \delta \phi , \tag{30.5}$$

where, for brevity, we have written

$$\frac{1 + \dfrac{h}{r} + \dfrac{h^2}{r^2}}{1 + \dfrac{h}{r}} = U . \tag{30.6}$$

Just what does (30.5) mean?

It is a recipe for calculating (as accurately as we wish) the increment

† 'Epsilon', the Greek e.

in ϵ, when you're given the increment in ϕ and the value of r (Figure 30.3). This gives us the tools for a step by step calculation of a table of ϵ against ϕ by essentially the same method as we used in Chapter 21 to calculate a table of ft against k. Maybe you should study §§21.18–23 again. At that stage we had not introduced the 'δ' notation for increments (§§21.39–41). So if you now translate the Chapter 21 material into the 'δ' language, you'll find it easier to cope with the next few pages.

30.6 We shall be concerned with light grazing the Sun on its way from a star to the Earth – both of which are at enormous distances compared with the radius of the Sun. So the initial and final values of ϕ will be indistinguishable from $-90°$ and $+90°$ (when it comes to practical calculations we revert from radians, §26.34, to degrees, minutes and seconds). But the two halves of the curve will be symmetrical, and so we need only do the calculation from $0°$ to $90°$ and then double the result. If your mathematical knowledge goes far enough, you may be able to skip to §30.11.

30.7 What size of steps shall we use? Just as in Chapter 21, smaller steps mean more accuracy *and* more work. But we're also up against the problem of calculating the value of h/r that corresponds to the value of ϕ at the start of each step. And unless you're acquainted with another branch of mathematics that I'm going to avoid, this is in general a very difficult calculation to make.

However, when ϕ has one of the values $0°$, $30°$, $45°$, $60°$ and $90°$, the calculation of h/r becomes easy (see §30.8). So we're in a position to do the working in either (*a*) two steps of $45°$, or (*b*) three steps of $30°$. Comparing the results will give a check on accuracy (§21.22).

For the same reasons as in §§21.19–20, we can get more accuracy if in place of (30.5) we use near-equation

$$\delta\left(\frac{h}{\lambda}\epsilon\right) \sim \bar{U}\delta\phi , \tag{30.7}$$

where $\bar{U}$ is the average of the values of U at the beginning and end of the step.

30.8M When $\phi = 0°$, SP coincides with SL (Figure 30.3), so that $h/r = 1$. As ϕ gets nearer to $90°$, r grows larger without limit, so that the fraction h/r grows smaller and smaller, till at $\phi = 90°$, $h/r = 0$.

If $\phi = 45°$ – Figure 30.8(i) – length LP is also h. And so Pythagoras' Theorem gives $r^2 = 2h^2$, from which $h/r = 1/\sqrt{2} = 0\cdot7071$.

In the large triangle of Figure 30.8(ii) all three sides have length r.

So all its angles are equal; and as they total 180°, each is 60°. If PL is perpendicular to SR, then length SL $= \frac{1}{2}r$ (by symmetry). So identifying the triangle SLP with the one similarly labelled in Figure 30.3, we find that when $\phi = 60°$, $h = \frac{1}{2}r$, so that $h/r = 0\cdot5$.

Obviously angle LPR is 30°. To find the corresponding value of h/r, we identify triangle PLR with SLP of Figure 30.3, so that length PL $= h$. Pythagoras' Theorem gives $r^2 = (\frac{1}{2}r)^2 + h^2$, so that $h^2 = \frac{3}{4}r^2$, whence $h^2/r^2 = \frac{3}{4}$ and $h/r = \sqrt{\frac{3}{4}} = 0\cdot8660$.

30.9 Then the calculation of the table in 45° steps will run:

(1)	(2)	(3)	(4)	(5)	(6)
ϕ	h/r	U	$\bar{U}$	$\delta\left(\dfrac{h}{\lambda}\epsilon\right)$	$\dfrac{h}{\lambda}\epsilon$
0°	1·0000	1·500			0·00°
			1·397	62·87°	
45°	0·7071	1·293			62·87°
			1·146	51·57°	
90°	0·0000	1·000			114·44°

See if you can understand the rationale of this calculation before reading the description that follows.

The values of h/r in column (2) are obtained as in §30.8, and from these we get U of column (3) by using (30.6). The column (4) entries, set on the half-lines, are the averages of the values of U in the lines immediately above and below. Then by (30.7) the (approximate) column (5) entry is the U from column (4) multiplied by $\delta\phi = 45°$. And each entry in column (6) is obtained by adding the column (5) increment to the previous column (6) entry. **How about trying the three-step calculation for yourself before looking at my version?**

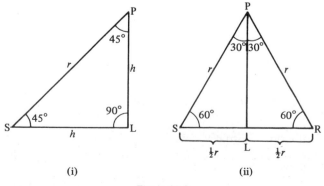

(i)   (ii)

Figure 30.8

ϕ	h/r	U	$\bar{U}$	$\delta\left(\dfrac{h}{\lambda}\epsilon\right)$	$\dfrac{h}{\lambda}\epsilon$
0°	1·0000	1·500			0·00°
			1·451	43·53°	
30°	0·8660	1·402			43·53°
			1·284	38.52°	
60°	0·5000	1·167			82·05°
			1·084	32·52°	
90°	0·0000	1·000			114·57°

30.10 By the same argument as in §21.22, the latter result will be accurate enough for our purpose. So, using ϵ^* to denote the final value of ϵ at 90°, we get

$$\frac{h}{\lambda}\epsilon^* = 114\cdot57° \approx 412\ 500 \text{ seconds}$$

(§26.34).

30.11 If you know about trigonometric functions and their increments (equivalent to their derivatives), you can avoid the above numerical computation. From (30.5) and (30.6) we have

$$\delta\left(\frac{h}{\lambda}\epsilon\right) \sim \left\{\frac{1}{1 + \cos\phi} + \cos\phi\right\}\delta\phi \sim \delta\left(\tan\tfrac{1}{2}\phi + \sin\phi\right).$$

So by §25.13, $(h/\lambda)\epsilon = \tan\tfrac{1}{2}\phi + \sin\phi$ (since $\epsilon = 0$ when $\phi = 0$). And $(h/\lambda)\epsilon^* = 2$ radians $= 412\ 530$ seconds.

30.12 The angle that concerns us (§30.6) is

$$2\epsilon^* = 825\ 000\frac{\lambda}{h} \text{ seconds.}$$

If we consider a ray just grazing the Sun's surface, then $h = 696\ 000$ kilometres, while $\lambda = 1\cdot477$ kilometres (§27.15), so that

$$2\epsilon^* = 1\cdot75 \text{ seconds.}$$

   In Figure 30.12 (drawn grossly out of proportion) the light from the star would, if the Sun were not there, reach the Earth in direction CE. With the Sun present, it comes in along the parallel line AD, but after gravitational deflection it approaches Earth along FE – as if it had originally come along BFE. So it will appear to an observer on Earth to be displaced by an angle of 1·75 seconds (1·75″) from where it should be in the heavens.

   That, then, is the prediction that must be tested by observation.

30.13 The most obvious alternative prediction is that light is not affected by gravity; there should be no deflection at all. Or if we retain the Newtonian theory of gravitation, there is also the possibility – which received considerable attention after Einstein had shown that light, having energy, must also have mass – that the Sun's attraction would act on light in the same way as on a material body. If we calculate how much something moving at 300 000 kilometres per second should be deflected by a Newtonian force of gravity, we get an answer exactly half of the Einstein prediction.

So what does observation say?

30.14 Using ordinary optical astronomy – all that was available till a few years ago – the observations are very difficult. First, the angle to be measured is extremely small. Secondly, the observation of stars whose light passes close to the Sun is only possible during a total eclipse† – so that

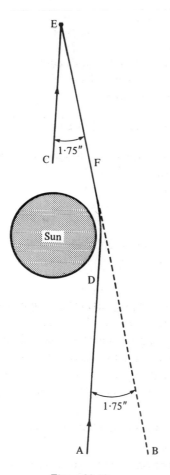

Figure 30.12

† Recently a technique has been suggested for doing it in daylight (*New Scientist,* 30 November 1978, p. 684).

the measurements have to be done on rare occasions and in a hurry. Then it is necessary to compare the apparent position of the star, as observed during the eclipse, with the position it would be seen in if the Sun were not there. So its position during the eclipse is compared with that of other stars well away from the Sun. And several months earlier or later, when the star is visible at night, this comparison is made again. The comparing is done by photographic methods of course. But one photograph was taken in the cold of the night, and the other during the day's warmth, so that allowance must be made for the expansion of the plate.

Without going farther, it will be obvious that the opportunities for error are manifold. We may expect large uncertainties in the results.

30.15 The first two observations, of the 1919 eclipse were by Eddington, who got $1 \cdot 98 \pm 0 \cdot 16$ seconds, and Dyson, $1 \cdot 61 \pm 0 \cdot 40$. (See §23.17 for this $\pm$ business.) The $\pm 0 \cdot 16$ indicates a low level of precision, since it says that even the large range from $1 \cdot 66$ to $2 \cdot 30$ seconds has only a 95 per cent chance of including the true value. Still, the predicted $1 \cdot 75$ seconds does lie inside that range; and so for a first attempt the result could be regarded as promising – and Dyson's also, despite its still lower precision.

One could hope that refinements would bring better results. But they never did. The best standard error ever obtained was $0 \cdot 1$: $2 \cdot 24 \pm 0 \cdot 10$ in 1929 and $1 \cdot 70 \pm 0 \cdot 10$ in 1952. The latter could be called 'not bad'. But the former, saying that there's a 95 per cent chance that the true deflection lies between $2 \cdot 04$ and $2 \cdot 44$ seconds, would – if we believed it – prove Einstein badly wrong. And while we've had $1 \cdot 72 \pm 0 \cdot 15$ in 1922 – tolerably good – we've also had $2 \cdot 73 \pm 0 \cdot 31$ in 1936 – shockingly bad. In addition, there is a considerable possibility that any or all of these observations involved large systematic errors.

According to S. Weinberg[†] we can conclude that the deflection is certainly greater than the $0 \cdot 875$ of the Newtonian prediction (§30.13), 'but as to its precise value we can say little more than that $2\epsilon^*$ is somewhere between $1 \cdot 6$ and $2 \cdot 2$ seconds'.

While only optical methods were available, the verdict had to be 'not proven'. And several rival theories still looked plausible.

30.16 Radio astronomy came to the rescue. Compact radio sources (quasars, for instance) will do as well as stars. As a result of the Earth's orbital motion some of these pass behind the Sun once a year, or very close to it. Their apparent shifts of position can be measured in relation to

† In *Gravitation and Cosmology* (1972) John Wiley & Sons, N.Y. p. 192. I've used $2\epsilon^*$ instead of his notation.

other radio sources. There is no waiting for eclipses and no difficult comparison of day and night measurements.

Five independent sets of observations were made in 1969–70 with equipment whose resolving power (ability to cope with small angles) was still very limited. And already the results were better than those given by fifty years of optical astronomy. Four out of the five clustered round the 1·75 seconds value; but they were still not precise enough to decide between Einstein and certain rival theories that I shall continue for the present to mention in a tantalisingly vague manner.

But methods continued to improve until in 1975 E. B. Fomalont and R. A. Sramek did a month-long series of observations at two frequencies on three radio sources that pass close to the Sun, using a 35 kilometre baseline. Their result† was that $2\epsilon^* = 1\cdot763 \pm 0\cdot016$. This represents an enormous improvement in precision, and Einstein's prediction is right on target. It makes things very difficult for any rival theory.

30.17 The prediction of a 1·75 second deflection was based on the assumption that $1/\alpha = 1 - 2\lambda/r$ (§§29.13–14). Rather more complicated calculations show that if we had taken $\alpha = 1$, the prediction would be a deflection of only half that amount. And $\alpha = 1 - 2\lambda/r$ would, surprisingly, give a zero deflection. The observations are inconsistent with either of these – or indeed with any plausible form other than (29.10). We can therefore take this last to be correct, and then it follows that (29.12) *is* the metric for the spacetime round the Sun.

The zero deflection predicted if α were $1 - 2\lambda/r$ does *not* imply that the light ray is straight. It bends away from the Sun as it approaches, comes momentarily back to its original line at closest approach, and then repeats the manoeuvre as it recedes. I'll sketch a proof of the above statements after we've dealt with the point deferred from §30.5.

It was my Derby student Roger Jackson who drew my attention to the oddities of the $\alpha = 1 - 2\lambda/r$ case, and set off the thoughts that emerge here and in §30.20. The last few chapters of this book owe a great deal to the constructively critical activities of that Derby class, which among other things saved me from committing one blunder so crass that it must remain a secret between us. And my Nottingham group three years later rounded off many rough edges. But then I've seldom met a student who couldn't teach me something. Roger and Stan (§26.39) must represent all the others.

† The published result is stated as $1\cdot007 \pm 0\cdot009$ of the expected deflection (*Physical Review Letters*, **36** (1976) 1475). So my figures may be 1 or 2 out in the last digit.

30.18M Repeated application of §16.12 yields another identity:

$$(u - v)(u^2 + uv + v^2) = u(u^2 + uv + v^2) - v(u^2 + uv + v^2)$$
$$= u^3 + u^2v + uv^2 - u^2v - uv^2 - v^3$$
$$= u^3 - v^3 . \qquad (30.8)$$

30.19 We have to prove (30.5). For convenience rewrite (30.2) and (30.3) as

$$\delta r^2 \sim r^4 X \delta\theta^2 \qquad (30.9)$$

and

$$\delta r^2 \sim r^4 Y \delta\phi^2, \qquad (30.10)$$

where

$$X = \frac{1}{h^2}\left(1 - \frac{2\lambda}{h}\right) - \frac{1}{r^2}\left(1 - \frac{2\lambda}{r}\right) \qquad (30.11)$$

and

$$Y = \frac{1}{h^2} - \frac{1}{r^2} . \qquad (30.12)$$

From (30.10) and (30.9) we get

$$\delta\phi^2 \sim \frac{\delta r^2}{r^4 Y} \sim \frac{X}{Y}\delta\theta^2 .$$

So

$$\delta\theta^2 - \delta\phi^2 \sim \delta\theta^2 - \frac{X}{Y}\delta\theta^2$$

§16.12

$$= \left(1 - \frac{X}{Y}\right)\delta\theta^2$$

§16.22

$$= \left(\frac{Y}{Y} - \frac{X}{Y}\right)\delta\theta^2$$

§21.12, footnote

$$= \frac{Y - X}{Y}\delta\theta^2 .$$

Then (16.10) gives

$$(\delta\theta + \delta\phi)(\delta\theta - \delta\phi) \sim \frac{Y - X}{Y}\delta\theta^2 . \qquad (30.13)$$

Now we make an approximation. Since $2\lambda/h$ and $2\lambda/r$ are very small fractions, the difference between X and Y is very small compared with either of them. So (30.9) and (30.10) tell us that the same applies to $\delta\theta$ and $\delta\phi$. Thus we shall not introduce serious error if we change $\delta\theta$ to $\delta\phi$ everywhere, *except* in the term $\delta\theta - \delta\phi$, where the difference between them obviously cannot be ignored. Then (30.13) becomes

$$2\delta\phi(\delta\theta - \delta\phi) \sim \frac{Y - X}{Y}\delta\phi^2 .$$

And dividing both sides by $\delta\phi$ and using (30.4) leads to

$$2\delta\epsilon \sim \frac{Y - X}{Y}\delta\phi . \tag{30.14}$$

Now from (30.11) and (30.12) – using §§15.23–4 and 16.12 several times – we get

$$Y - X = \frac{1}{h^2} - \frac{1}{r^2} - \frac{1}{h^2}\left(1 - \frac{2\lambda}{h}\right) + \frac{1}{r^2}\left(1 - \frac{2\lambda}{r}\right)$$

$$= \frac{1}{h^2} - \frac{1}{r^2} - \frac{1}{h^2} + \frac{2\lambda}{h^3} + \frac{1}{r^2} - \frac{2\lambda}{r^3}$$

$$= 2\lambda\left(\frac{1}{h^3} - \frac{1}{r^3}\right)$$

(30.8)

$$= 2\lambda\left(\frac{1}{h} - \frac{1}{r}\right)\left(\frac{1}{h^2} + \frac{1}{hr} + \frac{1}{r^2}\right) .$$

Again, (16.10) applied to (30.12) gives

$$Y = \left(\frac{1}{h} - \frac{1}{r}\right)\left(\frac{1}{h} + \frac{1}{r}\right) .$$

Substituting these expressions for Y and $Y - X$ in (30.14), cancelling out the factor that occurs in both numerator and denominator, and dividing both sides by 2 leads to

$$\delta\epsilon \sim \frac{\lambda\left(\dfrac{1}{h^2} + \dfrac{1}{hr} + \dfrac{1}{r^2}\right)}{\dfrac{1}{h} + \dfrac{1}{r}}\delta\phi ,$$

which, on multiplying numerator and denominator by h^2 (§16.22), becomes

$$\delta\epsilon \sim \frac{\lambda\left(1 + \dfrac{h}{r} + \dfrac{h^2}{r^2}\right)}{h\left(1 + \dfrac{h}{r}\right)}\delta\phi = \frac{\lambda}{h}U\delta\phi ,$$

by (30.6). Then multiplying both sides by h/λ and using §22.11 gives (30.5) as required.

Difficult? I haven't spelt out every detail for you. But you can cope – otherwise you'd have given up long ago.

30.20 Proving the statements of §30.17 needs rather more mathematical equipment than I've been assuming you have, and I propose to be brief. If you could cope with §30.11, you'll probably be all right here. Otherwise just take this article on trust. To cover all cases, assume that

$$\frac{1}{\alpha} = \left(1 - \frac{2\lambda}{r}\right)^{1-n} \approx \left(1 - \frac{2\lambda}{r}\right)\left(1 + \frac{2n\lambda}{r}\right).$$

Equation (30.1) for $1/H^2$ still applies, and so (29.9) gives

$$\delta r^2 \sim r^4\left(1 + \frac{2n\lambda}{r}\right)\left\{\frac{1}{h^2}\left(1 - \frac{2\lambda}{h}\right) - \frac{1}{r^2}\left(1 - \frac{2\lambda}{r}\right)\right\}\delta\theta^2$$

$$\approx r^4\left[\frac{1}{h^2} - \frac{1}{r^2} - 2\lambda\left\{\frac{1}{h^3} - \frac{1}{r^3} - \frac{n}{r}\left(\frac{1}{h^2} - \frac{1}{r^2}\right)\right\}\right]\delta\theta^2 \, ,$$

when we neglect terms in λ^2.

The reasoning of §30.19 (but with a different X) then leads to

$$\delta\left(\frac{h}{\lambda}\epsilon\right) \sim \frac{1 + \dfrac{h}{r} + \dfrac{h^2}{r^2} - n\dfrac{h}{r}\left(1 + \dfrac{h}{r}\right)}{1 + \dfrac{h}{r}}\delta\phi \, .$$

Unfortunately the method of §30.9 gives very inaccurate results, when $n \neq 0$, unless one uses smaller steps. But we can continue on the lines of §30.11, reaching the conclusion that

$$(h/\lambda)\epsilon = \tan \tfrac{1}{2}\phi + (1 - n)\sin \phi \, .$$

(Studying the relation between ϵ and ϕ when $n = 2$ reveals the strange behaviour sketched in §30.17, paragraph 2.) At $\phi = 90°$ we get $(h/\lambda)\epsilon^* = 2 - n$. Putting $n = 0$, 1 and 2 gives, respectively, the Einstein prediction and the cases considered in §30.17.

30.21 Let's return to the question of the speed of light under (almost) inverse square law gravitation (§§27.37–9). Substituting from (29.10) in (27.24), we find that

$$radial \text{ speed of light} = 1 - \frac{2\lambda}{r} \, . \tag{30.15}$$

Or you could get that directly by putting $\delta s = \delta\theta = 0$ in the (suitably rearranged) metric (29.12) and using §24.22. Similarly we find the transverse speed by putting $\delta s = \delta r = 0$ in this (rearranged) metric, which gives $r^2\delta\theta^2 \sim (1 - 2\lambda/r)\delta t^2$. By (26.37) $r\delta\theta$ is the transverse distance increment. So §24.22 tells us that

$$transverse \text{ speed of light} = \sqrt{\left(1 - \frac{2\lambda}{r}\right)} \, . \tag{30.16}$$

Thus the speed of light in these conditions depends on distance from the centre and on direction. **Any comment?**

30.22 We're guilty again of the naivety of interpreting (30.15) and (30.16) as if spacetime were flat (cf. §§28.15–16, 29.17). Constant speed motion in flat spacetime is analogous to a straight line in flat space (§8.23). But just as there are no straight lines in curved space,† so also there is no constant speed motion in curved spacetime. Since a signal line is a geodesic, the speed of light is as nearly constant as the curvature of spacetime permits.

Even if we return, as at the end of §29.17, to statements about how things will appear if we work on the assumption that spacetime is flat, it's still not obvious that (30.15) and (30.16) would be correct. Our co-ordinates t and r are related to time and distance, but not necessarily identical with them (cf. §27.39). It all depends on the practical question of how you decide to measure time and distance. If we agreed, for example, to put unqualified faith in radar once more, the speed of light would be constant by definition; and (30.15) and (30.16) would merely provide information about how our co-ordinates are related to distance so measured.

Astronomical distances are measured in several ways – by parallax (triangulation), from the apparent brightness or size of something whose actual brightness or size is known from other evidence, and so on. There is no guarantee that these methods all give the same answer, especially in curved spacetime (though practical astronomy can't detect any differences). And similarly there is no guarantee that our co-ordinate r (obtained in yet another way) will agree with any of them.

In any case, it doesn't really matter, as we noted way back in §22.5, since experimental tests necessarily refer to on-the-spot measurements only. For example, in the deflection-of-light test the prediction and observations concern *only* angles measured here on Earth. Co-ordinates t and r occur in the calculation of the prediction, but the logic of that calculation remains exactly the same, whether these symbols refer to 'real' time and distance measurements or not. Similarly, any statements the theory may make about the speed of light 'over there' are of purely academic interest, since we can only measure it on the spot. (Notice, by the way, that the speed of light will always be 1 according to an on-the-spot free-falling observer.)

---

† In certain curved spaces there may be some exceptional straight lines, but they do not affect the argument.

30.23 Independently of exactly how we interpret the equations of §30.21, they make clear that a radar signal passing close to the Sun on its way to and from another planet will take longer than it would have done if the Sun had not been there. It doesn't make any difference whether we say that the radar pulse was slowed down or that it had farther to travel. All that matters is *travel time*, which we can calculate along the following lines.

From (29.7) (which relates $\delta\theta$ and δt for a light signal) we get

$$r^4\delta\theta^2 \sim H^2\left(1 - \frac{2\lambda}{r}\right)^2 \delta t^2 = \frac{h^2\left(1 - \frac{2\lambda}{r}\right)^2}{1 - \frac{2\lambda}{h}}\delta t^2 \,,$$

by (30.1). And substituting for $r^4\delta\theta^2$ in (30.2) gives a near-equation connecting δr and δt. This would allow us to calculate (as accurately as we wish) the time δt that the radar pulse will take to travel along a portion of its path in which r increases by $\delta r - QQ'$ in Figure 30.2. And by the usual step by step process we could work out the total travel time, and hence the *delay* that the Sun's gravity imposes.

30.24 It would not be worth our while to do so, however, because the practicalities of the experiment demand the estimation (which we could not undertake) of corrections to allow for the effect of the Earth's motion, the electrical influences of the solar corona, the uncertainties arising from hills and valleys on the planet, and so on.

A yet more serious problem is that r has to be known to an accuracy far exceeding anything the astronomers can provide (even assuming that r does correspond to their distance). To circumvent these difficulties, a whole series of measurements is made as the planet approaches, passes behind, and recedes from the Sun – over a period that may be several hundred days. The now very complicated theory predicts how the delay due to gravitation should vary with time, rising from zero to a maximum of about 200 microseconds (when the target is Venus) and falling to zero again. And the real test concerns how well the day-to-day observations fit this prediction.

30.25 Tests have been made with passive radar reflections from Mercury and Venus, and with signals actively retransmitted by apparatus aboard *Mariner* spacecraft in orbit and *Viking* landers resting on Mars. According to Professor Irwin Shapiro, leader in this field, the *Viking* observations verify Einstein's predictions to an accuracy of about 0·1 per cent.

Altogether we can say that the General Theory of Relativity has done rather well in predicting how light behaves under the influence of gravitation.

30.26 *Postscript.* If you consider a ray going round the other side of the Sun in Figure 30.12, you will see that it would be possible, in certain circumstances, to see the same distant object twice – or even several times. Actually the Sun does not deflect light enough to bring this about, but in the right conditions a transparent galaxy might do so.

Observations made during 1979 and 1980 appear (though one must be cautious) to have yielded two instances of this actually happening. In one case, two quasars were found to be so similar and so close together that in all probability they are really two views of the same quasar, produced by the deflection of light by an intervening galaxy. The other seems to exhibit a galaxy deflecting the light from a quasar in such a way as to produce *three* close and similar images. (*Nature,* **279** (1979) 381; and *Nature,* **285** (1980) 641.)

31

The scandal about Mercury

31.1 In the region around the Sun there's only one other test to which we can subject our theory – we can ask how well it does in predicting the motion of a planet. Newtonian theory says that the orbit is an ellipse, and observation over the centuries has confirmed that this must be very near the truth. So we shall have to study some properties of ellipses.

If you're already well up in the subject, you may find it good enough to (1) note the definitions of the quantities, as summarised in Figure 31.2; (2) check that (31.7) is an old friend, lightly disguised by writing x/r instead of cos ϕ; (3) check that you know how to derive (31.12) from (31.7), and that you understand its significance; (4) check (as in §31.12 or by geometry) that the extreme values of r are given by (31.17); and (5) make sure you are familiar with (31.22) as a way of putting $\delta(\sin \phi) \sim \cos \phi \delta\phi$.

31.2M You probably know a practical recipe for drawing an ellipse. The ends of a piece of string are fixed to two pins stuck in a sheet of paper; if a pencil point (always in contact with the paper) is moved so that it keeps the string constantly taut, then it will draw an ellipse. To formalise that a little, we confine ourselves to a plane in which we have chosen two points S and S' (replacing the pins). An ellipse (Figure 31.2) consists of all points P which have the property that

$$\text{length SP} + \text{length S'P} = 2a , \tag{31.1}$$

where $2a$ is a constant length (replacing the string length of the practical method).

Each of the points S and S' is called a *focus* of the ellipse (plural *foci*). If O is halfway between S and S', and the line through these points meets the curve at A and A', **you can easily prove that**

$$\text{length OA} = \text{length OA'} = a .$$

31.3M The distance between S and S' is most conveniently specified as a fraction, e, of the 'string length' $2a$. That is length S'S = 2ae , (31.2)

where *e*–known for historical reasons as the *eccentricity*–is a handy measure of the extent to which the ellipse is elongated (check that by drawing a few curves with the same *a* and different values of *e*, using the pins-and-string procedure).

The curve has obvious symmetries–which Figure 31.2 emphasises–about the lines AOA' and BOB' (perpendicular to AA').

31.4M We shall discuss the ellipse's properties in terms of polar co-ordinates with S as pole–so that *r* stands for the length SP and φ for the size of the angle ASP (Figure 31.2). Also, with PM drawn perpendicular to AA', let *x* stand for length SM and *y* for length MP.

Then from (31.1),

length S'P = 2a − r , (31.3)

and from (31.2),

length S'M = 2ae + x . (31.4)

31.5M Applying Pythagoras' Theorem to the triangles SMP and S'MP, and using (31.3) and (31.4) for the latter, we get

$$x^2 + y^2 = r^2$$ (31.5)

and

$$(2ae + x)^2 + y^2 = (2a - r)^2.$$ (31.6)

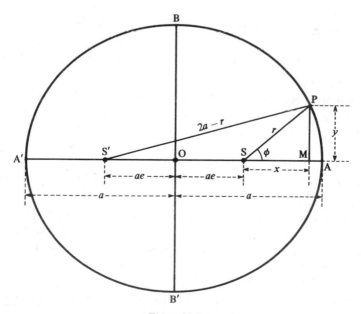

Figure 31.2

From these we can deduce by some dull manipulation (details in §31.8, but don't bother now) that

$$\frac{1}{r} - c = b\frac{x}{r},$$ (31.7)

where for brevity's sake we have written

$$b = \frac{e}{a(1 - e^2)} \quad \text{and} \quad c = \frac{1}{a(1 - e^2)}.$$ (31.8)

31.6M When we used the null geodesics to find the path of a light signal, this came out as a near-equation connecting the increments between two points of the path. And to study its shape we had to compare it with a similar near-equation for a straight line. Similarly, when we come shortly to employ ordinary geodesics to give the orbit of a planet, this will be expressed by a near-equation. And we shall want to compare it with the near-equation that connects the increments between two points of an ellipse – which we must now find.

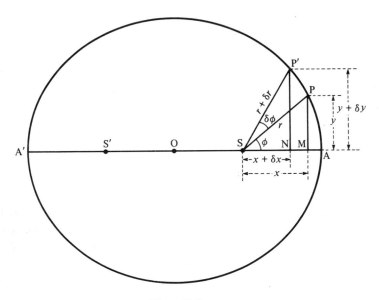

Figure 31.6

So we consider another point P′ on the ellipse, which has polar co-ordinates $r + \delta r$ and $\phi + \delta\phi$ (Figure 31.6). Let P′N be the perpendicular from P′ to AA′. Then length SN is $x + \delta x$ (so that δx is negative) and length NP′ is $y + \delta y$. Finding the near-equation connecting $\delta\phi$ and δr is easy enough in principle, but involves a longish calculation. I shall give only the main steps in §31.7, so that you can get yourself clear about the general line of advance, leaving the details to §§31.9–11, which you can consult later.

31.7M From (31.7) we can deduce (as in §31.9) that

$$\delta x = -\frac{c}{b}\delta r .$$ (31.9)

And from (31.5) we obtain (as in §31.10)

$$y\delta y \sim r\delta r - x\delta x .$$ (31.10)

Another relation arises from writing the metric in two different forms, given (with slight changes of notation) by (24.4) and (26.38), namely $\delta s^2 = \delta x^2 + \delta y^2$ and $\delta s^2 \sim \delta r^2 + r^2\delta\phi^2$. Since this is just the same δs expressed two different ways, we deduce that

$$r^2\delta\phi^2 \sim \delta x^2 + \delta y^2 - \delta r^2.$$ (31.11)

Now we combine (31.9)–(31.11)–with some help from earlier equations–in such a way as to eliminate x, y, δx and δy. By a tedious manipulation, which you can study *later* in §31.11, this leads to

$$\delta r^2 \sim r^4\left\{ b^2 - \left(\frac{1}{r} - c\right)^2 \right\}\delta\phi^2 .$$ (31.12)

And that is the near-equation we've been aiming at, describing the relation between the increments in the polar co-ordinates as we pass from one point of the ellipse to another. It plays the same role in relation to the ellipse that (26.43) plays for the straight line.

31.8M Articles 31.8–11 clear up details omitted earlier. First we have to prove (31.7). Using identities (26.26) and (26.27), (31.6) becomes $4a^2e^2 + 4aex + x^2 + y^2 = 4a^2 - 4ar + r^2$. From each side of this subtract the corresponding side of (31.5), and then divide both sides by $4a$, getting $ae^2 + ex = a - r$, whence (subtracting ae^2 from both sides and using §16.12) $a(1 - e^2) - r = ex$. And so, on dividing both sides by $ra(1 - e^2)$,

$$\frac{1}{r} - \frac{1}{a(1 - e^2)} = \frac{ex}{ra(1 - e^2)} .$$

And with b and c defined by (31.8), this gives (31.7).

31.9M From (31.7) we have

$$1 - cr = bx .$$ (31.13)

Then by §22.9, $\delta(1 - cr) = \delta(bx)$; and on applying (22.7) and (22.8) this yields $-c\delta r = b\delta x$, from which (31.9) follows immediately.

1.10M Read §27.42 if you skipped it previously. Applying (27.27) to each term in (31.5) and dividing both sides by 2 gives $x\delta x + y\delta y \sim r\delta r$, which leads to (31.10) when $x\delta x$ is subtracted from both sides.

31.11M Proving (31.12) is a big bore, though every step is simple. Equation (31.5) can be put in the alternative forms

$$r^2 - y^2 = x^2 \qquad (31.14)$$

and

$$y^2 = r^2 - x^2 . \qquad (31.15)$$

And (31.13) gives

$$x = \frac{1 - cr}{b} . \qquad (31.16)$$

Multiplying both sides of (31.11) by y^2 gives

$$r^2 y^2 \delta^2 \phi^2 \sim y^2 \delta x^2 + y^2 \delta y^2 - y^2 \delta r^2$$

(31.10)

$$\sim y^2 \delta x^2 + (r \delta r - x \delta x)^2 - y^2 \delta r^2$$

(26.27)

$$= y^2 \delta x^2 + r^2 \delta r^2 - 2rx \delta x \delta r + x^2 \delta x^2 - y^2 \delta r^2$$

(rearranging order)

$$= x^2 \delta x^2 + y^2 \delta x^2 - 2rx \delta x \delta r + r^2 \delta r^2 - y^2 \delta r^2$$

§16.12

$$= (x^2 + y^2) \delta x^2 - 2rx \delta x \delta r + (r^2 - y^2) \delta r^2$$

(31.5) and (31.14)

$$= r^2 \delta x^2 - 2rx \delta x \delta r + x^2 \delta r^2$$

(26.27)

$$= (r \delta x - x \delta r)^2$$

(31.9) and (31.16)

$$= \left(-\frac{cr}{b} \delta r - \frac{1 - cr}{b} \delta r \right)^2$$

§§16.12, 21.12 footnote, 15.24

$$= \left(-\frac{cr}{b} \delta r - \frac{1}{b} \delta r + \frac{cr}{b} \delta r \right)^2$$

$$= \left(-\frac{\delta r}{b} \right)^2 ,$$

since the first and last terms in the previous line cancel each other out. So summarising and using §24.20, we get $r^2 y^2 \delta \phi^2 \sim \delta r^2 / b^2$, whence

$$\delta r^2 \sim r^2 b^2 y^2 \delta \phi^2$$

(31.15)

$$= r^2 b^2 \{ r^2 - x^2 \} \delta \phi^2$$

§16.12

$$= r^2\{b^2r^2 - b^2x^2\}\delta\phi^2$$

(31.13)

$$= r^2\{b^2r^2 - (1 - cr)^2\}\delta\phi^2$$

§§16.12 and 26.37

$$= r^4\left\{b^2 - \frac{(1 - cr)^2}{r^2}\right\}\delta\phi^2$$

§17.25

$$= r^4\left\{b^2 - \left(\frac{1 - cr}{r}\right)^2\right\}\delta\phi^2$$

§21.12, footnote

$$= r^4\left\{b^2 - \left(\frac{1}{r} - c\right)^2\right\}\delta\phi^2 ,$$

which gives – at last! – the required proof of (31.12). It's very easy to miss shortcuts in work like that. If you find one, please let me know.

1.12M For a job like that of §30.2, we shall need to know the extreme values of r. Applying the Theorem of §29.20 to (31.12), we see that these are among the values of r which make

$$r^4\left\{b^2 - \left(\frac{1}{r} - c\right)^2\right\} = 0 .$$

Discarding the obviously irrelevant case of $r = 0$ (the ellipse doesn't pass through S), we want values of r which make

$$b^2 - \left(\frac{1}{r} - c\right)^2 = 0 ,$$

or

$$\left(\frac{1}{r} - c\right)^2 = b^2 .$$

There are two possibilities (§24.21): $1/r - c = b$ and $1/r - c = -b$. And so the extreme values of r are given by

$$\frac{1}{r} = c + b , \quad \text{and} \quad \frac{1}{r} = c - b . \tag{31.17}$$

Clearly the least and greatest values of r are the lengths SA and SA′ (Figure 31.2). So you can check that answer geometrically, using (31.8).

31.13m Articles 31.13–15 establish a very important near-equation concerning the quantities x, y, r and ϕ of §31.4. From (31.5) we have

$$\left(\frac{x}{r}\right)^2 + \left(\frac{y}{r}\right)^2 = 1 .$$

And if we write

$$\frac{x}{r} = X \quad \text{and} \quad \frac{y}{r} = Y , \tag{31.18}$$

this gives

$$X^2 + Y^2 = 1 . \tag{31.19}$$

Now we interpret X and Y as the co-ordinates of a point R in an ordinary rectangular co-ordinate system (§8.42), as in Figure 31.13. Using Pythagoras' Theorem, (31.19) tells us that length OR = 1. So no matter what the position of P may be on the ellipse, R lies on a circle whose centre is O and whose radius is 1.

From (31.18), the ratios y/x and Y/X are equal, and so angle LOR = angle MSP = ϕ.

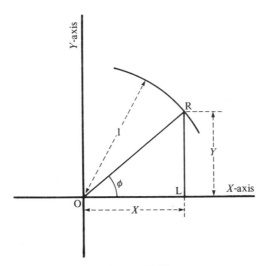

Figure 31.13

31.14m Figure 31.14 shows triangle OLR in the complete circle, and R' is another point of the circle with co-ordinates $X + \delta X$ and $Y + \delta Y$, so that δX (negative) and δY are the increments between R and R'. And the corresponding increment $\delta\phi$ is the angle ROR'. Let δs and s_i stand, respectively, for the length of the *arc* and *chord* RR' (§26.34). Then from (25.1) and (26.34) we have $\delta s \sim s_i$ and $\delta s = \delta \phi$, which combine to give $s_i \sim \delta \phi$.

Applying Pythagoras' Theorem to triangle KRR′, and using that last near-equation, we get $\delta \phi^2 \sim \delta X^2 + \delta Y^2$, whence

$$X^2 \delta \phi^2 \sim X^2 \delta X^2 + X^2 \delta Y^2. \tag{31.20}$$

1.15m From (31.19) we have $X^2 = 1 - Y^2$. And so – using successively §22.9, (22.7) and (22.8) with $c = -1$ – this leads to $\delta(X^2) = \delta(1 - Y^2) = \delta(-Y^2) = -\delta(Y^2)$. Applying (27.27) to this last gives $X\delta X \sim -Y\delta Y$, whence $X^2 \delta X^2 \sim Y^2 \delta Y^2$. Substituting this in (31.20) and then using §16.12 and (31.19) yields $X^2 \delta \phi^2 \sim Y^2 \delta Y^2 + X^2 \delta Y^2 = (X^2 + Y^2)\delta Y^2 = \delta Y^2$. That is $X^2 \delta \phi^2 \sim \delta Y^2$, from which it follows that

$$\delta Y \sim X\delta \phi \ . \tag{31.21}$$

But in view of §24.21, how do I know that this is the correct deduction, rather than $\delta Y \sim -X\delta \phi$? Well we can always decide to make $\delta \phi$ positive. In the case shown in Figure 31.14, δY and X are also positive – which agrees with (31.21) and not with the alternative.

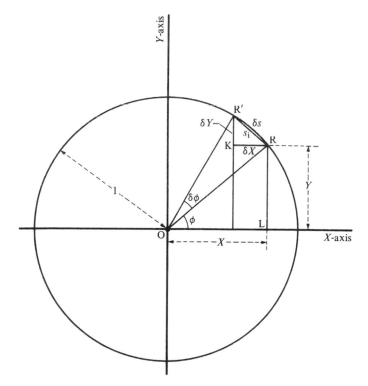

Figure 31.14

Check for yourself that, wherever you put R on the circle, δY and X will be both positive or both negative (provided you make $\delta\phi$ small enough) – so that (31.21) is always the correct choice.

Finally, using (31.18) to substitute for X and Y in (31.21) brings us to the near-equation we've been looking for, namely

$$\delta\left(\frac{y}{r}\right) \sim \frac{x}{r}\delta\phi . \tag{31.22}$$

31.16 And now we are ready to study the motion of a planet orbiting round the Sun. We use the same general strategy as we did when investigating the motion of light in Chapters 29 and 30, and you may find it helpful to look back occasionally.

To make the problem simple enough – even for expert mathematicians, not just you and me – we have to assume that the Sun has only the one planet, whose mass is so small that its effect on spacetime can be ignored. Then the metric is (29.12).

The planet's motion is inertial (§27.7), and so its world line is an ordinary geodesic (§25.28) – which we find by the recipe of §26.33. As usual (cf. §§26.42, 29.11–12) we identify r, θ and t with x, y and z of (26.30), so that

$$\alpha = -\frac{1}{1 - \dfrac{2\lambda}{r}} , \qquad \beta = -r^2 \quad \text{and} \quad \gamma = 1 - \frac{2\lambda}{r} .$$

Then using the second form of (26.31) and the first of (26.32), we find that the increments between two events on the planet's world line have to be related by the near-equations

$$\delta s \sim -hr^2\delta\theta \tag{31.23}$$

and

$$\left(1 - \frac{2\lambda}{r}\right)\delta t \sim k\delta s \tag{31.24}$$

(where h and k are constants) together with a third near-equation provided by the metric (29.12) itself.

31.17 We are interested only in the shape of the orbit – the relation between r and θ. So we combine these three near-equations in such a way as to get rid of δs and δt. Once more we'll relegate boring detail to §31.38. The resulting near-equation for the orbit can be put in the form

$$\delta r^2 \sim r^4\left\{B^2 - \left(\frac{1}{r} - C\right)^2 + \frac{2\lambda}{r^3}\right\}\delta\theta^2 , \tag{31.25}$$

where
$$B^2 = h^2(k^2 - 1) + h^4\lambda^2 \tag{31.26}$$
and
$$C = h^2\lambda . \tag{31.27}$$

31.18 Near-equation (31.25) describes the relation between the increments in the co-ordinates as we pass from one point of the planet's orbit to another – just as (31.12) did for the ellipse (using ϕ instead of θ for the vectorial angle has no significance here). Comparing the two, **what conclusion would you draw?**

They differ only by the term $2\lambda/r^3$ – which (with an r of more than 50 million kilometres for the innermost planet) is extremely small. **So you conclude. . .?**

. . . that the orbit is very nearly an ellipse, did you say? If so, **criticise yourself.**

Read again §§29.17 and 30.22. It's the same blunder once more. The correct conclusion is that if we treat the curved spacetime around the Sun *as if it were* flat, then the orbit of a planet will *appear* to be nearly, but not quite an ellipse.

31.19 Here follows a digression, which I think may be helpful. In spite of the discussion in §§28.15–17, you probably find it hard to imagine that a planet's complicated behaviour is described by a geodesic – that simplest of all world lines. I wonder if the following thoughts will help.

We have *not* said that the planet's path is a geodesic *in space* – only that it is the spatial aspect of a geodesic *in spacetime*. The latter is what you get by finding a geodesic world line and *then* eliminating δt to get the spatial path. The other is what you get by changing over to the curved space metric (29.13) and then finding a geodesic curve. **You can verify** that this last – the nearest possible thing to a straight line in the curved space round the Sun – is given by

$$\delta r^2 \sim r^4 \left(1 - \frac{2\lambda}{r}\right)\left(\frac{1}{h^2} - \frac{1}{r^2}\right)\delta\theta^2 .$$

This is only slightly different from (26.43), which refers to a straight line in flat space. But it is *very* different from (31.25). So unless you have an imagination that can cope with geodesics in curved spacetime (mine can't) and then extract their spatial aspects, you simply must not expect that the geodesic simplicity of a planet's motion will be in any way apparent to you.

Notice that even the path of a light ray, given by (30.2), is not a geodesic *in space* – just the spatial component of a spacetime null geodesic. It can be proved (and **you can probably do it**) that the spatial aspects of ordinary and null geodesics in spacetime – i.e., the paths of free-falling bodies and light rays – are always different from the geodesics of space as such if the δt^2 term in the metric is multiplied by an expression that involves the co-ordinates. They will be the same *only if* the metric has the form $\delta s^2 \sim \delta t^2 - \ldots$ (or $\delta s^2 \sim a\delta t^2 - \ldots$, where a is a constant).

31.20 Coming back to our main line of advance, let's apply to the last sentence of §31.18 the same sort of thinking as in the last paragraph of §29.17. We emerge with the prediction that if the planet's orbit is worked out from the observations using ordinary astronomical methods, we shall find it to be very nearly an ellipse (with the Sun at one focus). We talk in these flat-space terms from now on.

Now that is satisfactory, so far as it goes. Newton's theory predicts an *exact* ellipse, and observation has confirmed it to a very high degree of accuracy. If Einstein had predicted anything very different, he would simply have been wrong. But when the two predictions come out to be slightly different, the question naturally arises: which actually agrees more closely with the facts? To answer that we must first discover in what way the Einstein orbit differs from an ellipse, and by how much.

31.21 We shall do so by comparing the predicted orbit with an ellipse, just as previously we compared the light ray with a straight line (§30.3). And for this purpose we shall need to know the extreme values of r. Applying §29.20 to (31.25), we see that these are among the values of r which make

$$r^4\left\{ B^2 - \left(\frac{1}{r} - C\right)^2 + \frac{2\lambda}{r^3} \right\} = 0 \,,$$

or (since r can't be zero) which make

$$B^2 - \left(\frac{1}{r} - C\right)^2 + \frac{2\lambda}{r^3} = 0 \,. \tag{31.28}$$

31.22 That equation does not involve t. So the extreme values of r do not change with time. Now a little playing around with the pins-and-string recipe (§31.2) will convince you that if the maximum and minimum values of r (the lengths SA′ and SA, Figure 31.2) are fixed, then the size and shape of the ellipse are also fixed. So the orbit is very nearly the same as an ellipse (with one focus at the Sun's centre) which does not change in size and shape as time goes on. The only remaining possibility is that the orbit is what you would get by slowly *rotating* the ellipse around point S, while the planet continues to go round and round this moving ellipse.

Kindly imagine Figure 31.2 pivoted at S. Think of the whole figure as slowly rotating anticlockwise round that point, while the planet travels round and round the ellipse in the same direction but much faster. You can then visualise the combination of these motions giving the planet an actual path like that shown in Figure 31.22 (where the ellipse's elongation and the rate of rotation are grossly exaggerated). First time round it follows track 1, second time track 2, and so on.

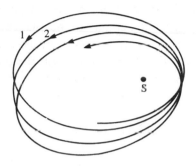

Figure 31.22

31.23 Astronomers meet motions of this type in other contexts. They normally avoid the complicated image of Figure 31.22, and think of the orbit in terms of a slowly rotating ellipse.

To specify the rotation of the ellipse as a whole, it is only necessary to say how fast the point A (Figure 31.2) goes round the Sun. Since A is called the *perihelion* (point nearest the Sun), this rotation is known as the *advance of the perihelion*. And we have to work out how fast it advances.

31.24 The method is very like that of §30.3–5 – I suggest another perusal. There we fitted the light ray and the straight line together at their nearest approach to the Sun. This time both nearest approach and farthest recession are involved. And so we have to arrange that *both* extreme values of r for the orbit shall be the same as the extreme values of r for a fixed ellipse that will be used as standard of comparison.

Now the extreme values for the ellipse are given by (31.17), while those for the orbit are the values of r which make (31.28) true. So to arrange coincidence at both extremes, we have only to say that substituting $c + b$ or $c - b$ for $1/r$ makes (31.28) true. These two substitutions lead, respectively, to

$$B^2 - (c + b - C)^2 + 2\lambda(c + b)^3 = 0$$
and
$$B^2 - (c - b - C)^2 + 2\lambda(c - b)^3 = 0$$

(31.29)

From these two we deduce (see §31.40 in due course) that

$$b^2 - B^2 = -(c - C)^2 + 2\lambda(c^3 + 3b^2c)$$ (31.30)

and

$$c - C = \lambda(3c^2 + b^2) .$$ (31.31)

31.25 We are now ready to do an operation akin to that of §30.3. Figure 31.25 shows the fixed ellipse and a portion AN of the orbit (the difference between the two being greatly exaggerated). Point Q on the orbit has polar co-ordinates (i) the distance r from S to Q, and (ii) the angle θ between SA and SQ. We now take the point P on the ellipse, which is at the *same* distance r from S. So its co-ordinates are (i) length SP, which is r again, and (ii) the angle ϕ between SA and SP.

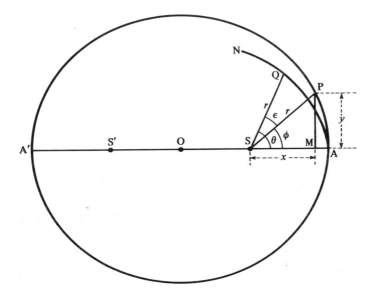

Figure 31.25

31.26 Following the tactics of §30.4, we put $\theta - \phi = \epsilon$, so that ϵ measures how much SQ has swung ahead of SP. Hence

$$\delta\theta - \delta\phi = \delta\epsilon .$$ (31.32)

Still using the method that worked for light (§§30.4–5), we combine (31.12), (31.25) and (31.32). And from a very long, very dull piece of manipulation (§31.41) there emerges the very simple result that

$$\delta\epsilon \sim 3\lambda c\delta\phi + \lambda b \frac{x}{r}\delta\phi \ , \tag{31.33}$$

where x is as defined in §31.4. Obviously this puts us in a position to calculate ϵ by a step by step process like that of §30.9. But this time there's an easier way.

31.27 Applying (31.22) to the last term of (31.33) gives

$$\delta\epsilon \sim 3\lambda c\delta\phi + \lambda b\delta\left(\frac{y}{r}\right)$$

(22.8)

$$= \delta(3\lambda c\phi) + \delta\left(\lambda b\frac{y}{r}\right) \ .$$

And so by (22.5),

$$\delta\epsilon \sim \delta(3\lambda c\phi + \lambda b\frac{y}{r}) \ .$$

Any ideas?

31.28 That near-equation has an increment *only* on each side. So the conditions of the Second Fundamental Theorem apply (§25.13), and using form (iii) we deduce that

$$\epsilon = 3\lambda c\phi + \lambda b\frac{y}{r} + K \ ,$$

where K is a constant. We have chosen to start P and Q off together from starting post A. So when $\phi = 0$, we also have $\epsilon = 0$ and $y = 0$; and to make this possible K must also be 0. So we now arrive at the equation

$$\epsilon = 3\lambda c\phi + \lambda b\frac{y}{r} \ . \tag{31.34}$$

This enables us to calculate ϵ – the angle by which SQ has got ahead of SP – in terms of ϕ and y, which are measurements associated with P (Figures 31.2, 31.25).

31.29 As SP swings round and round, the angle ϕ steadily increases,† and so also does the first term on the right-hand side of (31.34). But the second term has a more complicated history. It starts from zero at $\phi = 0$ (since P and M coincide at A, giving $y = 0$). As ϕ increases, it rises to a maximum and then falls again through zero (at $\phi = 180°$, where $y = 0$ again) to a negative minimum, and returns to zero when $\phi = 360°$ – and then repeats this oscillatory rise and fall every time SP does a complete revolution.

> † Don't think of ϕ as starting again from zero each time P returns to A. When it has been once round and done 13° more, it has reached 360 + 13 = 373° – and in that way it can grow as large as you like.

So the story of ϵ is one of steady increase (the $3\lambda c\phi$ term) modified by an up and down oscillation (the $\lambda by/r$ term).

31.30 We are only interested in completed orbits – when P has returned again to A (and Q has returned to perihelion, though no longer at A). In that case $y = 0$, so that (31.34) reduces to $\epsilon = 3\lambda c\phi$, or on using (31.8),

$$\epsilon = \frac{3\lambda}{a(1 - e^2)}\phi \, . \tag{31.35}$$

(a and e being the quantities that most conveniently specify the ellipse, §§31.2–3).

Let ϵ^* stand for the amount by which SQ gets ahead of SP during one complete orbit – in other words, the amount by which ϵ increases in one orbit, which is also the amount by which the perihelion advances in one orbit (§31.23). For a complete orbit, we have $\phi = 360° = 360 \times 3600$ seconds. And so from (31.35) we get

$$\epsilon^* = \frac{3\,888\,000\lambda}{a(1 - e^2)} \text{ seconds.} \tag{31.36}$$

So if we are given the dimensions of the ellipse to which the orbit closely approximates, this formula tells us how much the perihelion will advance every time the planet completes an orbit.

31.31 The smaller a is, the larger will be the effect predicted by (31.36). So the obvious thing is to work out how big it will be for the innermost planet, Mercury. In that case $a = 57\,910\,000$ kilometres and $e = 0.2056$, so that $1 - e^2 = 0.957\,7$, while $\lambda = 1.477$ kilometres (§27.15). So (31.36) gives

$$\epsilon^* = \frac{3\,888\,000 \times 1.477}{57\,910\,000 \times 0.957\,7} \text{ seconds.}$$

Now Mercury takes 87.97 Earth days to complete an orbit; and so does $(100 \times 365.26)/87.97$ orbits in a century. Combining this with the above value of ϵ^*, we predict that the advance of Mercury's perihelion per century should be

$$\frac{3\,888\,000 \times 1.477 \times 100 \times 365.26}{57\,910\,000 \times 0.957\,7 \times 87.97} = 43.0 \text{ seconds.}$$

So the orbit of Mercury (says General Relativity) is not a fixed ellipse, but one that rotates in a century through an angle of 43.0 seconds. This is a tiny, tiny amount. But measuring it is well within the powers of astronomers.

31.32 Now this prediction hits the Newtonian theory where it hurts. It strikes at the one notorious case in which that otherwise excellent theory blatantly failed to account for what the practical astronomers observed.

In the Newtonian scheme, if the Sun had only one planet, its orbit should be a fixed ellipse. Of course the Sun actually has several planets, which affect each other gravitationally, distorting the orbits. But these effects are small, and it turns out that you can always describe the actual situation by saying that the orbit of a planet is an ellipse which very slowly changes. Several sorts of change are involved – rotation of the type we've been discussing, changes in its size and shape, and tilting of the plane it lies in.

So the astronomical world spent centuries observing the motions of all the planets and describing them in terms of slowly changing ellipses. Then, since the changes are due to the influences of other planets, it was possible to eliminate these effects theoretically. In other words, starting from the observed motions, you could calculate what the orbit of a particular planet *would be* if the others didn't exist. According to Newton the answer ought to be an ellipse, an unchanging ellipse.

31.33 And when the tedious calculations have been completed, *do* we in fact get that answer? Yes we do – to a very high degree of accuracy – *except* in the one notorious case of the innermost planet, Mercury.

When the disturbing effects of the other planets have been eliminated, Mercury's orbit is still not a fixed ellipse. Its plane and its size and shape are fixed, as they should be. But it is found to be a rotating ellipse, one whose perihelion is advancing. Mercury is the black sheep of the Newtonian family – the one case in which the theory, despite persistent efforts, could not be made to account for the observations. And this scandal was known long before Einstein was born.

So Einstein's theory predicts the right *sort* of orbit for Mercury. Does observation confirm the *amount* by which it predicts the perihelion will advance?

Accurate observations have been carried out since 1765. When the full power of the theory is used to eliminate the influence of the other planets, it is found that Mercury's perihelion would still be advancing by $43 \cdot 11 \pm 0 \cdot 45$ seconds per century. The $\pm 0 \cdot 45$ indicates a very precise measurement (§23.17), a very small degree of uncertainty. **Compare that with the Einstein prediction of $43 \cdot 0$ seconds.**

One could not ask for better agreement. Einstein's theory is triumphantly vindicated. (I'll say a few words in my last chapter about a recent challenge – which seems to have failed.)

31.34 For other planets, the advance of the perihelion is much smaller. The figures are

| Planet | Advance of perihelion (seconds per century) | |
	Predicted	Observed
Venus	8·6	8·4 ± 4·8
Earth	3·8	5·0 ± 1·2
Icarus	10·3	9·8 ± 0·8

These give further confirmation – though with smaller effects and less refined measurements, not a great deal is added.

31.35 Till recently the evidence in favour of Einstein's gravitational theory depended on minute effects like this one or those of Chapter 30 – effects that are near the limits of what today's science can measure. But almost as a stop-press item in the making of this book comes news of a case like the one we've been discussing, but happening 35 000 times faster.

The new evidence is provided by a *pulsar*. A pulsar is an object, somewhere out there in space among the not very distant stars, from which we receive a rapid succession of short sharp pulses of radio waves. The period from one pulse to the next is commonly something like a second. But in one famous case it is as little as 0·033 second, and for the one that concerns us it is 0·059 second. These pulses are emitted with a regularity that rivals all but the very best man-made clocks (though every pulsar clock is in fact slowing down very very very gradually in a way that need not concern us here.)

According to the generally accepted view a pulsar is a 'neutron star' – a late stage in the life of what started off as a quite ordinary star, though considerably more massive than the Sun. Amid many adventures it may have had to get rid of surplus mass in a violent explosion, and in any case has exhausted its supply of nuclear energy and suffered a dramatic collapse, as a result of which it is now rather more massive than the Sun but compressed into a sphere something like 10 or 20 kilometres in diameter.

In the course of this contraction, the slow rotation that every star possesses has been transformed into a rapid spin (just as the whirling skater spins faster when she brings her arms close to her sides). It is emitting radio waves, which its magnetic field – enormously intensified as a result of the contraction – concentrates into narrow beams. And the rotation causes these beams to swing round like those of a lighthouse. If we happen to be placed where a rotating beam can hit us, we observe the neutron star as a pulsar. And the interval between successive pulses – typically about a second, as I've said – is the time it takes the star to go round once.

Even if that account is not entirely accurate (and it *does* depend on a lot of theorising derived from rather few facts), the rapidity with which the pulses grow and fade makes it quite certain that a pulsar must be small.† And in the case we are going to discuss there is observational evidence for the 'rather more massive than the Sun' statement also. I'm emphasising these points to show that even if further research changes the theory of how pulsars work, the following verification of General Relativity will probably still hold good.

31.36 The pulsar known as PSR1913 + 16 was discovered by Professor J.H. Taylor and colleagues in July 1974. And by the last week of August it was known to have an odd mode of behaviour. The interval between successive pulses was *not* regular. It kept on increasing to a maximum and decreasing to a minimum over and over again, repeating that cycle every $7\frac{3}{4}$ hours (27 906·981 72 ± 0·000 025 seconds to be exact). **Can you interpret that observation?**

The only likely explanation is a Doppler effect (§§15.4–10). In this $7\frac{3}{4}$ hour cycle the pulsar moves towards us – at about a thousandth of the speed of light, actually – stops, moves away from us, approaches again . . . We know of only one thing that could make it behave like that: it must be in orbit. It must have a companion star, even though that has not been observed. And the two are in orbit about their mutual centre of gravity. Such a pair is called a *binary*.

The optical equivalent of this phenomenon has been known for a long time – where the frequency of visible light from a star periodically increases and decreases, and one deduces that it is in orbit with a companion. Sometimes the companion can be observed, often it's invisible.

31.37 An intensive study of every detail of the signals coming from this binary pulsar can yield an immense amount of information. If I tried to state just what is known now, I should merely look foolishly old-fashioned, vague and inaccurate when you come to read my words. So a sketchy description of how things stood in early 1979 must do.‡ The masses of both stars are about 1·4 times that of the Sun – the best estimates at the moment being 1·39 for the pulsar and 1·44 for the companion. We'll again describe the orbit in terms of a rotating ellipse (§31.22). This time the ellipse is very elongated ($e \approx 0.617$) and very small ($a \approx 870\ 000$ kilometres – comparable to the radius of the Sun!).

---

† Think out for yourself what would be observed if a body's brightness were to fluctuate in a shorter period than it takes light to cross it from one side to the other.

‡ For a fuller summary, see J.H. Taylor and others, in *Nature,* **277** (1979) 437.

So (31.35)† predicts a very rapid rotation of the ellipse – a very rapid advance of the *periastron* (equivalent, for a star, of a perihelion). The observed advance is $4 \cdot 226 \pm 0 \cdot 001$ degrees per year. No, that's not a misprint: I do mean *degrees per year*. And (though various doubts still need investigating), Taylor and his colleagues seem to be justified in claiming that 'the observed rate of periastron advance is entirely caused by the general relativistic effect'. So (subject to the checking and cross-checking of the next few years) we appear to have another verification of Einstein's theory. And this time it is in terms of a really large-scale effect, not just a minor deviation from what Newton would have said.

31.38 And now for the dull work of checking up on mathematical details. First, to establish (31.25) we start by combining (31.23) and (31.24) to get

$$\left(1 - \frac{2\lambda}{r}\right)\delta t \sim -hkr^2\delta\theta . \tag{31.37}$$

Multiply both sides of the metric (29.12) by $1 - 2\lambda/r$, then add δr^2 to both sides and subtract $(1 - 2\lambda/r)\delta s^2$ from both – and we arrive at

$$\delta r^2 \sim \left(1 - \frac{2\lambda}{r}\right)^2\delta t^2 - \left(1 - \frac{2\lambda}{r}\right)\delta s^2 - \left(1 - \frac{2\lambda}{r}\right)r^2\delta\theta^2 .$$

Using (31.23) and (31.37) to substitute for δt and δs, we get

$$\delta r^2 \sim h^2k^2r^4\delta\theta^2 - \left(1 - \frac{2\lambda}{r}\right)h^2r^4\delta\theta^2 - \left(1 - \frac{2\lambda}{r}\right)r^2\delta\theta^2 .$$

$$= r^4\left\{h^2k^2 - h^2 + \frac{2\lambda h^2}{r} - \frac{1}{r^2} + \frac{2\lambda}{r^3}\right\}\delta\theta^2 ,$$

using §§16.12, 15.24 and 26.37 to make that last step. Next we modify the form of the expression inside the curly brackets without changing its value, by (a) both adding and subtracting $h^4\lambda^2$, and (b) slightly altering the order, and so obtain

$$\delta r^2 \sim r^4\left\{h^2k^2 - h^2 + h^4\lambda^2 - \frac{1}{r^2} + \frac{2h^2\lambda}{r} - h^4\lambda^2 + \frac{2\lambda}{r^3}\right\}\delta\theta^2$$

$$= r^4\left\{h^2\left(k^2 - 1\right) + h^4\lambda^2 - \left(\frac{1}{r^2} - \frac{2h^2\lambda}{r} + h^4\lambda^2\right) + \frac{2\lambda}{r^3}\right\}\delta\theta^2$$

$$= r^4\left\{h^2\left(k^2 - 1\right) + h^4\lambda^2 - \left(\frac{1}{r} - h^2\lambda\right)^2 + \frac{2\lambda}{r^3}\right\}\delta\theta^2 ,$$

† Modified to cover the case where two bodies of roughly equal mass are involved. But if you apply (31.35) crudely, you'll get an answer of roughly the right size.

where the step from the first to the second line uses §§16.12 and 15.23–4, and the last step employs (26.27) with $a = 1/r$ and $b = h^2\lambda$. And introducing B and C according to (31.26) and (31.27), this reduces to (31.25).

31.39M By repeated application of §16.12 (cf. §26.28) you can prove that

$$(c + b)^3 = c^3 + 3c^2b + 3cb^2 + b^3 \tag{31.38}$$

and

$$(c - b)^3 = c^3 - 3c^2b + 3cb^2 - b^3. \tag{31.39}$$

31.40 A change of order gives

$$(c + b - C)^2 = (c - C + b)^2$$

(26.26)

$$= (c - C)^2 + 2(c - C)b + b^2;$$

and similarly $(c - b - C)^2 = (c - C)^2 - 2(c - C)b + b^2.$

Applying the last four equations to (31.29), we get

$$B^2 - (c - C)^2 - 2(c - C)b - b^2$$
$$+ 2\lambda(c^3 + 3c^2b + 3cb^2 + b^3) = 0 \tag{31.40}$$

and

$$B^2 - (c - C)^2 + 2(c - C)b - b^2$$
$$+ 2\lambda(c^3 - 3c^2b + 3cb^2 - b^3) = 0 . \tag{31.41}$$

Here – and frequently from now on – we must keep in mind the rules for adding and subtracting negative quantities (§15.23–4); I can't remind you every time. Next we add corresponding sides of (31.40) and (31.41) (using a slight elaboration of the methods of §15.28) and divide by 2, which gives

$$B^2 - (c - C)^2 - b^2 + 2\lambda(c^3 + 3cb^2) = 0 .$$

And when $b^2 - B^2$ is added to both sides, this gives (31.30).

Again, subtracting each side of (31.40) from the corresponding side of (31.41) and dividing both sides by $4b$ gives

$$c - C - \lambda(3c^2 + b^2) = 0 ,$$

whence (31.31) follows immediately.

31.41 Proving (31.33) involves perhaps our most tedious bit of manipulation, even if every step is simple in principle. We rewrite (31.25) and (31.12) as

$$\delta r^2 \sim r^4 X \delta\theta^2$$

and

$$\delta r^2 \sim r^4 Y \delta\phi^2 ,$$

where

$$X = B^2 - \left(\frac{1}{r} - C\right)^2 + \frac{2\lambda}{r^3} \qquad (31.42)$$

and

$$Y = b^2 - \left(\frac{1}{r} - c\right)^2 \qquad (31.43)$$

(31.7)

$$= b^2 - b^2\frac{x^2}{r^2}$$

§16.12

$$= b^2\left(1 - \frac{x^2}{r^2}\right) . \qquad (31.44)$$

We can then deduce, just as we did in §30.19, that

$$2\delta\epsilon \sim \frac{Y - X}{Y}\delta\phi . \qquad (31.45)$$

From (31.42) and (31.43),

$$Y - X = b^2 - B^2 - \left(\frac{1}{r} - c\right)^2 + \left(\frac{1}{r} - C\right)^2 - \frac{2\lambda}{r^3}$$

(31.30)

$$= -(c - C)^2 + 2\lambda(c^3 + 3b^2c) - \left(\frac{1}{r} - c\right)^2 + \left(\frac{1}{r} - c\right)^2 - \frac{2\lambda}{r^3}$$

changing order and §16.12

$$= -(c - C)^2 + \left(\frac{1}{r} - C\right)^2 - \left(\frac{1}{r} - c\right)^2 + 2\lambda\left(c^3 + 3b^2c - \frac{1}{r^3}\right) . $$

(31.46)

But, $-(c - C)^2 = -(c - C)(c - C) = (c - C)(C - c)$. And (16.10) gives

$$\left(\frac{1}{r} - C\right)^2 - \left(\frac{1}{r} - c\right)^2 = \left(\frac{1}{r} - C + \frac{1}{r} - c\right)\left(\frac{1}{r} - C - \frac{1}{r} + c\right)$$

$$= \left(\frac{2}{r} - c - C\right)\left(c - C\right) . $$

Then applying these last two equations to (31.46) gives

$$Y - X = (c - C)(C - c) + \left(c - C\right)\left(\frac{2}{r} - c - C\right)$$

$$+ 2\lambda\left(c^3 + 3b^2c - \frac{1}{r^3}\right)$$

§16.12

$$= 2\left(c - C\right)\left(\frac{1}{r} - c\right) + 2\lambda\left(c^3 + 3b^2c - \frac{1}{r^3}\right)$$

(31.31)

$$= 2\lambda\left(3c^2 + b^2\right)\left(\frac{1}{r} - c\right) + 2\lambda\left(c^3 + 3b^2c - \frac{1}{r^3}\right).$$

At this point B and C have finally disappeared from our considerations. The b and c that remain refer to the ellipse, and so we can use (31.7) to substitute bx/r for $1/r - c$ and $(c + bx/r)^3$ for $1/r^3$. This gives

$$Y - X = 2\lambda(3c^2 + b^2)b\frac{x}{r} + 2\lambda\left\{c^3 + 3b^2c - \left(c + b\frac{x}{r}\right)^3\right\}$$

§16.12, (31.38)

$$= 2\lambda\left(3c^2b\frac{x}{r} + b^3\frac{x}{r} + c^3 + 3b^2c - c^3 - 3c^2b\frac{x}{r}\right.$$
$$\left. - 3cb^2\frac{x^2}{r^2} - b^3\frac{x^3}{r^3}\right).$$

Now we cancel out terms that occur with both $+$ and $-$, and use §16.12 yet again to get

$$Y - X = 2\lambda b^2\left(b\frac{x}{r} + 3c - 3c\frac{x^2}{r^2} - b\frac{x^3}{r^3}\right)$$

changing order and §16.12

$$= 2\lambda b^2\left\{3c + b\frac{x}{r} - \frac{x^2}{r^2}\left(3c + b\frac{x}{r}\right)\right\}$$

§16.12

$$= 2\lambda b^2\left(1 - \frac{x^2}{r^2}\right)\left(3c + b\frac{x}{r}\right)$$

(31.44)

$$= 2\lambda Y\left(3c + b\frac{x}{r}\right).$$

So substituting for $Y - X$ in (31.45), and dividing both sides by 2, we have

$$\delta\epsilon \sim \lambda\left(3c + b\frac{x}{r}\right)\delta\phi,$$

which leads directly to (31.33) – at last! I repeat the appeal with which §31.11 ended.

32

How Einstein did it

32.1 I have a confession to make. In the last few chapters I've led you to Einstein's more important results, but I've been using methods very different from his. Everything we did on the *Special* Theory of Relativity was Einsteinian in principle, though our techniques were different in detail. In working out the General Theory, we are still in step with him in using the Principle of Equivalence to connect gravitation with acceleration, in describing the geometry of spacetime via the metric, in taking free-fall world lines to be geodesics, and in thinking in terms of curved spacetime. But in the central problem of working out the metric, the need to keep the mathematics simple has forced me to use a method which is in principle quite different from his – and far inferior.

We reached the metric by a series of fairly easy plodding steps. Einstein, after several unsuccessful attempts, got there by one gigantic leap of intuition – rather like the leap we met in §19.10, but taking mankind across the Grand Canyon compared to that previous brook. I'm afraid I can only give you a vague idea of how he did it.

The struggle towards the new theory began in earnest in 1907, when Einstein introduced the Principle of Equivalence. After that he tried many wrong roads before travelling, from about 1913, along the right one. The theory finally took shape during 1915, and he published a definitive summary in 1916. The following description does not attempt to be an accurate history of his progress – only a synopsis of the type of thinking that led to his triumph.

Remind yourself as necessary of the contents of Chapter 28, particularly the definition of curved spacetime (§28.9).

32.2 You'll remember that even in gravitational conditions, it is always possible to choose a system of co-ordinates in the neighbourhood of any event such that the flat metric of Special Relativity will apply *locally* (see §§23.12, 27.3–5). To do so we have only to take as co-ordinates the measurements of an on-the-spot free-falling observer.

But this flat metric is essentially local. Another free-falling observer some distance away will be accelerating relative to the first, and so the same flat-spacetime co-ordinates will not do for him. Any co-ordinate system that serves for more than purely local measurements will necessarily have the more complicated geometry of curved spacetime. Einstein had reached that point by 1913.

32.3 When you know the metric of a curved spacetime, you can deduce (via geodesics) how free-falling bodies will move. We want the resulting motions to be ones that could be produced by the action of gravity. So that implies some sort of restriction on possible types of metric. And the key problem is to discover just what this restriction is.

32.4 The obvious thing is to start with a metric in the most general form possible. And this involves not only squares of co-ordinate increments like δt^2 and δx^2, but also products like $\delta t \delta x$ and so on. So (using what any self-respecting relativitist would condemn as a bastard notation) let's write the metric as

$$\delta s^2 \sim g_{11}\delta t^2 + g_{12}\delta t \delta x + g_{13}\delta t \delta y + g_{14}\delta t \delta z$$
$$+ g_{22}\delta x^2 + g_{23}\delta x \delta y + g_{24}\delta x \delta z$$
$$+ g_{33}\delta y^2 + g_{34}\delta y \delta z$$
$$+ g_{44}\delta z^2 . \tag{32.1}$$

You'll have spotted the subscript scheme: 1, 2, 3 and 4 are related, respectively, to the first, second, third and fourth co-ordinates – t, x, y and z as I've written them – so that g_{23}, for example, is attached to the product $\delta x \delta y$. However, in spite of appearances, I do *not* intend the t, x, y and z of that metric to stand for a time co-ordinate and three ordinary space co-ordinates; they are to stand for *any* four co-ordinates that will specify an event in spacetime. I'm just using the familiar symbols to avoid the 'official' notation, which would be an unwelcome complication for us. And incidentally, from here on I'll usually speak of *points* when I really mean events – when you're discussing spacetime in a thoroughly geometrical manner, the word 'event' can be confusing.

The $g_{11}, g_{12}, \ldots$ ('the gs', as I'll call them for brevity) stand for expressions involving the co-ordinates – expressions like the $1 - 2\lambda/r$, etc., of (29.12), but usually more complicated and involving all the co-ordinates. So the value of each g varies from one point of spacetime to another.

32.5 It's helpful to think of the gs as set out in an array like

$$\begin{array}{cccc} g_{11} & g_{12} & g_{13} & g_{14} \\ & g_{22} & g_{23} & g_{24} \\ & & g_{33} & g_{34} \\ & & & g_{44} \end{array}$$

—where the first row refers to gs connected with δt, the second row and column to gs connected with δx, and so on.

32.6 When you change the co-ordinate system, then the gs change also. For example, the Special Relativity metric for a plane in polar co-ordinates is $\delta s^2 \sim \delta t^2 - \delta r^2 - r^2 \delta \theta^2$ (§26.41), whereas in rectangular co-ordinates it is $\delta s^2 \sim \delta t^2 - \delta x^2 - \delta y^2$. So g_{33} is $-r^2$ in one case and -1 in the other. You can get another example by comparing (24.3) and (24.8).

Collectively the gs work as a single entity, whose job is to specify the geometry of the spacetime we're discussing (§24.10); and clearly the nature of spacetime does not depend on what co-ordinate system we choose to describe it in. Yet *individually* the gs depend on what co-ordinate system you're using. Change the co-ordinate system and you automatically change the gs.

Further investigation shows that there is a set of precise mathematical rules governing these changes. If you are told the equations that describe the change of co-ordinate system – the equations that express the new co-ordinates in terms of the old – then by using these rules you can deduce the corresponding set of equations that express the new gs in terms of the old ones.

32.7 The set of gs taken as a whole is an example of what is called a *tensor*. A tensor is a set of quantities describing some feature of the geometry we're dealing with (or some feature of physical reality that we're approaching through the geometry). The tensor exists independently of the co-ordinate system. But when you set out to describe it in terms of the co-ordinate system, the tensor appears as a set of numbers or quantities like the gs of §32.5. These are called the *components* of the tensor. And the components depend on what co-ordinate system you're using.

What makes it a tensor, however, is that when you change the co-ordinate system, you can deduce – by means of a set of not *very* difficult rules – the corresponding change in the components. The change in the components is related in a well-defined manner to the change in the co-ordinates.

And of course the tensor will usually vary from point to point within the spacetime. That is to say, the values of the components will usually depend on the values of the co-ordinates.

32.8 For example, the set of gs in §32.5 form collectively a tensor, which is called the *metric tensor*. And individually the gs are the components of the metric tensor.

When we want a shorthand symbol for 'the metric tensor' we simply write $g_{\mu\nu}$. Here it is understood that† μ stands for one of the numbers 1, 2, 3 and 4, and ν stands for the same or another of these numbers. So $g_{\mu\nu}$ is a sort of abbreviated catalogue of the metric tensor's components and can therefore serve to symbolise the tensor itself.

32.9 Tensors are divided into different types, and we shall be particularly concerned with the ones called *covariant tensors* – distinguished by the precise way in which the change of components is related to the change of co-ordinate system. The metric tensor is covariant.

32.10 Before the present chapter we have always known (pretty well) what co-ordinate system we are using. But in Einstein's approach we have to allow the use of any conceivable system of co-ordinates (subject to a few mathematical restrictions that only a purist need worry about) – co-ordinates resulting from the measurements of any observer, inertial or accelerated in any way; and more generally even co-ordinates that could not possibly come from any single observer's measurements, so long as they will serve to specify events. But the truth about nature cannot depend on the system of co-ordinates that we choose for describing it. And so Einstein was led to formulate what he called the

Principle of General Covariance
The general laws of nature are to be expressed by equations which hold good for all systems of co-ordinates.

For example, if there were a law of nature (which there isn't) that could be expressed by saying that the sum of all the gs of §32.5 is zero when one particular co-ordinate system is used, then it would have to say that the sum of all the gs is still zero whatever co-ordinate system you use.

It was proved a little later that the Principle of General Covariance actually doesn't say anything at all! – because starting with any equation you like, you can (with a little ingenuity) twist it into covariant form. But when we add the idea that important truths can usually be stated in simple elegant form, then the general covariance idea proves in practice to be a powerful instrument for suggesting useful equations, whose validity must eventually be established in other ways.

32.11 Now it turns out that the Principle of General Covariance will automatically be obeyed if the laws of nature are expressed entirely in terms of *covariant* tensors (§32.9). This is a consequence of the particular way in which changes in the components of a covariant tensor are related to changes in the co-ordinate system.

† These are 'mu' and 'nu' – the Greek m and n.

So to make progress along Einstein's road one has to learn the branch of mathematics concerned with manipulating tensors – the 'tensor calculus' as it is called. And the task of mastering that subject is not lightly to be wished on one's bitterest foe. Not that great conceptual profundities are involved – it is actually a shallow subject. But it is technically extremely complicated. Twice in my life I've learnt enough of it to follow Einstein's thought. And each time I've quickly forgotten it again – for the complexity of its interconnections is more than my poor brain can retain except when I'm actually using it. I refuse to learn it a third time. And so I'm writing this chapter on the basis of vague memories of what I twice temporarily knew, plus hasty consultations of books to compensate (I hope) for huge gaps in my memory. And you ought to be deeply grateful to me for having given you the gist of Einstein's theory without requiring you to learn the tensor calculus.

32.12 When Einstein realised that spacetime must be curved, he seems to have conceived this curvature in a very concrete, almost pictorial way, and made abundant use of it as an aid to the imagination. Gravitation is a consequence of curvature (§§28.13–17). The laws of gravitation must be equations that restrict the curvature – choose particular types of curved spacetime out of all the possibles.

But gravity is caused by matter – it's the Sun's presence that makes the planets behave as they do. The presence of matter distorts the spacetime, gives it curvature; then the curvature produces the accelerations we call gravity. As put aphoristically by today's Grand Old Man of gravitational theory, Professor John A. Wheeler, 'Space tells matter how to move and matter tells space how to curve.' So the law of gravity must somehow connect up the curvature of spacetime with the position and motion of the matter within it. Or in more modern terms, it must make a connection between the curvature and the distribution and flow of energy.

32.13 Now the curvature of spacetime is naturally a good deal more complicated than the curvature of a surface. To describe it completely you need *twenty* quantities – 20 quantities at each point, of course. These will usually vary from point to point, and they can be calculated from the gs of the metric tensor.

Curvature is obviously a feature of spacetime as such, and does not depend on what co-ordinate system you decide to use in describing it. Yet the 20 quantities do depend on the co-ordinate system. As we'd expect, they turn out to be components of a tensor bearing the grand name of the *Riemann–Christoffel curvature tensor*. (Actually it has far more than 20 components, but as a result of several symmetries there are only 20 *different* components – which is what matters.)

32.14 And now at last we can clear up a difficulty that arose in relation to the definition of curved spacetime (§28.9) and was briefly discussed in §28.5. A spacetime is flat – it can be proved – if and only if all the components of the Riemann–Christoffel curvature tensor are zero everywhere in it.

32.15 The Riemann–Christoffel curvature tensor is not covariant (§32.9), and so will not serve our purpose of establishing the law of gravitation (§32.11). But from it one can derive, by a standard technique, another tensor – called the *Ricci tensor,* and denoted by $R_{\mu\nu}$ (using the scheme of §32.8) – which has ten components. Putting it rather crudely, the Ricci tensor does a job equivalent to *half* of the Riemann–Christoffel tensor's task of describing the curvature.

And the Ricci tensor *is* covariant. So the answer to our problem might be something to do with this.

32.16 As we said in §32.12, the problem is to connect up the curvature with the distribution and flow of energy (mass) that provides the source of the gravitation. So how does the tensor calculus deal with energy?

If the theory is to cover all possibilities, it can't just refer to how the energy is distributed in space. It must also describe how the distribution is changing – how the energy is moving about. When you pursue that line of thought, you discover that you'll need a set of ten quantities altogether – describing energy density (energy per unit volume), energy flux density (flow of energy across unit area), momentum density and momentum flux density. It comes to ten quantities in all. And of course the values of these quantities at any point will depend on the co-ordinates.

These ten quantities form a tensor – a 10-component covariant tensor, called the *Energy–Momentum tensor* and denoted by $T_{\mu\nu}$. **Any ideas?**

32.17 We've got two 10-component covariant tensors: the Ricci tensor (§32.15) concerned with curvature, and the Energy–Momentum tensor concerned with the distribution of energy (and therefore of mass). Could the law of gravitation consist of some simple relation between them? The simplest guess would be that $R_{\mu\nu}$ is proportional to $T_{\mu\nu}$. (That just means that each component of one would be equal to the corresponding component of the other multiplied by the same constant.) In this way we might hope to find the connection between curvature and energy distribution (§32.12). At some stage Einstein must have had that thought.

32.18 Alas, it won't work. The Energy–Momentum tensor has a particular property ('zero divergence') which is in effect a mathematical formulation of the Principles of Conservation of Energy and Momentum (§§18.8, 19.5). The Ricci tensor does not possess this property. And so it is not possible for them to be proportional.

We can easily imagine Einstein's disappointment. It had looked so promising . . . Is there no way of making the idea work? Worry, worry, worry . . .

32.19 . . . till at last came the flash of inspiration. There is a certain quantity, denoted by R, which acts as a sort of one-number summary of the curvature. This R varies from point to point of the spacetime, but does not change when you change the co-ordinate system. From each component of $R_{\mu\nu}$ subtract $\frac{1}{2}R$ times the corresponding component of $g_{\mu\nu}$ (§§32.4–5). The result is a new tensor

$$G_{\mu\nu} = R_{\mu\nu} - \tfrac{1}{2}g_{\mu\nu}R,$$

which is known as the *Einstein tensor*. This also has ten components and is covariant. *And* it has the desired property of zero divergence. In fact the mathematical properties of $G_{\mu\nu}$ are in every significant way the same as those of $T_{\mu\nu}$ (§32.16).

32.20 So Einstein modified the guess of §32.17. It is not $R_{\mu\nu}$ but $G_{\mu\nu}$ which is proportional to $T_{\mu\nu}$. The constant of proportionality can be fixed by saying that when gravity is weak and speeds are slow, the results must agree with Newton. It turns out to be $-8\pi G$. (For π see §27.12. And G is the *gravitational constant* – a universal constant whose value is found by experiment and which fixes the strength of gravity. In fact the λ of our inverse square law metric is equal to G times the mass of the Sun – or whatever central body we are dealing with, §27.14).

32.21 So, putting together the substance of the last two articles, Einstein's bold suggestion was that the law of gravity, the relation between energy–matter and curvature, is given by the statement that

$$R_{\mu\nu} - \tfrac{1}{2}g_{\mu\nu}R = -8\pi G T_{\mu\nu} . \qquad (32.2)$$

As we give μ and ν their possible values, this provides an equation for each component. And these are collectively known as the *Einstein Field Equations*.

32.22 The left-hand side of (32.2) can be expressed entirely in terms of the $g_{\mu\nu}$, while the right-hand side stands for whatever matter–energy distribution we wish to consider. So when you start from any particular matter–energy distribution, (32.2) tells you how to calculate the $g_{\mu\nu}$ and discover how they vary from point to point (event to event). And that fixes the geometry of spacetime. So the geometry of spacetime is completely determined, through the Field Equations, by the distribution of matter–energy.

Unfortunately the Field Equations, when you write them in full detail, turn out to be extremely complicated. And so in practice the problem of deciding the geometry from the energy distribution can only be solved in a few simple cases.

32.23 There once was an Arab philosopher who wrote out all the knowledge and wisdom of the world in a hundred volumes. When friends pointed out that few would find time to read so much, he condensed it to one volume. But people urged still more brevity, and he compressed it to a chapter, and then – having caught the habit – to a paragraph, to a sentence. And at last he ended up with one word: Allah. If you know what 'Allah' means, you know everything.

Similarly, if you know the meanings of the separate symbols in the Field Equations, you know everything about gravity. This is one of mankind's supreme examples of compressing a lot of knowledge into a tiny compass.

The trouble is that it takes more than a lifetime to learn what 'Allah' means. Understanding $R_{\mu\nu}$, $T_{\mu\nu}$ and the rest is not quite so hard. But it's a biggish task even for a mathematics graduate. That's why I've chosen another way.

And I draw your attention once more to the great intuitive leap, the mighty act of imagination by which Einstein reached his goal. What a contrast with the plodding steps by which we have laboriously got there.

32.24 In empty space the components of $T_{\mu\nu}$ are all zero; and so (32.2) tells us that the same applies to the Einstein tensor. But remember that the Ricci tensor, and hence the Einstein tensor, only does half of the job of specifying the curvature (§32.15). The other half of the Riemann–Christoffel curvature tensor can still be non-zero. And when you insist that the Field Equations are to hold good both where there is matter ($T_{\mu\nu}$ non-zero) and where there is not, you find that they give a non-zero Riemann–Christoffel tensor even in empty space. Spacetime *between* the stars is curved, even though it's the matter *in* the stars that produces the curvature.

32.25 When he had hit on the Field Equations, Einstein used approximate numerical methods to deduce the predictions of Chapters 30 and 31. But within a few months Karl Schwarzschild had deduced that in the empty space round a single body like the Sun the metric must be (29.12). To reach this conclusion he started from a few reasonable assumptions to the effect that the metric does not change with time, is spherically symmetrical (i.e., the same in all directions from the centre), and so on. But it has since been shown that only one assumption is needed.

That assumption is simply that the matter which is producing the gravitational effect has a spherically symmetric distribution. It need not be static – it could, for example, be pulsating radially. And with that single assumption applied to the Field Equations, G. Birkhoff proved in 1923 that the metric in the empty space outside this matter must be (29.12). (Naturally it is taken for granted that any other matter in the Universe is of such small mass or so far away that its effects are negligible.)

Notice again the superiority of the Einstein method. To get (29.12) *we* had to put the inverse square law into our workings (§27.24) – and then we got it out again in slightly modified form in §§27.45–6. But in the Einstein theory, the metric (29.12) is an inevitable consequence of the spherically symmetrical situation; and so (nearly) inverse square law gravitation is the only possibility.

32.26 On the whole, then, our last few chapters have done no more than take us by a simpler and duller route (and with the help of rather more assumptions) to some of the results that Einstein attained far more elegantly. Yet there is one respect in which our results are not identical with Einstein's. Way back in §§22.19–20 we had to make an approximation that is only valid if fx is small compared with 1 (f being an observer's acceleration and x his distance from an observed event). As a consequence of that approximation, our later reasoning about inverse square law gravity is valid only if $2\lambda/r$ is a small fraction – that is, if r is very much larger than λ.

Thus in our way of doing things, the metric (29.12) is only an approximation, whereas in Einstein's theory it is an *exact* consequence of the Field Equations and spherical symmetry. So far this has not mattered, since we've only applied the theory in situations where the approximation is certainly good enough.

32.27 But suppose we were concerned with the state of affairs near a very very very much more compressed body (assuming one to exist) whose radius is actually less than its own appropriate value of 2λ – the Schwarzschild radius, as it's called (§27.14). It would be possible to approach so close to such a body that $r = 2\lambda$. Then $1 - 2\lambda/r = 0$.

And in (29.12) the δr^2 term would be infinite†–which rouses one's curiosity.

Furthermore (30.15) and (30.16) tell us that at that distance the speed of light would be zero–so an outsider could never *see* anything happening there. The maximum possible speed (Chapter 10) would now be zero; in other words, nothing at (or inside) the Schwarzschild radius could ever get away. And (27.33) says that any body at that distance would be falling inwards with infinite acceleration–though in honesty we ought to do that calculation for u of (27.26) rather than v of (27.31). It can also be shown that a clock placed at the Schwarzschild radius would seem to us to be running infinitely slow–that is, to have stopped. Even remembering that the co-ordinates may not correspond to 'true' times and distances (§§27.39, 30.22), there is clearly something very queer going on.

Following up that line of thought–in a way that we can't pursue here–leads to the concept of a *Black Hole*, a region of space round a very highly compressed star that things can fall into but nothing can come out of. Most, though not quite all, relativitists believe today that Einstein's theory does predict the existence of black holes (unless some unknown phenomenon prevents stars–or even larger masses of matter–from being compressed to that extent). Some think there is observational evidence of their existence.

These predictions are all consequences of the metric (29.12), which is *exact* according to Einstein's theory, but would only be an approximation if you were to put your faith in the method you and I have followed. So suppose we hadn't made the approximation in §§22.19–20. What metric should we get, and what would be the consequences? One more bit of mathematics is needed to answer that.

32.28M Let's go back to explore the mathematics of the relation between k and t that we worked out in Chapter 21. Forget what quantities k and t stood for–it's only their mathematical relationship that interests us. We'll simplify things by taking $f = 1$. Mathematicians have two brief ways of talking about this relation. One way‡ is to say that k is the *exponential* of t–which is written symbolically as

$$k = \exp t . \qquad\qquad (32.3)$$

So Tables 21.23 and 21.27, if we changed the column headings to exp t and t, would become tables of this exponential relation–from which

---

† 'Infinite' just means 'bigger than any quantity you can name'.

‡ For the other, see §21.43.

you could read, for example, that exp $0\cdot41 = 1\cdot5$. And all the relations between k and t that we found in Chapter 21 will still hold good when expressed in the new language. For example, the proposition of §21.5 translates into

$$\exp t \times \exp \tau = \exp(t + \tau), \tag{32.4}$$

(You *can* fill in the steps I've omitted.) And so

$$\exp t \times \exp(-t) = \exp 0 = 1$$

(since the starting point of the Chapter 21 calculation was that $k = 1$ when $t = 0$, §21.18). So

$$\exp(-t) = 1/\exp t. \tag{32.5}$$

32.29M By inspecting Tables 21.23 and 21.27 and imagining them extended as far as you like by the use of (32.4) you can see that as we increase t, exp t also keeps on increasing steadily (and much faster than t). So by (32.5), exp $(-t)$ keeps on steadily decreasing. By taking t large enough we can make exp $(-t)$ as near to zero as we like. But no matter how large we take t to be, we can never make exp $(-t)$ *equal* to zero – it will always be some positive quantity.

32.30m Using (32.3) we see that in our new language (21.20) becomes

$$\frac{\delta(\exp t)}{\exp t} \sim \delta t. \tag{32.6}$$

Along with the starting condition that exp $t = 1$ when $t = 0$, this can be taken as the *definition* of exp t – since this near-equation in the cruder form of (21.13) was the basis for working out the relation between $k = \exp t$ and t.

32.31 Now turn to (22.22) and (22.23) – the near-equations that we replaced by approximations a page or two later. The second of these has the alternative form (22.24). Furthermore $fX = 1$ when $fx = 0$ (§22.19). So the definition of §32.30 gives $fX = \exp(fx)$. And substituting in (22.22) and (22.23) we arrive at near-equations (22.29) and (22.30), which I formerly stated without proof. So these are the accurate near-equations that should replace the approximate (22.27) and (22.28).

32.32 If we now repeated the reasoning of later chapters, using these exact near-equations wherever we formerly used the approximations, we should get the same results as before, *except* that every time $1 - 2\lambda/r$ occurs we should have to change it to exp $(-2\lambda/r)$. Thus our version of the metric would be

$$\delta s^2 \sim \exp\left(\frac{-2\lambda}{r}\right)\delta t^2 - \exp\left(\frac{2\lambda}{r}\right)\delta r^2 - r^2\delta\theta^2, \tag{32.7}$$

using (32.5) for the δr^2 term. And (29.12) would merely be an approximation to this that does well enough when r is large compared with λ.

Let's exclude the extreme case $r = 0$, which would require the central mass to have shrunk to a point. Then with this new metric, none of the terms ever becomes infinite. Furthermore a modified (30.15) would now give the radial speed of light as $\exp(-2\lambda/r)$, which does not become zero at any (non-zero) distance. And (27.33) would put the acceleration at $-\lambda\exp(2\lambda/r)/r^2$ which does not become infinite.

All the oddities of §32.27 would disappear. And in fact there would be no black holes!

32.33 Presumably Einstein is right and I am wrong – the theory we've worked out is only an approximation which is good enough in all conditions where empirical testing can be done at present. But there is just an outside chance that (32.7) is the correct metric and (29.12) is the approximation that does well enough in testable conditions. Then there would be no black holes – which would upset a good deal of current thinking.

There is no vanity in a humble follower hoping to improve slightly on the work of a great pioneer – that's how most of scientific progress is made. The clumsy and inefficient methods used in this book leave too much room for error. At the best they could suggest this idea as perhaps worth thinking about; they certainly couldn't establish it. But could (32.7) perhaps be shown to arise from some modification of Einstein's own methods? Would some acceptable amendment to the Field Equations lead to this exponential form? After the confession of §32.11, you won't expect answers from me. But the matter might merit the attention of any real Relativity expert who has done me the honour of reading so far. Meanwhile you'd better ignore this brief excursion into unorthodoxy, and assume that Einstein is right.

33

A few conclusions

33.1 It's time to attempt some sort of summing up. We've left the Special Theory so far behind that I'm not going to return to it now. Anyhow, since it's so generally acknowledged as a well-established theory that lies at the roots of so much of today's scientific thought, what is there to discuss? The General Theory is a more interesting topic – still subject to active controversy.

Let's begin with a limitation. It never lived up to its title, with the suggestion of a great all-round generalisation of the Special Theory. It turns out, as I said in the Introduction, to be only a theory of gravitation, and as such we must judge it. So let's start by comparing Einstein's theory of gravitation with Newton's at a philosophical level. You'll know by this time where my preference lies.

33.2 Please reread §§26.46–7. There we have one reason for preferring Einstein's outlook: at the most basic level it involves one assumption fewer than Newton's. And so (although technically more complicated) it is simpler in principle. And most of us would agree (though it's hard to say precisely why) that simple explanations are philosophically preferable to complicated ones.

33.3 We can't, of course, *blame* Newton for being more complicated – the Einstein approach was inconceivable in his time. But we can note what caused him to be so. And that is simply that his theory begins by saying, in essence, 'If there were no gravity', things would behave in such and such a way. That's unsatisfactory, because you can never find true no-gravity conditions in which to test its truth. And it leads inevitably to Newton's further assumption, via 'But there always is gravity, and so. . .'

Naturally in our search for simple theories we want to consider how things will move when there are no forces acting on them. So we say 'If there are no strings attached, and nothing pushing, and no electric or magnetic attractions at work, and so on', then the body will do so and so. That makes sense because we can test it by cutting the strings,

neutralising electric charges and the like. But Newton's fatal (if histori-
cally inevitable and excusable) error was to treat gravity as just another
force which could also be cut off (in some sort of theoretical sense at
least).

So greater simplicity results from Einstein saying that when you've
cut off all the forces that you can cut off, a body will follow a geodesic;
and of course the shape of the geodesic will be decided by whatever
gravitational influences are at work.

33.4 Newton's theory involves what we call 'action at a distance'. The existence of the Sun in one place affects the motion of a planet somewhere
else, without any medium or mechanism that transmits the effect from
Sun to planet. There are other phenomena – electrical or magnetic attractions and repulsions, for instance – that can also be described
mathematically in action-at-a-distance terms. Such descriptions have
led to very good quantitative predictions. And of course there is no
logical reason why action at a distance should not exist – why causes
here should not produce effects over there.

Yet we have a tendency to distrust action-at-a-distance explanations.
When the conjuror waves his hands and the ball floating in the air at
the other side of the stage moves in response, we don't believe he is
acting at a distance – we try to worry out what invisible wires or other
connections he's using.

And that's how most people have felt about electric and magnetic
attractions. Distrusting action at a distance, they invented various theories about connections through space between cause and effect. At
first they used the aether (§1.11); and when that had to be abandoned,
they put their explanations in terms of the electromagnetic properties
of space. An electrically charged body disturbs the electromagnetic
balance in the immediately adjacent bit of space – creates, as it were,
an electromagnetic *stress* there. This stress passes on the disturbance to
the next bit of space . . . and so on, till the disturbance (which is
spreading in all directions) exerts its influence on the distant body.
Theories of this sort are called Field Theories.

33.5 For many purposes the phenomena of electricity and magnetism can be
explained equally well by either action-at-a-distance or field theories.
But when rapid change is involved – for example, a high frequency
alternating current having effects on things at some distance from the
wire – then the problem arises of *how fast* the influence will be transmitted. Action-at-a-distance can give no answer – it might even be instantaneous. But the field theory says that the changing influence will

be transmitted by a wave motion. It predicts that the waves will travel at 300 000 kilometres per second; and the predictions fit the facts (radio!). So far as electricity and magnetism are concerned, the field theories win.

33.6 But no successful field-theory version of Newton's law of gravitation has ever been developed. If you want Newton, then you must accept action at a distance, and all that this implies philosophically.

(I don't relish having to knock Newton like this. But it doesn't seem so bad when you realise that many of the criticisms are ones that he actually made himself. His correspondence shows him aware of the difficulties of the action-at-a-distance concept. But, he said, it gives the right answers, and so we've got to use it, no matter what philosophy may say. Unfortunately continued success after success dulled the critical faculties of his followers, so that the renewed questioning this century came as a shock.)

33.7 On the other hand, Einstein's theory is a field theory from the word Go. The presence of the Sun distorts the space around it, and the distortion spreads outwards, producing the curved-space metric. Then this curvature decides the planetary orbits. For most people – though not for all, as many an argument in class has shown me – this concrete connection between Sun as cause and planet's curved orbit as effect is more satisfactory than the action-at-a-distance abstraction.

33.8 You'll remember that we've had the much more serious objection (§10.16) that the Newtonian theory will only work as an *instantaneous*-action-at-a-distance theory (or at any rate requires cause–effect transmission much faster than light), which is impossible according to Special Relativity.

So what does the General Theory of Relativity have to say about the transmission of gravitational influences? Suppose (to take an absurd illustration) the Sun's mass were to double. The spacetime round it would have to change its shape, become even more curved. How would this change be transmitted? That's a line of research that we can't follow. But the theory answers that such changes are transmitted as waves, which travel *at* the speed of light.

The predicted waves would be very weak; and claims to have detected them experimentally have not been generally accepted. But the binary pulsar (§31.36) provides indirect evidence for their existence. According to the theory, such a system should send out gravity waves – one wave for each completed orbit. These carry away energy, which has to be provided by the two components of the binary falling towards each other. Kepler's Third Law (§27.12, but modified to cope with two bodies of comparable masses) says that as the size of the orbits decreases, the time to go round will also decrease. Putting in the observational data, one can deduce how fast this decrease should take place – close to a ten-thousandth of a second per year, as it turns out. And the ratio of the observed rate of decrease to that predicted by the theory turns out to be $1·3 ± 0·15$. Obviously more precise measurement is desirable – and will come in the next few years. Meanwhile this comparatively crude observational result agrees as well as we should expect with the prediction based on the assumption that energy is being lost by gravitational radiation only. While other possible causes cannot yet be definitely ruled out, this is evidence for the reality of the expected gravitational waves.

33.9 At any given spot, all bodies fall with exactly the same acceleration. This has been known from the time of Galileo and verified to extremely high accuracy (§23.10). On the Einsteinian theory this uniqueness of gravitational acceleration presents no problem. All free-falling bodies have geodesics as world lines. So at any given point they all have the same motion, including the same acceleration.

33.10 But if gravity is to be considered as a force, then its power to produce the same acceleration in all bodies makes it so different from other forces that a special explanation is required. Being well acquainted with (18.5), Newton explained it by postulating that the gravitational attraction acting on a body is necessarily proportional to its mass. Or, as later workers tended to put it, a body has two distinct masses: its *inertial mass,* which is the one defined in §§18.7–9, the one that measures the resistance that it puts up to having its motion changed; and its *gravitational mass,* which is a measure of the strength with which Newton's force of gravity acts on it. These are defined as two distinct properties, and yet the Newtonian theory only works if they are *always equal* (or proportional if awkward units are used).

The fact that electrical influences produce different accelerations in different bodies is simply explained by saying that the electrical force acting on a body is proportional to a quantity that is usually called its electric charge, but could equally well have been named its 'electrical mass'. This electrical mass is not proportional to inertial mass – and indeed can be changed without appreciably changing the inertial mass. The same goes for the magnetic case. One would expect – in the absence of reason to the contrary – that gravitational mass would also be independent of inertial mass. But Newtonian theory requires them to be the same, and yet can give no explanation of this amazing coincidence.

33.11 Weight – the force required to stop something from falling – is found to be proportional to mass. This is really the same phenomenon looked at from another angle. Newtonian theory again fails to provide an explanation – except to say that it is another way of expressing the unexplained identity of inertial and gravitational mass.

But in the Einsteinian view we think of a free-falling body as behaving naturally. By the Principle of Equivalence, a body that does not fall is to be considered as accelerating upwards. Its weight is the force that is required to give it this upward acceleration. That puts it on a par with the force that gives it any other acceleration. Only inertial mass is involved, and the proportionality of mass and weight is an obvious consequence.

33.12 So in all these respects Einstein's theory is more satisfactory than Newton's. It uses fewer assumptions; it avoids the traps of action at a distance; it steers clear of unexplained coincidences like that of the two types of mass. A purely philosophical critic would find it distinctly preferable to Newton's theory.

But we can't be purely philosophical. In the long run what matters is how a theory fits the facts. Newton's theory may have been philosophically suspect, but it worked! That was why it had to be accepted till a better one arrived.

So how well does Einstein's General Theory of Relativity stand up to empirical testing? (We already know that the Special Theory does exceedingly well.) In a sense it is one of the best confirmed theories of all time – because everything predicted by Newton's theory is also predicted by Einstein's, except that when gravity is strong, the predictions begin to diverge. Newtonian theory has been confirmed by a host of patient and accurate observations over a period of three centuries. And every one of these confirmations is also a confirmation of the General Theory of Relativity.

33.13 But of course the real interest lies in the few places where the theories give different predictions. If Einstein didn't do significantly better than Newton in these cases, then we'd prefer the latter – in spite of philosophy – because the calculations are easier.

In this Einstein versus Newton contest we have to confess that the amount of empirical evidence is rather small. Let's first discuss how things stood around 1970. Only four types of test had been possible so far. In each case Einstein's prediction was better than Newton's, but the differences were minute. Let's recap.

33.14 First there's the gravitational shift (§§23.14–18, 27.30), a prediction that has been very well confirmed. But it only tests the Principle of Equivalence, not the wider ramifications of General Relativity. And indeed one can show that this prediction would follow from any theory of gravity that is consistent with the *Special* Theory of Relativity.

33.15 Even the other three tests have a more limited range than one would like. They only test the correctness of the Schwarzschild metric (29.12). The validity of the Field Equations (32.2) in other conditions is not tested. So they fail to distinguish between General Relativity and any other possible theory that might yield (29.12) as the spherically symmetrical metric.

And even the Schwarzschild metric is tested only for cases in which r is very much larger than λ. Hence the queries at the end of Chapter 32. If the astronomers could give unchallengeable proof that black holes exist, that would be the verification we want for small values of r. But it's in the nature of astronomy that few of its interpretations of observational fact are unchallengeable.

These two limitations still apply today.

33.16 Consider the prediction about the deflection of light passing near the Sun (§30.12). For many years the observations gave ambiguous results (§30.15). They showed that Newton's theory is definitely wrong, even with the added assumption of §30.13. And the Einstein prediction was obviously of roughly the right size. But on the basis of these optical observations one could not regard the theory as anywhere near to being confirmed. It might qualify as the best attempt at a gravitational theory so far, but no better than that. And several rivals could not be excluded.

Radio observations (§30.16) looked more promising, but around 1970 they still had not attained enough accuracy to distinguish Einstein from his rivals. And the same could be said for the third type of test – radar echo delay (§30.23–5) – as it stood at that time.

And so we come to the prediction about the advance of Mercury's perihelion. The observational result of $43 \cdot 11 \pm 0 \cdot 45$ seconds per century (§31.33) agreed so exceedingly well with the prediction of $43 \cdot 0$ seconds (§31.31) that one could say that this test fully and unambiguously confirms the Einstein theory.

33.17 Or rather that's what one *would have* said until the 1960s, when C.H. Brans and R.H. Dicke proposed their rival 'scalar–tensor theory'. This was supposed to incorporate Mach's Principle – a suggestion put forward by Ernst Mach in 1872. He was impressed by the fact that bodies which are inertial in our sense (§5.2) are also observed to move in a straight line at constant speed relative to a frame of reference defined by the fixed stars (or as we should now think of it, defined by the totality of the galaxies). As an explanation he suggested that inertial mass (§33.10) is not just a property of a body as such, but is produced by the action of the totality of matter in the Universe on the body concerned.

Philosophically Mach's Principle looks attractive. But attempts to state it in precise scientific form have had little success; and no empirical verification has been found.

33.18 I don't feel competent to outline the Brans–Dicke theory. So let's have it in Dicke's own words. He says that it

> ... can be thought of as Newton's law of gravitation treated relativistically within the geometrical framework of Einstein's curved space. Alternatively, it can be described as a theory for which the gravitational force on an object is due partially to the interaction with a scalar field† and partially to a tensor interaction. The third equivalent description of the theory is that it is a relativistic theory of gravitation for which the gravitational constant‡ is not constant but is determined locally by a scalar field variable.

The Brans–Dicke theory works out just like Einstein's, but with a modified metric, which is very nearly the same as what you would get by substituting for α in (29.2) one of the forms that we considered in §30.20. This time, however, n would be a smallish fraction, which can be adjusted to some extent to fit the evidence.§ If $n = 0$, the theory becomes identical with Einstein's (§30.20). And by the way, if you don't know what $(1 - 2\lambda/r)^{1-n}$ could mean when n is a fraction, use the §30.20 form $(1 - 2\lambda/r)(1 + 2n\lambda/r)$, in which the difficulty does not arise.

33.19 This theory gives the same prediction as Einstein's about the gravitational shift (§23.14–15, 27.30). On the question of the deflection of light passing near the Sun, the calculation of §30.20 shows that it will give a slightly different prediction from Einstein's. But the difference is too small for the poor optical observations (§30.15) to give any decision. And the radar delay experiments (§§30.23–5) were only just beginning at that time.

† A scalar is a quantity that does not change when you change the co-ordinates (contrast tensor, §32.7). 'Scalar field' refers to a scalar that varies from point to point of spacetime in a manner that is related to the distribution of mass–energy.

‡ The G of §32.20.

§ My n is $1 - \gamma = 1/(\omega + 2)$ of the usual notation.

33.20 So the crucial test was the advance of Mercury's perihelion. And with any plausible value of the adjustable n (or γ or ω) of §33.18, the best that the Brans–Dicke theory could predict was 39 or 40 seconds instead of 43.

At first sight that would seem to be the end of it. But Dicke is a man of resource. The Einstein calculation, he pointed out, assumes a perfectly spherical Sun (§32.25). Of course the Sun's rotation makes it slightly oblate (flattened, orange-shaped). But with its slow rotation rate – roughly once a month, as sunspot observations show – the deviation from a perfect sphere would be too small to matter.

However, said Dicke, the interior might be rotating much faster than the surface. (It's difficult to see, physically, how this differential rotation could persist; but it's conceivable.) If it were going 10 or 20 times faster, the resulting oblateness would distort the gravitational influence in such a way as to make Mercury's perihelion advance – this was a well known result of Newtonian theory. A flattening of about 5 parts per 100 000 would give an advance of about 4 seconds per century. Then Dicke's theory, accounting for the remaining 39 seconds, would give the right prediction. And Einstein would be wrong!

33.21 Dicke is a fine experimenter as well as an original theorist – a rare combination these days. With colleagues he carried out a series of very refined observations, which did indeed appear to show a flattening of about 4·5 parts per 100 000. Things looked bad for Einstein – whereas, with a suitable value for n, these observations fitted the Brans–Dicke theory perfectly.

However, later observers have not confirmed Dicke's results. The Sun's oblateness appears to be too small to account for anything like that 3 or 4 seconds that the scalar–tensor theory requires. Meanwhile improved measurements of the deflection of radio waves by the Sun's gravity (§30.16) had come down strongly in favour of Einstein and against Dicke; and so did the radar delay experiments (§30.25).

33.22 Another hammer blow came, curiously enough, from the Principle of Equivalence. In Einstein's theory this Principle permits no exceptions. Putting it in Newtonian terms, gravitational and inertial mass (§33.10) must be exactly equal in all circumstances. But in the Brans–Dicke theory this is not quite true.

Part of the mass of a body comes from its 'gravitational binding energy' – the energy arising from the fact that its parts are attracted to each other gravitationally, the energy we should have to supply to overcome gravity in pulling them apart. According to Brans–Dicke this gravitational binding energy contributes differently to the inertial and gravitational masses. The effect would be too small for laboratory detection, but it predicts a measurable distortion of the Moon's orbit – a stretching in the direction of the Sun, which would show up as variations in the Earth–Moon distance.

Measurements were carried out over the years 1969–75 by timing laser beam pulses reflected from fixed reflectors on the Moon (and Dicke was one of the team). These showed to an accuracy of about 30 centimetres that the predicted stretching did not take place.

33.23 To make the Brans–Dicke theory fit the observations that were available by early 1979, the adjustable n of §33.18 would have to be less than about 0·002. So it had become virtually indistinguishable from Einstein's, for which $n = 0$ (§§30.20, 33.18). And there is no point in adopting a more complicated theory if a simpler one gives essentially the same predictions.

There seems to be little chance that any new observations or refinement of old ones will go in favour of Brans–Dicke. And the same goes for other rivals – about which I'm going to say almost nothing, for the very good reason that I know almost nothing about them, except that their challenges are generally agreed to be even weaker.

33.24 Actually it is difficult to make really conclusive decisions between Einstein and his rivals by means of solar system observations – for the weak gravity and slow speeds lead to predictions that only differ slightly from one theory to another. The much higher speeds and much stronger gravity of the binary pulsar (§31.36) should reveal bigger differences, allowing the question to be settled beyond reasonable doubt. The gravity-wave results of §33.8 already weigh heavily against most of Einstein's rivals, since their theories predict a much more rapid dissipation of energy by gravitational radiation (unless the masses of pulsar and companion turn out to be virtually identical, in which case a cancelling-out effect would reduce their predictions to about Einstein size).

The next few years will still be characterised by a certain amount of nail-biting. But I think it is fair to say that the General Theory of Relativity, after a period of considerable doubt, is more strongly confirmed than ever.

33.25 Shall we end by trying to assess the status of General Relativity as a scientific theory? Let's set up a few standards for comparison.

(a) *Newtonian Dynamics*, but excluding gravitation theory – the subject that began on the lines of Chapters 18 and 19 (but with restriction to low speed and therefore constant mass) and flowered and ripened through three centuries to give nearly complete mastery of motion, force, energy and the rest at non-relativistic speeds. One could not say that the relativistic dynamics of Chapters 18 and 19 proved Newton wrong – merely that it revealed limits to the territory in which Newtonian dynamics can be applied. His theory remains valid, provided we watch these limits.

(*b*) *The Special Theory of Relativity* – not just the dynamics, but the whole theory that spreads out from what we did in our first 21 chapters. This has stood firm in a multiple ordeal of experimental tests. Applied in other branches of science, it has illuminated whole fields of phenomena that were previously cloaked in darkness. Philosophically it has a satisfactory 'feel' about it. And it has passed what for me is the most important test of all: it has enabled us to *do* things that we couldn't have done without it (§§13.16, 18.20, 19.24, 19.33–4).

One would guess that this theory – like Newton's dynamics – will never be thought of as *wrong*. Some day, presumably, we shall find limits to its range of applicability. But I think we shall always regard it as valid within these limits.

(*c*) *The Newtonian Theory of Gravitation.* When it comes to the crunch, this theory gives the wrong predictions and General Relativity gives the right ones. And there is no sense in which the Einstein view of gravity can be regarded as an extension or modification of the Newtonian. On the contrary, they say such different things – one that gravity is a force, the other that it is a distortion of spacetime geometry – that if one is true, the other must be false. Einstein fits the observed facts, and so we are forced to say that the Newtonian theory is not just limited in its range of validity, but is actually wrong. Rather curiously, it gives right answers in all but extreme conditions, and yet it must be *wrong*. So in cynical mood, one could even classify it as a stop-gap that kept the world going for 200 years or so until Einstein's theory arrived.

33.26 Now how shall we classify the General Theory of Relativity? The new views of spacetime have surely given a deeper appreciation of the Universe we live in. But then one thinks of how the Special Theory has entered into so many branches of science, scattering light all over the place. By contrast one feels, on pessimistic days, that General Relativity has done no more than illuminate a few small distant corners that Newton left in the dark – though many experts would maintain that it promises far-reaching consequences for the rest of physics which they are only beginning to explore.

At any rate – and in strong contrast with the Special Theory – it can make few claims on grounds of practical usefulness. There are rumours that it is used for accurate missile navigation systems, but naturally I cannot confirm that. In space travel of the distant future, maybe, it will be needed for navigational calculations in regions of intense gravitation. But a theory can hardly claim practical justification on the basis of a problematic future. And so we must assess it mainly in aesthetic, philosophical and empirical terms.

We've already given it much philosophical praise, though it's not philosophically beyond criticism. For example, many feel that its inability to incorporate Mach's Principle (§§33.17) is a serious defect. That was one reason why the Brans–Dicke theory was given very serious consideration. And on the empirical side we've said that far too few tests are possible, but that it has passed these with head erect.

33.27 And so there remains the aesthetic question. Maybe you find it difficult to apply ideas of beauty to mathematical–physical theories. If so, you must just ignore this article. But many people genuinely feel that General Relativity *is* beautiful, in the sense that a Beethoven symphony or a Rembrandt painting is beautiful (or should I have cited Picasso and Stockhausen?). *I* feel like that about it. Some theories may be good and useful, but dull, ordinary and merely utilitarian. Others have a beauty of structure. General Relativity has this to a high degree.

Part of the beauty comes from its simplicity (§33.2) – a very real simplicity of principle, despite the technical complexity. And this is much more impressive when it is approached through the Field Equations (§§32.21, 32.23). But even using the clumsy step by step methods of this book, there surely comes a moment (perhaps around Chapter 28 or 29) when a barren ridge is crossed and a splendid new panorama opens up in front.

33.28 Yet when we've drunk our fill of the beauty, we're recalled with a jolt to the remembrance that the theory has achieved remarkably little – a few minor improvements on Newton (with the binary pulsar hinting that bigger improvements may be on the way); some help towards understanding cosmology (a subject I can't start on) but less than was once hoped; and some future prospects about black holes or gravitational waves.

So how shall we classify it? A well-confirmed theory that we'd expect always to remain valid within its limitations (whatever they turn out to be) – like (*a*) and (*b*) of §33.25? Or a promising speculation that might make the grade? Or a stop-gap, on the lines of my cynically phrased last sentence of §33.25 – good enough to serve till a better comes along? (And if so, will this better theory be so different that we are eventually compelled to say of Einstein's gravitational theory what we now say of Newton's: that it may have given good answers, but still it's wrong?)

My last bold-type request for action: **Would you please think about these questions?**

33.29 In the early 1970s, when observations of the deflection of light were still very uncertain, when the radar delay work was little more than an idea and the Brans–Dicke challenge looked promising – at that time a good many eminent scientists would have taken the last-mentioned point of view: General Relativity provides the best gravitational theory so far; maybe it will last, maybe it won't; but it will have to do till a better comes along. But since then everything seems to be going Einstein's way. I think the consensus would now put it down as a well-confirmed theory – one which we'd expect to survive as valid, even when its limitations have become clear.

What is not in doubt is its magnificence as an intellectual construct. Whatever its ultimate fate in other respects, it will remain a monument to the powers of the human mind. That is something which Einstein shares with other giants like Newton – though he is now proved to be wrong (on gravitation) – and Aristotle – though he is known to be *very* wrong (on a lot of things). Rightness and wrongness is such a tiny matter compared with creativeness!

I have always felt a sense of privilege when I'm allowed in my humble way to pass on some of Einstein's thought to the ordinary women and men who are my students – and now my readers. I hope that you, at the receiving end, share something of this feeling of privilege – and awe.

INDEX

absolute value, 148
acceleration
    in circular motion, 216, 272, 323–5
    constant, *see* constant acceleration
    effect on time measurement, 215–19,
        221–6
    of 1*g*, 189–90; *see also* space travel
    relation to mass and force, 197–8
    speed–distance relation, 335–6
    status, 182
    see also motion, accelerated
advance of
    periastron, 390
    perihelion, 383, 386, 387–8; Mercury's,
        386, 387, 411, 413
aether, 13–14, 17, 24, 60
ageing, 53, 227–8, 243–4
allowable operations
    definition, 105
    examples, 40–1, 104–5, 166–7, 170
angles, measurement, 311–12, 385n
approximations, 143–4, 191, 192, 258–9,
    280
arc, 311
arguments against Relativity Theory, *see*
    counterarguments
as-accurately-as-we-wish technique, 186–
    7, 219–23, 224–5
    see also Fundamental Theorem; incre-
    ments; near-equations
Assumption, Additional, 221, 222, 226
axes, 87–91, 94–5, 132–3
    rectangular, 99–100
    rule for slopes, 86–7, 94–5, 99

BIRKHOFF, G., 402
black hole, 403, 405, 411
BONDI, HERMANN, *x*, 342, 343
brackets
    'curly', 253
    general rule, 173
    particular instances, 39, 136, 142, 169,
        170, 247
BRANS, C. H., 411

causality, 114–21
Causality Principle, 114, 116

chain calculation, 204–5
chord, 311
CLIFFORD, WILLIAM KINGDON, 201
clock
    atomic, 63, 217–18
    nuclear, 226
    standardised, 63
    see also acceleration, effect on time
    measurement
clock paradox
    approached tentatively, 52–3, 159–60,
        215
    firm calculations, 215–16, 223–5, 226–7
    from traveller's viewpoint, 264–5
    verifications, 216–18, 226
    see also space travel; space-twin paradox
clock rate factor, 261
combining Doppler factors, 179–81
combining speeds, 102–5, 107–8, 109–11,
    179–80
constant acceleration, 111–12, 184–5,
    188, 317–18
    equations describing, 188, 191–2
    of 1*g*, 189–90
    observer with, *see* observer with con-
    stant acceleration
    peculiarities of motion, 192–3
    size, 186–7
    speed variation, 185, 192
    traveller with, 183, 184–5, 188–93; time
        experience of, 229–36, 239–46
    world line, 182–3, 185, 188
    see also space travel
constant gravity, 268, 270, 272–3, 320, 321
    see also under free fall; metric; space
    time; speed of light in vacuum
co-ordinates, 89
    distance and time, 89; relation to prac-
        tical measurements, 250, 334, 369;
        relation of two observers', 181
    polar, 312; straight line in, 314–16
    radar, 161, 251; relation of two ob-
        servers', 165–7, 181
    relation between radar and time–dis-
        tance, 172–3, 251, 253–4
co-ordinate systems
    distance-time, 89–90, 93–6